CINEMATIC SOCIOLOGY

To
Savannah Sutherland Bindas
for enduring my asides through gazillions of kid movies and sitting beside me anyway
—jas

Zachary Feltey
for living your truth and being the change
—kmf

CINEMATIC SOCIOLOGY

SOCIAL LIFE IN FILM

JEAN-ANNE SUTHERLAND
University of North Carolina at Wilmington

KATHRYN FELTEY
The University of Akron

PINE FORGE PRESS
An Imprint of SAGE Publications, Inc.
Los Angeles • London • New Delhi • Singapore • Washington DC

For information:

Pine Forge Press
A SAGE Publications Company
2455 Teller Road
Thousand Oaks, California 91320
E-mail: order@sagepub.com

SAGE Publications Ltd.
1 Oliver's Yard
55 City Road
London EC1Y 1SP
United Kingdom

SAGE Publications India Pvt. Ltd.
B 1/I 1 Mohan Cooperative
Industrial Area
Mathura Road, New Delhi 110 044
India

SAGE Publications Asia-Pacific
Pte. Ltd.
33 Pekin Street #02-01
Far East Square
Singapore 048763

Printed in the United States of America.

Library of Congress Cataloging-in-Publication Data

Cinematic sociology : social life in film / edited by Jean-Anne
Sutherland and Kathryn Feltey.
p. cm.
Includes bibliographical references and index.
ISBN 978-1-4129-6046-5 (pbk.)
1. Motion pictures—Social aspects. 2. Sociology. I. Sutherland, Jean-Anne, 1962-
II. Feltey, Kathryn, 1954-

PN1995.9.S6C543 2010
302.23'43—dc22 2009033425

This book is printed on acid-free paper.

09 10 11 12 13 10 9 8 7 6 5 4 3 2 1

Acquisitions Editor: David Repetto
Editorial Assistant: Nancy Scrofano
Project Editor: Karen Wiley
Copy Editor: Trey Thoelcke
Proofreader: Andrea Martin
Typesetter: C&M Digitals (P) Ltd.
Cover Designer: Gail Buschman
Marketing Manager: Jennifer Reed Banando

Contents

Preface

Cinematic Sociology: Social Life in Film began with many conversations about how to teach a course called Sociology Through Film. We both taught the course, and knew what we wanted to assign the students to read, but had never found a complete textbook featuring the right combination of theory, methods, and substantive material. Sharing our teaching strategies, film lists, and assignments, we developed a list of goals for the book, including:

1. Creating a book of readings that is accessible, engaging, and addresses central issues in the discipline of sociology, such as class, race/ethnicity, gender and sexuality, social institutions, and social change.
2. Compiling readings that creatively apply sociology, using film(s) as the subject matter.
3. Framing the readings with an overview of how to think about film as a cultural product that both reflects and shapes the social world in which we live.
4. Incorporating social constructionist and critical perspectives to uncover the multiple stories within film that might not be readily seen or understood.
5. Integrating the sociological imagination throughout to demonstrate how individual experiences occur in cultural and social contexts where structures constrain and enable individual action.

This book can be used alone as a text/reader, as a supplement to a text, or in combination with other articles and monographs. It is designed for undergraduate students and can be used as a stand-alone text/reader for courses on sociology through film, or as a supplement for Introduction to Sociology, Social Problems, or Social Inequality.

Sociologists at the Movies

When we teach the Sociology Through Film course, we find that a good starting point for exploring social life through film is to consider our personal relationship to film. We often ask the students in our respective Sociology Through Film classes to recall the movies they watched as children. What were their favorite movies? What were some of the major themes in these films? Are there scenes that stand out? Can they recall the circumstances of their early film viewing? With whom did they watch movies? Our earliest memories of film might involve family and close friends. Perhaps they led us to games and fantasies that were integral to our young lives. Those films might have captured ideals we clung to in our childhoods, such as princesses falling in love or action heroes saving the world. The first notes of a soundtrack might elicit memories that touch us deeply and personally, taking us back to a particular place and time, or memories of significant loved ones from our pasts.

The influence of the movies as agents of socialization, along with all forms of media, cannot be underestimated. In asking our students to take a trip down movie memory lane, we are helping them to begin thinking like sociologists, to make the connection between their own lives and the social worlds to which they belong. Our own stories of growing up with the movies serve as illustration.

Jean-Anne

When I was a kid, the small Central Florida town in which I lived had one movie theater on Main Street, which showed but one movie at a time. A few years later, a second theater opened in town, this one with fancy red rocking chairs. It gave us a second option for Friday night movies. Because that's what the kids did in our town—we went to movies on Friday night. I can recall my growing-up years by reflecting on movies that I saw and with whom I saw them. I remember when I was old enough to go to a movie without a parent: a friend and I saw *Young Frankenstein,* and I still remember how hard we laughed. I remember seeing *Tommy* and looking down the aisle as several classmates hissed "Jean-Anne, Jean-Anne, *move*!" because I didn't want to move down a seat. I loved Elton John and got there early so I would be situated perfectly in the middle to see his Pinball Wizard. I *did* move down (albeit grudgingly) in part because Friday nights at the movies meant more than opportunities to see our heroes and idols on the big screen—we were also there to be together. We experienced our first dates, our first kisses, our many fights, and make-ups. We experienced moments of popularity and/or social exclusion. When I reached high school age, I got the best job in the world: at the movie theater. On Friday afternoons, several of us would gather and preview the newly arrived films. The theater was ours alone until evening when it would fill with the faces of those we'd known for years. One night I was alone in the small concession area when a man quickly left the theater in tears midway through a showing of *Apocalypse Now.* It moved me then, though I understood it more clearly later. I grew up watching movies, but the significance of those movies came from the community in which I experienced them.

Kathryn

My childhood memories of going to the movies include standing at attention, hand over heart, watching a screen size American flag waving in the breeze as the national anthem played prior to the beginning of the feature film. My siblings and I were Air Force "brats" (Wertsch, 1996) raised, for the most part, on military bases. But the first movie I remember seeing was off-base at the drive-in on a warm summer night when I was 10 years old. My parents and the four kids (soon to be five) were packed in our 1963 Mercury station wagon with the fake wood paneling on the side. We parked backwards, opened the back, and I laid side-by-side with my siblings as we entered a world where a nanny brought magic, wonder, and order into the lives of Jane and Michael Banks. I fell in love with the idea of Mary Poppins, a woman who flew in with the east wind, took charge, brought order, and left when her job was done. I most identified, of course, with Jane Banks, but I was keenly aware of her mother in the background organizing and rallying for women's rights. For my birthday I asked for the book, and over the next couple of years read all eight books in the series by P. L. Travers. To this day, I remember all of the words to the songs from the film, and when I am with my siblings during holidays we sing those and other songs from childhood movies we saw together.

From the Personal to the Sociological

As we reflect back on our early movie memories, we look at the experiences through a sociological lens. While our stories appear to be very personal, like all stories, they are best understood in social context. As sociologists we experience what C. Wright Mills (1959) described as "the urge to know the social and historical meaning of the individual in the society." The way we view film and the meanings we derive are affected by our locations and engagement with the social world.

For Jean-Anne, growing up in a small town, the theater was a pivotal site in building and sustaining social networks. In Goffman's (1959) terms, it was the setting for the front-stage of adolescence in her community. Jean-Anne's memory of the man leaving *Apocalypse Now* in tears is part of her understanding of the ongoing legacy of Vietnam in the lives of those affected. For Kathryn, going to the movies, like all activities on a military base, was structured by patriotism and nationalism. Because she was an avid reader, movies provided visual representation of the texts she read and directed her attention to social events outside of school curriculum and family life, such as the women's suffrage movement that she had never heard about before seeing *Mary Poppins*.

Going through this exercise with our students, we learn how their lives have been affected by the films they have watched, and invite them to think about films and film viewing differently than they have in the past. Many students tell us that their friends and family do not necessarily appreciate their sociological approach to watching movies; they complain, "You're not fun to go to the movies with anymore!" But overall the students feel they have been enlightened and given a set of tools that will be useful throughout their lives. In any event, they say they will never see movies in the same way!

Acknowledgments

We would like to thank the many students who have taken our Sociology Through Film classes over the years. They have brought to this relatively new course ideas and insights (and film suggestions!) that have contributed to our approach to the course and informed the pedagogical framework we are developing.

We appreciate the support provided by our families and friends. Jean-Anne would like to thank Mark Cox for his advice, insight, and loving patience on those many occasions that she trapped him in conversation about this project. Thanks also to Daren Painter for his keen perception of film and social life, Kenneth Bindas for many years of advice and unwavering support, and DeeAnna Merz Nagel—from that Central Florida town to here. Kathryn would like to thank Diane Moran for sharing life, love, and the Cleveland International Film Festival. Thanks also to Robert Marcum for sharing his wealth of knowledge about film, and Cheryl King for sharing her insight, wisdom, and, best of all, laughter.

We thank the contributors to this book for their work using film to teach about the social world we share with one another. We appreciate Pine Forge Press, our Acquisition Editor, David Repetto, and Nancy Scrofano, Sociology Editorial Assistant, for guiding us and shepherding the book through the publication process.

Many thanks to our friends and colleagues at the University of North Carolina, Wilmington (UNCW), and the University of Akron (UA) for their support, including Kim Cook, John Rice, Diane Levy, Dave Monahan, John Zipp, Becky Erickson, Matt Lee, and Bill Lyons. We also thank our graduate assistants Katie Gay (UNCW) and Virgil Russell (UA) for helping to compile and edit the film index.

We gratefully acknowledge the following reviewers: Ingrid E. Castro, Northeastern Illinois University; Alexander M. Hicks, Emory University; Jacqueline Carrigan, California State University, Sacramento; Jacqueline Clark, Ripon College; Tricia M. Davis, University of Wisconsin, River Falls; Robert E. Kapsis, Queens College of CUNY; Donald P. Stewart, II, University of Nevada, Las Vegas; Marcie Goodman, Weber State University; Shirley Jackson, Southern Connecticut State University; Elizabeth Jenner Gustavus, Adolphus College; Eric J. Krieg, Johnson State College; Christine Von der Haar, Indiana University, Bloomington; Lisa Rashotte, University of North Carolina, Charlotte; Tricia M. Davis, University of Wisconsin, River Falls; Mathieu Deflem, University of South Carolina; and Laura L. Hansen, University of Massachusetts, Boston.

We are especially grateful to our friend, colleague, and mentor, Michael Kimmel, who encouraged, evaluated, advised, and ultimately empowered us to write from the heart. Michael, just as you encouraged us to work from our hearts, you daily modeled for us a colleague who worked unselfishly from his. Thank you for your relentless cheerleading and support.

Jean-Anne Sutherland
Wilmington, North Carolina

Kathryn Feltey
Akron, Ohio

References

Goffman, E. (1959). *The presentation of self in everyday life.* New York: Anchor Books.

Mills, C. W. (1959). *Excerpt from C. Wright Mills,* The sociological imagination *(originally published in 1959).* Retrieved May 8, 2009, from http://www.camden.rutgers.edu/~wood/207socimagination.htm

Wertsch, M. E. (2006). *Military brats: Legacies of childhood inside the fortress.* St. Louis, MO: Brightwell.

Foreword

Jean-Anne Sutherland and Kathryn Feltey's exciting volume brings together many strains of the new sociological scholarship on film. Incorporating documentary films into our teaching has been standard practice in the field. Yet, increasingly sociologists bring narrative films into their classrooms using them in a range of ways, but also encouraging students to learn to read them as texts. Narrative films with compelling stories are powerful in communicating issues to audiences. Films take us to places we have never been, in terms of geography and time. They also put us into the shoes of various people and ask us to think about a different point of view. This volume can help readers identify the social issues explored in films, develop a critical analysis of how film production meets specific goals, and what those constructs mean for how audiences frame issues.

In building this book, Sutherland and Feltey were inspired by Lewis Coser's 1963 volume *Sociology Through Literature*. Rather than reading fiction, the editors ask us to view films to analyze social experiences, institutions, and the theoretical perspectives within them. Films are ways of looking at the very social constructions that pattern our lives and provide us with glimpses into the forces that shape the lives of others. The editors encourage readers to interrogate those representations. The chapters in this volume are ideal for tackling many of those profound issues that the powerful medium of film frames for viewers.

I teach a Social Inequality and Film course, where students learn sociology through film, and also the sociology of film, as they think critically about what they view. These are major agenda items for the editors in pulling together readings that help explore not only what we see on the screen, but why these images are on the screen. A few weeks ago, my students followed Norma Rae into the textile mill as she confronted the challenges of a single mother doing low-wage work, as was the tradition of her family. As Ruben Warshawsky, the labor organizer, comes to Henleyville, students see how having the right to organize is different from the reality of organizing a union in a town dominated by one industry with a history of paternalism that "favored" white workers. We discussed the role of unions for working-class people and how unionization was successful in many southern textile plants in the wake of the civil rights movement because black workers brought their perspective and organizing skills into these mills. While *Norma Rae*, the film directed by Martin Ritt in 1979, was based on the life of a real woman, Crystal Lee Jordan, it follows the Hollywood formula as a narrative of an individual's transformation, with a "major" star (Sally Fields), as key to both the unionization effort in the film and the film itself. We do not hear the voices of other workers in the struggle to unionize. We also do not see how Norma Rae brings people into the union other than with her charm. While the film is a compromise, it is one of few Hollywood productions about working-class people in the 1970s. Viewers can read between the lines and still look at the collective struggle even through the prism of Hollywood individualism. In terms of the individual transformation, we can see how Norma Rae took learning seriously when the lessons were connected to her empowerment.

Sutherland and Feltey's chapter introductions and selections explain how representations in films have been contested throughout their history. We learn that even within studios, the money often followed a simple narrative with a positive message. Yet, audiences have also challenged various representations as biased and supportive of a status quo. Many audiences wanted films that represented new visions of racial and ethnic groups, women, and sexualities. Many viewers have learned to read between the lines, celebrating visibility of their group and issues, yet often crafting a different interpretation than the one intended. Viewing films is complex, as the readings here encourage viewers to think critically about the images and stories offered in them.

Sutherland and Feltey structured this volume to provide the tools for exploring films as rich texts that teach us about core areas of social life and prompt readers to question the images produced for their consumption. Their introduction provides a brief history of film production and viewing in the United States that stresses the connection between the state of the society and the nature of representations and social relations on the screen. Films were essential in socializing immigrants and the children of immigrants early in the 20th century, just as they now teach us about other groups in a society characterized by a high degree of racial and social class segregation. The book is ambitious, covering many areas of social life with offerings that can push the readers to think about what we view and what those viewings mean in terms of our thinking about social issues.

Sutherland and Feltey solicited and wrote sections/pieces that explore social identity, interaction, inequality, and institutions. The contributors include many of the usual suspects who have been writing about film representations of identity and institutions, theoretical treatments in film, and how films frame our perspectives on the wider society. The editors also have contributions from some new voices revealing the innovative ways that sociologists are tackling discussions of social class, sexualities, relationships, and social change.

The films we can watch in multiplexes, on flat-screen televisions, on laptops, and all manner of electronic devices come from many sources. We still have Hollywood productions and independent film producers, even though the lines are getting blurred between these two sources. We also have foreign films—now many of them hybrids in this global economy. Films from these streams all offer ways to frame identity issues for various groups, the nature of interaction in this global economy, and conceptions of inequality and the institutions that shape our lives and make negotiating the lives we want complex. A film can make social classes and inequality visible, even to a privileged audience. Then we have to ask if the film frames social mobility as easy, if one follows the appropriate gender routes. Or does a film expose the rigidities to the system offering the subject the choices of either compromise or struggle?

In the past 30 years we have had Hollywood productions that explore the historical legacy of race and ethnicity. But do these films encourage viewers to critically investigate the mechanisms that establish various racial and ethnic hierarchies or do they leave the legacy of race unquestioned? Race scholars have identified how Hollywood films avoid an analysis of power while offering scenarios that depict white people in a positive light (Vera & Gordon, 2003). The writers in Sutherland and Feltey's book challenge ideologies that do not ground our viewings in the social, political, and economic realities of group interactions, while questioning the power to represent others. We have to question how a film such as *Crash* avoids any analysis of power differentials among the various individuals colliding with each other in Los Angeles. The readings suggest that as viewers we can interrogate the ideologies presented in film and ask if different films prepare us to explore the current configurations of social class, gender, race, ethnicity, and sexuality in our society and around the world.

Films help us think about the major institutions that shape our lives and our interactions within them. In this volume, attention is given to films about work and systems of production, family, and schools—all worthy of considerable discussion, since these treatments are windows into different spheres of social life. Yet, films vary in how they present the structures that shape our lives, offering either simple narratives or suggesting that one's social class, race, gender, sexuality, and citizenship require complex negotiations to secure the basics of life. In this fast-paced world, films also frame our thinking about changes within and outside of our borders in this new global enterprise—as well as the possibilities for the future.

The readings in Sutherland and Feltey's book demonstrate that filmmakers vary in their presentations, but they remind us that as viewers we are decoding films from our own unique social positions. What we see on the screen, regardless of the size, is very much about who we are. If we are advantaged, we might be comforted by stories that celebrate the status quo that does not push us out of a comfort zone. In fact, Hollywood productions have long celebrated whiteness, a particular brand of masculinity and femininity that changes with the times, as well as heterosexuality and middle-class status. Those viewers who lack privileges along some dimension of inequality might search for different role models or restructure narratives to celebrate their own point of view. For example, a viewer might identify with the monster or mobster who is wreaking havoc on the community rather than the hero. Thus, this book is important in opening up viewers' eyes to think more critically about how we view images and identify the ideologies that are encoded in films.

In the end, films are still dream makers, products of our culture that seek to satisfy some human needs. Popular films might offer simple solutions to problems at a time when there is great complexity of any social issue. We might want happy endings for romances at a time when gender and sexuality images and expectations are contested and often hard to negotiate. Yet, we can also set these needs aside to think about the visions offered to us. As viewers we can sharpen our tools for evaluation of the stories and images offered to us. Sutherland and Feltey's volume is an important contribution to that endeavor, preparing viewers with a critical eye for thinking about how issues are framed for us by different directors and producers, and then deciding what we want to do with the meanings of what we see on the screen.

Elizabeth Higginbotham
University of Delaware

References

Coser, L. (1963). *Sociology through literature.* New Jersey: Prentice Hall.

Vera, H., & Gordon, A. M. (2003). *Screen saviors: Hollywood fictions of whiteness.* Lanham, MD: Rowman & Littlefield.

CHAPTER 1

Introduction

Davis: This is life. . . . You haven't seen enough movies. All of life's riddles are answered in the movies.

Grand Canyon (1991)

American Society and Film

In U.S. society, "going to the movies" has been a favored pastime since the opening of the first movie theaters more than 100 years ago. It is hard to imagine a small town, Anywhere USA, without envisioning the movie theater on the town square, complete with marquee and ticket booth with moviegoers lined up to pay the admission fee. The movie theater also has been an integral part of the creation of community in the movies. Indeed, the image of the town movie theater is a staple in movies, and it is often used, along with other institutions such as the pharmacy and barbershop, to convey a sense of community, a *gemeinschaft* haven in a changing world. For example, in the classic *It's a Wonderful Life* (1946), we see from the Bijou Theater marquee that *The Bells of Saint Mary's* is playing in Bedford Falls, a community protected from capitalist and political greed by George Bailey's (James Stewart) goodness. In a similar vein, in the first *Back to the Future* (1985) when Marty (Michael J. Fox) travels 30 years back in his hometown, Hill Valley, he orders a soda at the downtown drugstore and observes that the nearby Essex Theater is playing a Ronald Reagan film.

The antithesis of this communal setting is what towns could/would become if small-town values were sacrificed to modernization. When George Bailey is rescued from his Christmas Eve suicide attempt by his guardian angel, Clarence, he is shown the town that Bedford Falls would have been without his influence. Pottersville is a tawdry town with a main street dominated by pawnshops and sleazy bars. The theater marquee now boasts "Georgia's sensational striptease dance with Girls! Girls! Girls!" Community values have been replaced by individualism, cynicism, and distrust. In *Back to the Future,* in an alternate 1985 we see a run-down Essex Theater, now an adult movie house showing "Orgy, American Style." The run-down condition of these theaters, along with the commodification of sexuality (in a place formerly reserved for family entertainment), symbolizes the alienation that accompanies modernization.

The process of modernization, however, included the rapid expansion of the motion picture industry in the first half of the 20th century. Between 1914 and 1922, there were 4,000 new theaters built in the United States (Starker, 1989). Many of these were built in residential areas, signaling a new trend in the industry, the "neighborhood theater." During the 1920s, around 100 million people attended movie theaters each week, twice the number that attended church (Library of Congress, n.d.). The popularity of the movies could, in part, be attributed to the low admission prices and unreserved "democratic" seating, making this leisure-time activity widely accessible to diverse audiences.

One of the functions of the theater in the early 20th century was to aid in the assimilation of immigrants into the community, ensuring that a shared way of life was protected and passed on. For example, in some communities, theaters ran special features on Sundays (when theaters were not ordinarily open), "to educate and familiarize foreign speaking people with the customs, principles, and institutions of our American life" (Abel, 2004, p. 107). Many films during this period focused on the immigration story and life in the new country. For example, in movies like *Making an American Citizen* (1912), *The Immigrant* (1917), and *The Making of an American* (1920), immigration status is resolved when the immigrant has "melted into the American mainstream" (Library of Congress, n.d.).

Assimilation into the American mainstream was, of course, reserved for immigrants who most resembled the dominant group. Theaters, like other public institutions, reflected the racial, ethnic, and class-based divisions of the larger society in the content of the films shown and the treatment of the attending audiences. Until the latter part of the 20th century, the policy of racial segregation was enforced at movie theaters by time (showing films for African American audiences late at night), by section (seating African American viewers in the balcony), by entrance (requiring African Americans to enter through a side alleyway), and by neighborhood, with black-only theaters serving patrons in African American neighborhoods, especially in northern cities (Hearne, 2007). This history of racial segregation in theaters has been portrayed in recent movies. In *The Secret Life of Bees* (2008), set in 1964 South Carolina, Lily (Dakota Fanning) goes to the movies with Zack (Tristan Wilds), and she follows him to the balcony so they can sit together. Zach is dragged out of the theater and beaten by white men, outraged at the transgression. In the film *Ray* (2004), Charles (Jamie Foxx) refuses to perform at a segregated concert after encountering young civil rights activists protesting outside the theater. In a lesser known movie, *Hope* (1997), a young African American boy dies in a theater fire when he is trapped in a balcony that lacks an emergency exit.

In southwestern states, schools, public facilities, and movie theaters were segregated, with Asian and Mexican immigrants and citizens forced to sit in the balcony, even if seating was available downstairs (Ross, 1998). In Kansas, Mexican Americans were restricted from some sections of city parks, churches, and other public facilities, and they were routinely segregated in movie theaters through the mid 20th century (Oppenheimer, 1985). This institutional segregation, visible across communities, was based on an ideology of racism reflected in the official policies of organizations, businesses, and homeowners associations. Two of the most common examples of segregation, in addition to schools, were movie theaters and swimming pools (Montoya, 2001).

Moving through the decades of the 20th century, changes in the film industry reflected societal and cultural changes. The location and structure of venues for movie viewing changed from the picture palaces of the early 20th century, to the art-deco architecture of the Depression era, to the drive-in movies of the mid 20th century, to the muliplex, cineplex, and megaplex of the mid to late 20th century. Just as the content of film celebrated the culture of consumerism and individualism, the way we watched movies became less public with the expansion of in-home movie viewing services and options. However, while 75% of Americans say they would rather watch a movie at home, between 1995 and 2006 there was only a 5% decline in those reporting they go out to see a movie at least once a month, from 31% to 26% (Pew Research Center, 2006). While the decline in monthly theater attendance was modest among the full adult population, it was more substantial among the age group most coveted by the theater industry—younger, better educated, and higher-income consumers (from 56% to 43%). Regardless of where we are watching, seven in ten Americans are viewing movies at least once a week; we are still "going to the movies" in our leisure time.

Sociology and Film[1]

Paul D: I ask him, "How come you didn't run with Sethe?" He never answered me. I seen him once more, the day Schoolteacher sold me to Brandywine. Takin' me away in the wagon. There Halle was, in the yard, sittin' by the churn. They didn't need to put no chains on him no more. His mind was gone. . . . You see, you the only one of us who made it out that night. The only one.

Beloved (1998)

All over the world, people go to the movies—to escape, to be enlightened, to be entertained. Indeed, movies are among the world's most popular social experiences. In 2007 there was a decline in movie attendance; however, two years later, with a "teetering" economy, there has been a "box-office surge that has little precedent in the modern era" (Cieply & Barnes, 2009, para. 1). With a 16% increase in movie attendance, it appears that "people want to forget their troubles, and they want to be with other people" (para. 4).

Going to the movies is a social event; we view movies with significant others and are affected by the experience in a social context. Consider the young dating couple who share a soda (or something stronger) after watching a movie together, talking earnestly, seriously, about their feelings for the characters. Or the child, seated between his or her parents, munching popcorn and watching *March of the Penguins*

[1]This section was developed largely through conversations with Michael Kimmel. We appreciate his contributions that made it possible to write about sociology and film.

(2005) and deciding to become an oceanographer. Or families who share their anguish and deepen their understanding about social issues such as AIDS (*Philadelphia,* 1993), sexual harassment (*North Country,* 2005), personal heroism and sacrifice (*Saving Private Ryan,* 1998), or the cold cruelty of the Holocaust (*Schindler's List,* 1993). Or even the everyday people who, hit hard during troubled economic times, watch a musical or comedy and momentarily laugh their troubles away. As the leader character in *Sullivan's Travels* (1941) said in the final lines of the movie, "There's a lot to be said for making people laugh. Did you know that that's all some people have? It isn't much, but it's better than nothing in this cockeyed caravan."

Reading a Film Sociologically

Every field of study offers a slightly different way of seeing film. An anthropology of film class might integrate ethnographic films as a means of "seeing" culture. Older films might be seen as historical artifacts, offering a window onto the culture at particular points in time. Many documentaries are themselves products of anthropological interest in documenting the practices and ceremonies of other peoples and cultures. A film course in the history department would take a look at films of historical significance and question authenticity, accuracy, and interpretation. For example, you might take a film like *Glory* (1989) and discuss the cinematic treatment of the Civil War. A film such as *Sense and Sensibility* (1995) might enhance discussions of class and gender relations during different historical periods.

In a film studies class, attention would be placed on film production, cinematography, *mise-en-scène,* film genres, film history, and film theory. In mass media or cultural studies departments, perhaps the "cousin" to sociology in terms of film analysis, you might be asked to consider social and cultural "powers" at work in the film and in the film industry that remain largely unchallenged as social disseminators of cultural ideas. How does cinematic "form" work with (or, sometimes, against) the cinematic content?

Sociological perspectives draw from these traditions. In this text, we are less concerned with the technical aspects of film production and more concerned with the stories told through film and how these stories are told. Is there something specific that a sociological perspective brings to the study of film? More importantly for our inquiry, sociology *through* film, how can film be used by sociologists to better understand the society in which we live?

The core of any sociological curriculum revolves around four interrelated themes: (1) *identity,* (2) *interaction,* (3) *inequality,* and (4) *institutions.* Sociologists can use films as social texts to explore these core themes.

Identity: Many films involve the development of an individual character as a central theme: how they come to know themselves, how they try out identities until they find one that fits, how they adapt their identities for strategic purposes. The sociological concept of the "looking-glass self" and the dramaturgical metaphors made popular by Goffman and others are amply illustrated in films such as *In & Out* (1997) in which a closeted gay teacher is outed in the national media and comes to accept his homosexuality, just as the small town in which he lives comes to accept him. In *Forrest Gump* (1994), Forrest (Tom Hanks) and Jenny (Robin Wright) come to know who they are in very different ways. Forrest moves through his life, unconcerned as to whether others see him as heroic, iconic, or mentally ill. In fact, we do not learn until late in the film that Forrest is aware of his "difference," when he asks Jenny about their son, . . . "is he smart or is he . . . ?" Jenny responds, "He's very smart. He is one of the smartest in his class." Jenny, on the other hand, is very aware of how others view her, and her sense of self develops in the context of unloving and abusive relationships beginning in her family of origin.

Interaction: All film involves social interaction, even when the story is one of separation or isolation from others. For example, in *Cast Away* (2000), Chuck (Tom Hanks) develops a friendship with Wilson the volleyball, who becomes his companion and confidante. As the viewing audience, we come to know Chuck through his interaction with Wilson. By watching film, we learn about social interaction in the context of relationships, everything from friendship (*Sisterhood of the Traveling Pants,* 2005), to romance and dating (*The Breakup,* 2006), to family life (*The Joy Luck Club,* 1993). Indeed, many people probably first know what sex and intimacy looks like from having seen it in the movies.

Inequality: A common theme in movies is inequality based on social class, race, gender, and nation. Many people who live in relative comfort encounter the stark reality of people oppressed in systems of inequality through the movies. A student spoke of her political apathy until she saw the film *Hotel Rwanda* (2004) at the university cinema. The film so moved her that she became involved in campus political organizations that confronted inequality, eventually changing her major to Public Sociology to make her newfound awareness and politics her future.[2] A friend told of his first memory of learning about the Holocaust, watching *Exodus* (1960) with his parents and grandparents as a young child, and for the first time, his family began to talk about what had happened to the Jews of Europe. They had been silent; the film enabled them to give words to the unspeakable. How many films reveal the brutal injustices of the prison system (*Cool Hand Luke,* 1967; *Green Mile,* 1999), or the Jim Crow South (*To Kill a Mockingbird,* 1962; *Driving Miss Daisy,* 1989), or slavery (*Amistad,* 1997; *Beloved,* 1998), or any other type of structured inequality? Films can engage us viscerally; make inequality palpable and real, and thus far less tolerable.

Institutions: Finally, film enables us to locate ourselves in the institutions that shape our lives. Films such as *Patton* (1970) and *A Few Good Men* (1992) illustrate more than the atrocities of war—they provide a lens into the military as organization and institutional force. In the same way, movies about work and the workplace (*Working Girl,* 1988; *Michael Clayton,* 2007) reveal the often invisible structural barriers encountered in institutions, whether it's the glass ceiling encountered by women or unethical business-as-usual practices. Institutions such as the workplace, schools, or religious establishments shape our identities, and films, usually without

---

[2]MacDonald, J. (2008). Presentation in Introduction to Sociology Class, University of North Carolina, Wilmington.

our knowledge, provide us with understandings of the social structure of these "spaces" where we live, work, learn, pray, and are entertained.

Sociologists explore these four themes in the film content—the stories that are presented in films. Sociologists also might look for patterns in viewership—the ways some movies target specific audiences by class, race, gender, and sexuality. Or they might examine the centrality of film in childhood socialization; many Americans might say that some of their most tender and powerful memories of childhood involve going to the movies. Sociologists might also consider the ways that films, like other media texts, elicit, structure, and facilitate the expression of different emotions to which we, as humans, need access. We consider the social dynamics of audiences, the effects of images on social life, the complex interaction of cinematic technique with the experiences of viewers. We consider the ways in which social problems and identities are represented in film—the manner in which films both reflect and create culture.

Reading films with a sociological eye makes us conscious of ourselves as we watch movies. In a sense, we move from our seats in the theater to the projectionist booth. From there we can look at the audience as they "see" the film. Also, we can appreciate the film as *film;* a strip of material that produces images and ideas. From this angle we can recognize the social nature of the movie experience—the collective of people coming together for entertainment, for storytelling, for a glance at our culture. Movies are social experiences, and we often remember family events or our childhoods through associations with movies. Baby boomers (including the editors of this book) can probably still recall when small theaters sold pickles from a big jar at the concession stand, or biting through the outer layer of a Good & Plenty to get to the chewy licorice inside, or the sticky-sweet smell of black Jujubes that got stuck to the soles of their shoes as children. For many, a social life in high school meant meeting friends at the movies on a weekend night, where they would see dozens of classmates—friends and foes.

The dynamics of social interactions—homophobia, racial awareness, class interaction—are evident as we take our seats to watch the movies, just as much as they are in the movies themselves. For example, when two women go to the movies together, how many seats do they take? The answer is invariably two. But ask men how many seats they take. The typical answer is three; they leave the middle seat open, or put their coats there, because they don't want anyone to think that they are "together."

The Film as a "Text"

But it is mostly what is on the screen that rightly preoccupies the sociologist of film. Like other forms of art—music, literature, paintings—films "speak" to us, exposing us to the ideas of writers, actors, and directors, who use various techniques to explain, explore, or exploit our experiences. We can look at the history of cinema—the ways that a film will reference other "texts"—or the actual technology and technique, but only insofar as they enable the director, screenwriter, and actors to make a particular case to viewers. Applying the sociological imagination to film involves awareness of the economic, political, and social forces at the point in history when a particular film or set of films is produced. In fact, this is one of the strengths of film as a pedagogical tool; it can provide access to multiple sociohistorical contexts. However, it also reflects the (biased) view of the filmmaker(s) telling the story. Our interests are sociological—social—not aesthetic. We can appreciate the aesthetic beauty of, say, a Merchant Ivory production,[3] but our sociological imaginations are animated by the actual story and the moral lessons they present to us.

At the same time, reading films sociologically does not mean that we drain the fun out of the experience. While a sociological reading asks us to look beyond commonsense thinking and long-held interpretations (in short, to be critical thinkers), we can still appreciate the entertainment value of movies. We can critique expressions of gender in film without dismissing Disney, a central piece of our childhood movie memories! "Seeing" films through a sociological lens does not make movie watching laborious. Rather, it should enhance our view of film, the society in which it is produced, and the ones it portrays.

This book is about the ways in which films "speak" to us. It's about film as a "text"—in the same way that a novel is a text, not only telling a story, but providing moral instruction, social observation, social context, and political judgment. In 1963, Lewis Coser's edited volume *Sociology Through Literature: An Introductory Reader* sought to fuse social science and art—specifically literature. Coser made visible sociological thought within literary works and explored how classic works of literature expressed various classical preoccupations of sociologists. More current writing continues this tradition through analyses of literature and group identities, decoding systems, and the impact of life experiences on reader response (Griswold, 1993). Works of literature are seen not only as products of the authors' imaginations and examples of shared meanings, but also as *social* texts of culture, socialization, identity, inequalities, and social structures.

Literature is appreciated not only for the telling of a good story, but for the ways in which authors go about capturing social context, social and psychological struggles, and social problems. The great novels leave us with commentaries on social class (*The Great Gatsby, Pride and Prejudice*), race (*The Bluest Eye; Huckleberry Finn*), and gender (*The Awakening, The Scarlet Letter*). Narratives such as these provide frames through which we can understand a social class to which we do not belong, *feelings* associated with racial tensions, and social constraints associated with being male or female. Before motion pictures, these were the stories on which we relied.

In today's world, Americans attend movies far more often than they read fiction. In this sense, films have become a new kind of text through which we are provided stories, frames, and representations of social life. Films, of course, require a different kind of "read" than a novel. For instance, while the author of a novel might also be encouraged to consider production and sales—there are fewer investors to which the author must answer. With film, there is the ever-present issue of currency—films cost an enormous amount of money to produce and, thus, must earn an even

---

[3]Merchant and Ivory were filmmakers known for producing visually beautiful films. Their company's double-barreled name has entered the popular lexicon to refer to a particular way of life or social standing.

greater amount upon release. The issue of money has the potential to change the telling of the story so that it "satisfies" a much larger community. The novel can be as controversial as the author intends, as he or she has more autonomy. A film, on the other hand, must often play down controversy to reach a broader market. The novel is one voice, that of the author. Film is the product of many voices—the screenwriters, the director, the actors, and so forth—thus potentially diluting the original "voice" (i.e., the author of the text on which the film is based). When one reads a novel, the interpretations lie mostly in the active imaginations of the reader. A film can utilize technology—using visuals and sound to guide viewer response. Thus, while film is a modern "text," we must read it conscious of the ways in which symbols, language, and author intentions are specific to this medium (as opposed to reading literature as text). Not unlike film studies' entree into "film as text," we think of a film as "comprising visual language, verbal systems, dialogue, characterization, narrative, and 'story'" (Shiel, 2001, p. 3). Films reflect our culture at the same time as they are an element that constitutes it. We see our society through film and we see film through the prism of our social norms, values, and institutions.

Sociology of Film—Sociology Through Film

Early social science approaches to popular culture were steeped in deterministic thinking. For example, the Frankfurt School warned that the "liberatory" power of jazz—antiauthoritarian, sensual, rebellious—actually served to further authoritarian domination because it temporarily "freed" individuals, and thus muted or siphoned off anger at an unjust system. Any system that gives us such pleasure can't be all bad—thus the systemic domination is, ironically, legitimated by embracing its own resistance. Popular culture was an "opiate of the masses," to use Marx's famous phrase about religion, numbing us to the pain of inequality. "Keep you doped with religion and sex and TV/So you think you're so clever and classless and free," sang John Lennon in his devastatingly plaintive song "Working Class Hero."

Historically, sociology's approach to film was also deterministic. From the origin of the moving picture, social scientists worried that film would instruct people to passively follow the normative instructions contained in the film. Somewhat like contemporary anxious parents, they worried that a film about crime would inspire a young person to commit crimes. Surely, pornography would cause rape, and cowboy movies would inspire violence. Ultimately, sociologists renounced this conception of the audience as a gullible mass, somehow incapable of deciphering the descriptive from the normative.

In contrast, today, sociologists have come to see the film audience more as decoder of symbols and meanings, actively engaged with the film itself, and as "contextualizer," constantly interpreting images within the contexts of social life. This notion of the active audience lies as the foundation of current sociological interest in film—both sociological studies of popular culture as they have gained broader acceptance, and in the college courses on film and sociology that have become available to students. Some call the course "Sociology *of* Film." They do so because it is a class *of* film. It is likely the course will include topics such as filmmaking, film production, and distribution, as well as the content, form, and impact of film. Certainly when a class is *of* something, that *something* is the unit of analysis—gender, for example, in a sociology of gender class.

Others call the class "Sociology *Through* Film." In these classes, *film* is not the unit of analysis per se. Rather, film is used as a concrete means of illustrating principals in the study of social life. The goal of the course involves learning to recognize sociological concepts in our life experiences (in movies, rather than solely relying on textbooks). We are concerned with the social construction of reality (what do different films suggest about the meaning of gender, race, and social class status for people in society today?). How does the "language" of film produce different meanings? We are concerned with the repetition of images in our culture (how is a group or a political system presented to us and does this change over time?). We consider the socializing effects of film (how and what do films teach us?). Our goal is to "see" movies differently.

We see this of/through distinction in studies of sociology and literature. Coser's (1963) text was called *Sociology Through Literature,* not *of* literature. Coser compiled an impressive (and abundant) collection of excerpted literature from Shakespeare to Norman Mailer, arguing that literature was yet another source from which we gain knowledge about society and, hence, ourselves. He said, "if a novel, a play, or a poem is a personal and direct impression of social life, the sociologist should respond to it with the same openness and willingness to learn that he [*sic*] displays when he interviews a respondent, observes a community, or classifies and analyzes data" (p. xvi). This was quite a bold stance during a time when sociology tended toward more "serious" matters of scientific study (and typically relied on functionalist theory). What is of importance here is that Coser set out to explore literary works, not for purposes of literary analysis, but as data to be analyzed about social experiences for sociological inquiry. He wrote:

> This book is not meant to be a contribution to the sociology of literature. Sociology of literature is a specialized area of study which focuses upon the relation between a work of art, its public, and the social in which it is produced and received. . . . The attempt here is to use the work of literature for an understanding of society, rather than to illuminate artistic production by reference to the society in which it arose. . . . This collection, then, should help to teach modern sociology through illustrative material from literature. (p. xvii)

Coser then went on to "teach" such substantive sociological areas as stratification (through George Orwell's "The Lower Classes Smell," from *The Road to Wigan Pier*); sex roles (Virginia Woolf's *A Room of One's Own*); race relations (Mark Twain's "Huck Breaks the White Code," from *Huckleberry Finn*); poverty (Charles Dickens's "I Want Some More," from *Oliver Twist*); and family (Lev Tolstoi's "The Perils and Dilemmas of Marital Choice" from *Anna Karenina*).

This book follows the model set forth by Coser. Thus, we consciously call it sociology *through* film. This book will teach you about the field of sociology, using film as the source. Thus far we have delineated the sociological perspective used in the analysis of film. But how exactly do sociologists do it? How do we decode a film as text? What tools do we have available to us?

The Sociological "Toolkit"

Andrew: Everyone is born with blinders on knowing only that one station in life to which they are born. You, on the other hand, Madam have had the rare privilege of removing your bonds for just a spell to see life from an entirely different perspective. How you choose to use that information is entirely up to you.

Overboard (1987)

The sociological viewing, analysis, and interpretation of film require an understanding of the relationship between historical context, social structure, and individual experience (Shiel, 2001). In Mills's (1959) terms, the sociological imagination enables us to grasp history and biography, and the relations between the two. Engaging our sociological imaginations allows us to see how those events that seem extraordinarily personal are in fact produced and shaped by social forces.

Take, for example, divorce. This life event seems to those involved as the utmost private and personal decision that two individuals make, based on their personal lives. But as sociologists, we note trends in divorce (e.g., divorce rates rise when the economy is bad; those with lower incomes are more likely to divorce; those from volatile homes are more likely to divorce). Thus, divorces *are* personal and individual, but they also occur in societies with structured social orders, at a particular moment in history, with different types of men and women inhabiting this place and time. Consequently, when we view *Kramer vs. Kramer*, a 1979 film about a couple who separate, leaving the dad to learn child care and housework until the working mom returns and the custody battle begins, we must place it into context. Rather than viewing this as one couple's struggle with divorce, or even as *representing* divorce, our sociological imaginations allow us to see the impact of the historical context of the 1970s: the second wave of feminism was barely a decade old, so how might gender politics have impacted the telling of this story? We also consider social structure: how were "working moms" and "caregiving dads" represented considering "appropriate" gender roles at the time? And individual experience: what did the characters' emotional states and priorities reveal about divorce and child custody?

One of the benefits of studying sociology is the "toolkit" it provides for analyzing and understanding the social world. In this section we present theoretical perspectives and methodological approaches that can be used as a frame for "reading" films sociologically. As we have noted, film can be thought of as text, in the sense that it provides material (data) subject to observation and analysis.

Theoretical Perspectives

The theoretical perspective you use depends on the types of questions you intend to ask. If your questions have to do with issues of representation (the "meanings" a film seems to provide) such as socialization, construction of reality, or interaction, *symbolic interactionism* would provide the framing necessary. If you intend to explore power relations, including political power, patriarchy, or racial dominance, then a *conflict perspective* would be more appropriate.

In the symbolic interactionist or constructionist traditions, knowledge, truth, and reality are determined by "the context in which they are practiced" (O'Brien, 2006, p. 9). We engage in an interpretive process in which schemas and culturally specific "commonsense" realities help us to arrive at meanings. Helping to shape those meanings are the values, norms, and ideologies of a culture that inform what is "real."

Constructionists remind us that race, ethnicity, gender, sexuality, and social class have no meaning until we give them meaning. Watching film and engaging with the content is part of the interpretive process in which repetitive images take on meaning and become "real," often turning into "commonsense knowledge." Through films we are told what it means to be man, woman, gay, lesbian, black, white, Asian, Latino, middle or working class. Whether the images and information are "objectively" real is all but irrelevant in terms of the consequences. Mediated images and information become part of the material that supports or contradicts our notions of what is "real" in the social world.

Using a conflict perspective reminds us that film is not a documented recording of social events or alternative worlds; film is carefully crafted and produced within industrial and economic relations of power (Giroux, 2002). A continuous critical approach (Lehman & Luhr, 2003) allows for a view of film that shifts the center (Collins, 2000) from dominant discourse to the matrix of domination along the axes of race, class, gender, sexuality, and geopolitical location. These interlocking systems operate from the individual to the social structural level. Critical theories provide the best vehicle for this approach. Feminist theory allows you to ask about prevailing attitudes and assumptions concerning gender, as well as the structure of patriarchy. Critical race theory and critical white studies frame questions concerning racism, racial discrimination, and the acceptance of whiteness as the norm. Marxist theory is concerned with issues of power specific to capitalism, exploited workers, and class divisions. Marxist theory can also address power in the making of movies; who has the power to (re)produce ideologies? Using Pierre Bourdieu (1977) would allow an analysis of power via cultural capital; how do tastes and knowledge of culture "locate" people in social classes?

Methods

There are many methodological approaches to the study of film. We recommend the two processes outlined below.

The first approach is based on the process for conducting research, revised for film analysis.[4]

---

[4]Bulman, R. (2007). Handout presented in Teaching Sociology Through Film session. American Sociological Association, New York.

1. Develop a Research Question
    a. What is the sociological question motivating the research?
    b. What is significant about your question?
    c. What is your argument?
2. Literature Review
    a. How have others approached this question?
        i. Keep in mind, others may or may not have used film in addressing this question.
    b. What were the contributions regarding those findings?
    c. What theories and methods have been used?
    d. What questions are left to be asked?
    e. Did reading this literature shape the way you are approaching your question?
    f. Is it important that you ask your question in another way?
3. Methods and Data
    a. How did you select your sample of films?
    b. How did you collect data?
        i. How did you watch the film(s)?
        ii. How did you take notes while watching the film?
        iii. Did you code the data, looking for patterns and trends?
        iv. What did you do with your notes when you were finished watching the films?
        v. Did these data help you to answer your question?
4. Analysis of Data
    a. What are your findings?
    b. What do the data tell you regarding your research question?
    c. Use your data to make your argument—convince the reader.
    d. Relate your findings back to the literature review—is this something new?
5. Conclusion
    a. Sum up your findings.
    b. What is the significance of this work?
    c. Are there implications for future research?

Using this model has several advantages. First, it does not deviate from traditional research models. Thus, there may be familiarity with the structure of the project. Second, it specifically guides not only the kinds of questions one might ask, but also the order in which to address them.

The second approach involves a five-step process to analyzing film drawn from the work of Norman Denzin (1989). The goal is to help students orient to the content of film in the research role of nonparticipant observer (Tan & Ko, 2004). In addition to honing skills of observation, this process incorporates an ethnographic tool called "thick description" (Geertz, 1973). The key to thick, as opposed to thin, description is that even the simplest act can mean different things depending on the social and cultural context. Thus, it is the responsibility of the researcher to report not just actions, but the context of the practices and discourse within the society being observed. This means reading the text of the film at multiple levels, attending to implied as well as explicit content. The guiding question is how the film creates multiple meanings through language, action, and "what-goes-without-saying" (Barthes, cited in Denzin, 1989). This process should help you to "bracket" taken-for-granted assumptions about the social world in order to observe first and then analyze the content of the film.

1. Select the film and view it multiple times. While there are disadvantages to viewing a film outside the context for which it was made (e.g., large screen, particular sound system), one benefit to viewing a film at home is the ability to stop, rewind, and review throughout the process.
2. Outline the narrative themes of the film. Themes are patterns or constructs that, in qualitative research, are induced from texts in a process of open coding. Two strategies are worth mentioning here: the "social science query" and the "search for missing information" (Ryan & Bernard, 2003). In the first, the text is examined for topics important to social scientists, such as social conflict, cultural contradictions, methods of social control, and setting and context. In the second, the attention is on what is not represented, including the topics of this text, such as class, culture, race and ethnicity, and sex and gender.
3. Conduct a realist reading of the hegemonic interpretations of the text in terms of their dominant ideological meanings. This step involves a close reading of the films' characters, content, and dialogue. What universal features of the human condition are addressed? What are the stories told that reproduce existing relations of power and inequity?
4. Develop a subversive or oppositional reading of the text such that the taken-for-granted hegemony is revealed by making visible and explicit the connections between particular lives and social organization (Ewick & Silbey, 1995). Shifting away from characters, content, and dialogue, this analysis "focuses instead on how the film creates its meanings through the organization of signifying practices that organize the film's reality" (Denzin, 1989, p. 46). In this step, it is critical to attend to the individual-as-subject bias that structures much of film ideology, in other words to apply the sociological imagination.
5. Compare the realist/hegemonic and oppositional readings of the film. In this step the analysis is further developed as the results of the two previous steps are evaluated in relationship to one another. What is the story being told through the film and how is it supported explicitly in how the characters are portrayed, what they say (and how they say it), and the events that construct a story line? How does this compare to the "underlying ideological forces of the film" (Denzin, 1989, p. 46)?

In This Book

Atticus Finch: If you just learn a single trick, Scout, you'll get along a lot better with all kinds of folks. You never really understand a person until you consider things from his point of view . . . until you climb inside of his skin and walk around in it.

To Kill a Mockingbird (1962)

As Atticus Finch (Gregory Peck) explained to his daughter, the best way to understand others is to suspend what you think you know and to see the world through their eyes. In sociology we call this the practice of *verstehen,* the term Weber used to refer to the social scientist's attempt to understand both the intention and the context of human action and interaction (Munch, 1975). In this chapter, we have provided you with a toolkit (theory and method) and a data source (film) to put sociology into practice. Films can be effectively used in this way because they (1) give viewers access to social worlds beyond their own; and (2) create the opportunity to understand the relationship between individual experience and "the broader social context which structures one's actions and choices" (Prendergast, 1986, p. 243).

Chapter 2 addresses social class and inequality. Chapter 3 addresses race and ethnicity, and Chapter 4 covers gender and sexuality. While the readings in these chapters focus on social class, race/ethnicity, or gender and sexuality, these are not separate dimensions in the lives of individuals, or in larger social structures; they are intersecting systems of oppression (Collins, 2000). Individuals have intersecting identities and locations; social organizations and institutions are structured in terms of intersecting inequalities. However, there are many methodological challenges to observing and analyzing intersectionality.

The readings in these three chapters use a "relational model" (Lucal, 1996) that includes both oppression and privilege, those who benefit (directly or indirectly), as well as those who are disadvantaged by inequality. In Chapter 2, Michael Kimmel discusses the central themes of classical sociological theories of class as they appear in film, including the centrality of class and class conflict, the importance of culture in shaping the social and personal experience of class dynamics, the influence of ethnicity and gender on class relations, and the effects of rationalization into what Weber described as the "iron cage." James Dowd introduces the concept of social mobility and cultural loyalty to the idea that upward mobility is not only possible, but certain if individuals are motivated and industrious. Robert Bulman directs our attention to the tension between middle-class perspectives on education and upward mobility, and the realities of living and going to school in poor urban areas.

In Chapter 3, Carleen Basler points out stereotypes of Latinos (in film and life) that are used by the dominant group to exclude or marginalize the "Other." Thus, the dominant group's privilege and power are dependent on successfully "othering." This process operates in "trickle-down" fashion, as Ed Guerrero illustrates in his discussion of the film *Rosewood.* The film ends with the beating of a lower-class white woman by her husband, offering resolution (payback?) for genocide by displacing the responsibility on what Guerrero calls "yet another Hollywood outgroup, disenfranchised women." Both Basler and Guerrero claim that one hope for change lies in shifting the source of the story from the dominant group, who historically has been seen as better able to tell the story of the Other. However, the knowledge produced from a privileged perspective cannot encompass the stories of those at the margins, excluded from film writing, filmmaking, and often even acting (as in films where people of color are played by white actors in blackface). Basler notes a gradual change in film images, attributing it to the presence of Latino/a filmmakers in the industry. Guerrero describes the need for "wave after wave of new black filmmakers" with diverse experiences and standpoints. For example, gender differences in the filmmaking of black women and black men are apparent in the focus on character and drama versus action and violence.

Susan Searls Giroux and Henry Giroux go further than opening the field of film production to "others." They argue that the fight against racism on the screen and in the streets requires the commitment of all citizens to "critically engage and eliminate the conditions" that both produce structural racism and prevent true democracy. Further, the political arena is not the only site of action, since it is the responsibility of all citizens to create alliances across perceived racial differences in the community settings where they meet (and sometimes collide).

In Chapter 4, the readings on gender and sexuality provide new frameworks for thinking about power, patriarchy, and the social construction of gender and sexuality. In fact, the sex/gender/sexuality binary based on a two-category model where people are defined as male (masculine) and female (feminine) is called into question. As Betsy Lucal and Andrea Miller point out, the result of enforced categories (e.g., homosexual, woman) is the blinding erasure of identity where people's experiences, history, interactions, and emotions are collapsed into narrowly scripted roles. Bree (Felicity Huffman), in *Transamerica,* is offered as a sign of hope in that she "queers the gender binary," not by going through sex reassignment surgery, but by "comfortably mixing masculinity and femininity."

Bree's story is one of personal transformation: finding her place in a binary social world. She is persistent in reaching her goal of completing the transition from male to female, and in the process discovers the power she has to create change. Jean-Anne Sutherland tells us that this would be the "power-to" model of social change for women. Like Bree, Ana in *Real Women Have Curves* and Celie in the *Color Purple* become increasingly aware of their objective conditions, as well as ways to get free. However, unless this form of empowerment is linked to collective action ("power-with"), gender as structure remains intact and unchanged. The real challenge to patriarchy and gender inequality is represented in films like *North Country* where the oppressed come to see their circumstances as shared.

As Josie (Charlize Theron) explained, sexual harassment is not a personal problem, and as a social issue it can only be effectively challenged by collective action. In a similar vein, Michael Messner calls for collective action based on an ideal not of "power-over" where might is the means to achieving desired ends, but rather care and compassion. He challenges the masculine hegemonic, embodied in the film roles of Schwarzenegger (as the Terminator, for example) and Stallone (as Rambo).

In Chapter 5, we turn our attention to the social institutions of work and family. Drawing from Bourdieu (1977), we can understand social context in terms of the structured social relations between individuals, groups, and social institutions. Accordingly, social structure is embedded in everyday events that take place in social spaces. Social spaces in western culture are thought of as private or public domains of social action. The activities of the marketplace, as Karla Erickson points out, take place in public. She encourages us to ask how film versions of work compare to our own experiences as workers and consumers.

Similarly, Janet Cosbey suggests that we can use films about family life to think about our private lives and to understand ourselves and others better. Perhaps more important, as social observers we can view films about families, comparing hegemonic and subversive readings of the text. In *Sleeping With the Enemy,* Laura (Julia Roberts) is able, through her own resourcefulness and personal agency, to escape a life-threatening abusive marriage and find true love in another town. We might wonder, sociologically, how Laura can accomplish this without any social support. If we watch another film with the same story line, *Enough,* Slim (Jennifer Lopez) fights her way to freedom from abuse in equally spectacular ways. However, Slim receives emotional and material support from a close network of friends who are with her throughout her ordeal. Turning back to Sutherland's article, we might argue that these two films represent a power-to rather than a power-with model of social change. Where in the film world is the battered women's movement that wrought significant social changes in the latter half of the 20th century, establishing safe houses and battered women's shelters, influencing legislative changes in defining and criminalizing marital rape, and helping to shape national policies on violence against women?

In Chapter 6, the readings address the global context of work, immigration, race/ethnicity, and intergroup conflict. Terrorism in the 21st century is a global reality and L. Susan Williams and Travis Linnemann use films made before and after the terrorist attacks on the United States on September 11, 2001, to explore the "scripting" of an enemy. Tying together the themes of racism, xenophobia, and the social construction of "otherness," the authors argue that Arabs have become the scapegoat for our generalized fears of external threat. In Chapter 3, Searls Giroux and Giroux suggest that individual racist acts are made obvious in the film *Crash,* while institutionalized racism is rendered invisible. In films focusing on Arabs, we find that they are categorically presented as "evil" with only treatment of Native Americans rivaling the vilification of Middle Easterners in Hollywood (Sheehan, 2001). With the designation of enemy status, the question of individual versus institutional racism is muted, along with the historical context of anti-Arab sentiment in the United States.

The theme of intergroup relations is continued in Chapter 6 by Roberto Gonzales as he turns to low-wage migrant workers in the global economy. Gonzales argues that the new economies of the global market have created a new "serving class" to meet the needs of a time-strapped professional class. Immigrant workers fill this need as a flexible, low-wage source of labor. From a functionalist perspective, immigrant labor keeps society and economy running smoothly; conflict theorists would argue that they are the exploited labor. In *Pretty Dirty Things*, the effect of complex global economic systems on the individual lives of immigrant workers is made apparent. The intersection of the social institutions of economy (labor needs) and politics (regulatory power of the law) in the lives of individuals is seen in the ways that immigrant workers (barely) survive day-to-day.

The topics of Chapter 7 are social change and the environment. Both articles in this chapter are concerned with social change rooted in social relations (between social actors) and involving social structure, the environment, tradition, and the anticipated future (Sztompka, 1994). Both articles consider questions related to the survival of humanity, be it intergroup violence or destruction of the environment. What are the alternatives? Kathryn Feltey explores film stories involving nonviolence, both as an organizing principle in social movement activism (India's campaign for home rule; the civil rights movement in the southern United States) and in a *gemeinschaft* subculture (the Amish).

Christopher Podeschi emphasizes that nature and society's relationship with nature are socially constructed, and therefore can be changed. He provides a vehicle for exploring potential future scenarios in the form of science fiction films. Both articles are ultimately concerned with questions of power—and leave us with important questions to consider about the future: will we create a social world where domination and exploitation of people and nature are the status quo, or will we engage in "progressive self-transformation" (Sztompka, 1994) of ourselves, communities, and societies?

In sum, *Cinematic Sociology: Social Life in Film* provides a way to explore our social world and the lives of those who share it with us. It is our belief that sociology offers the tools to interrogate and change the conditions that produce discord in relationships—individual, communal, national, and global. It is our hope that by turning the sociological lens on the movies we watch, we will be better equipped to contribute to these changes and together "dream better futures" (Feagin, 2001, p. 17) and create a world in which happy endings are not found only in the movies.

References

Abel, R. (2004). History can work for you, if you know how to use it. *Cinema Journal, 44*, 107–112.

Bourdieu, P. (1977). *Outline of a theory of practice.* Cambridge, UK: University of Cambridge Press.

Cieply, M., & Barnes, B. (2009, February 28). In downturn, Americans flock to the movies. *New York Times*, p. A1. Retrieved March 8, 2009, from http://www.nytimes.com/2009/03/01/ movies/01films.html

Collins, P. H. (2000). *Black feminist thought: Knowledge, consciousness, and the politics of empowerment.* New York: Routledge.

Coser, L. (1963). *Sociology through literature: An introductory reader.* New Jersey: Prentice Hall.

Denzin, N. K. (1989). Reading Tender Mercies: Two interpretations. *The Sociological Quarterly, 30*(1), 37–57.

Ewick, P., & Silbey, S. (1995). Subversive stories and hegemonic tales: Toward a sociology of narrative. *Law & Society, 29*, 197–226.

Feagin, J. (2001). Social justice and sociology: Agendas for the twenty-first century. *American Sociological Review, 66,* 1–20.

Geertz, C. C. (1973). *The interpretation of cultures: Selected essays.* New York: Basic Books.

Giroux, H. (2002). *Breaking in to the movies.* Malden, MA: Blackwell.

Griswold, W. (1993). Recent moves in the sociology of literature. *Annual Review of Sociology, 19,* 455–467.

Hearne, J. (2007). Race and ethnicity. In B. K. Grant, J. Staiger, J. Hillier, & D. Desser (Eds.), *Schirmer encyclopedia of film,* Vol. 3 (pp. 369–378). Farmington Hills, MI: Thompson Gale.

Lehman, P., & Luhr, W. (2003). *Thinking about movies: Watching, questioning, enjoying* (2nd ed.). Malden, MA: Blackwell.

Library of Congress. (n.d.). *Making Americans. Program notes.* Retrieved November 19, 2008, from http://www.alamotheatre.org/nhfWeb/alamotheatre/MakingAmericans.htm#Introduction

Lucal, B. (1996). Oppression and privilege: Toward a relational conceptualization of race. *Teaching Sociology, 24,* 245–255.

Mills, C. W. (1959). *The sociological imagination.* New York: Oxford University Press.

Montoya, M. (2001). A brief history of Chicana/o school segregation: One rationale for affirmative action. *La Raza Law Journal, 12,* 159–172.

Munch, P. A. (1975). "Sense" and "intention" in Max Weber's theory of social action. *Sociological Inquiry, 45,* 59–65.

O'Brien, J. (2006). *The production of reality* (4th ed.). Thousand Oaks, CA: Pine Forge Press.

Oppenheimer, R. (1985). Acculturation or assimilation: Mexican immigrants in Kansas, 1900 to World War II. *The Western Historical Quarterly, 16,* 429–448.

Pew Research Center. (2006). *Increasingly, Americans prefer going to the movies at home.* Retrieved April 20, 2009, from http://pewresearch.org/assets/social/pdf/Movies.pdf

Prendergast, C. (1986). Cinema sociology: Cultivating the sociological imagination through popular film. *Teaching Sociology, 14*(3), 243–248.

Ross, S. (1998). *Working-class Hollywood: Silent film and the shaping of class in America.* Princeton, NJ: Princeton University Press.

Ryan, G. W., & Bernard, H. R. (2003). Techniques to identify themes. *Field Methods, 15,* 85–109.

Sheehan, J. (2001). *Reel bad Arabs: How Hollywood vilifies a people.* Portland, OR: Interlink.

Shiel, M. (2001). Cinema and the city in history and theory. In M. Shiel & T. Fitzmaurice (Eds.), *Cinema and the city: Film and urban societies in a global context.* Oxford, UK: Blackwell.

Starker, S. (1989). *Evil influences: Crusades against the mass media.* New Brunswick: Transaction.

Sztompka, P. (1994). *The sociology of social change.* Cambridge, MA: Blackwell.

Tan, J., & Ko, Y. (2004). Using feature films to teach observation in undergraduate research methods. *Teaching Sociology, 32,* 109–118.

CHAPTER 2

Social Class and Inequality

Professor Higgins:	You mean, you'd sell your daughter. . . . Have you no morals, man?
Alfred Doolittle (Eliza's father):	No, I can't afford 'em, Governor. Neither could you if you was as poor as me. Look at it my way. What am I? I ask ya, what am I? I'm one o' the undeserving poor, that's what I am. Think what that means to a man. It means he's up against middle-class morality for all the time.

My Fair Lady (1964)

Joe Kenehan:	You think this man is the enemy? Huh? This is a worker! Any union keeps this man out ain't a union, it's a goddam club! They got you fightin' white against colored, native against foreign, hollow against hollow, when you know there ain't but two sides in this world—them that work and them that don't. You work, they don't. That's all you get to know about the enemy.

Matewan (1987)

In the classic film *My Fair Lady,* Professor Higgins (Rex Harrison) tests the hypothesis that social class position can be learned, particularly through the practice of "proper" English and comportment. To prove his point, he takes on the challenge of transforming Eliza Doolittle (Audrey Hepburn) into a lady. Eliza's father (Stanley Holloway) shows up, shocking Higgins by asking for compensation for the use of his daughter, explaining that poor people cannot afford morality, it is a luxury of the middle class (who use it as a standard for the lower class). Ultimately, Higgins is successful in his endeavor, and in the end Eliza transcends her social class when the two fall in love.

The movie *Pretty Woman* (1990) is a modern version of this fairy tale. In this film, prostitute Vivian Ward (Julia Roberts) and wealthy businessman Edward

Lewis (Richard Gere) fall in love and Vivian leaves the streets for the life of an upper-class wife-of-a-businessman. On the way, Vivian is transformed by a new wardrobe, a makeover, learning appropriate dinner manners, and Edward's appreciation for her frank honesty and basic goodness.

Many movie plots include the rags-to-riches theme for both men and women, though men usually ascend through a combination of hard work, luck, and ingenuity, while women marry upward to exit the lower class. Celebration of the American Dream, based on the values of individualism, success, and hard work, is a constant in American film, reflecting our belief in an open class system and reinforcing attitudes that failure is only possible in the absence of effort or desire.

Understanding and explaining social inequality is central to the discipline of sociology. Social inequality refers to differential access to and distribution of resources and rewards in society based on statuses such as social class, race, sex, and age, and the intersection of these statuses in people's lives. In this chapter, the focus is on *social class.* How sociologists define the concept of social class varies, depending on the theoretical perspective and the questions asked, but the overarching concern is with systems of economic inequality.

In the first reading, Michael Kimmel begins with the rise of the market and the social contract, using the movie *Burn,* to illustrate what capitalism creates and what it destroys in the process. This is followed by a consideration of the works of Karl Marx and Max Weber, which have shaped the way sociologists think about and study class to the present day. Kimmel uses the epic masterpiece *1900* to illustrate a Marxist perspective on the changing relationship between the ruling and working class in the transition from feudalism to capitalism.

To Marx, social class, and in particular class conflict, is the engine that drives historical development. The conflict between classes is focused on control and use of the means of production with struggle between those who labor for a wage and those who derive benefit from the labor of others (Giddens, 1971). As Joe Kenehan (Chris Cooper), a labor organizer for the United Mine Workers in the movie *Matewan,* explains to the miners and the would-be African American and Italian immigrant scabs brought in by the mine company to break the strike, "there ain't but two sides in this world—them that work and them that don't."

While the economic order was critical in the work of Weber, he also theorized that structures such as religion and bureaucracy were influential in shaping class dynamics (Hadden, 1997). Kimmel uses the film *Babette's Feast* to explore the meaning of the Protestant ethic and cultures of self-denial as foundational to economic and social systems under capitalism. He then turns to the process of rationalization, which, for Weber, was the most important characteristic of the development of western society and capitalism. Drawing on the movies *The Godfather* and *The Godfather: Part II,* Kimmel follows the Corleone family and the transition from traditional authority to the rationalized "iron cage" in which power is concentrated at the top, but not without the loss of meaning, connection, and family.

In the second reading, James Dowd uses three films, *Maid in Manhattan, Lady and the Tramp,* and *Good Will Hunting,* to explore the American Dream of upward social mobility. The cultural belief that individual motivation and effort is the key to economic success (and, conversely, economic failure in the absence of motivation and effort) in American society is one that is widely held by white Americans, and is increasingly embraced by African and Hispanic Americans (Hunt, 2007). This belief is reflected in and reinforced by the movies we watch in childhood (*Lady and the Tramp*) and adulthood (*Maid in Manhattan* and *Good Will Hunting*). Interestingly, this belief may be waning; in a 2006 opinion poll, over half of the Americans surveyed said the American Dream is no longer attainable for the majority of their fellow citizens (Sawhill & Morton, 2007).

In the third reading, Robert Bulman explores social class and education, focusing on the experiences of poor students in urban schools, in contrast to middle-class students in private or public suburban schools. According to sociologist Adam Gamoran (2001), students from privileged backgrounds have more success in school and can translate their education into economic success in the marketplace. In this way, economic, cultural, and social differences combine to preserve the position of the privileged from one generation to the next. Americans tend to think of education as the means for achieving upward mobility and public education as the mechanism for creating a level playing field. However, economic inequality, racial segregation, and inequalities created by inherited wealth result in public schools that are separate and unequal, a direct contradiction to the ideology of the American Dream (Johnson, 2006).

Bulman argues that the Hollywood version of urban high schools reveals middle-class values based on rationality and individual achievement. This is exemplified by the idea that poor students have choices, a point made by Louanne Johnson (Michelle Pfieffer) in *Dangerous Minds*—they just need to make the right ones to be successful. The formula for the central story in the films Bulman analyzes is that an urban school plagued by poverty, drugs, gangs, violence, and a rejection of middle-class values is saved by a lone teacher or principal who single-handedly rescues the students from a life of crime or early death. Ignoring the structural inequalities that have created the urban school, these films are a celebration of individualism, might (often expressed through violence) as right, and conformity.

Exploring social class, the readings in this chapter invite you to consider the ways that inequality is structured economically, socially, and politically. As sociologists, we examine social class in terms of historical trends, global politics, and ideologies. We are particularly interested in the effect of economic inequality on life chances, including education and material well-being. Going to the movies is one way to learn about cultural constructions of social class across the years.

References

Gamraon, A. (2001). American schooling and educational inequality: Forecast for the 21st century. *Sociology of Education,* Extra Issue, 135–153.

Giddens, A. (1971). *Capitalism and modern social theory: An analysis of the writings of Marx, Durkheim and Max Weber.* Cambridge, UK: Cambridge University Press.

Hadden, R. W. (1997). *Sociological theory: An introduction to the classical tradition.* Peterborough, ON: Broadview Press.

Hunt, M. O. (2007). African American, Hispanic, and white beliefs about black/white inequality, 1977–2004. *American Sociological Review, 72*(June), 390–415.

Johnson, H. B. (2006). *The American dream and the power of wealth: Choosing schools and inheriting inequality in the land of opportunity.* New York: Routledge.

Sawhill, I. V., & Morton, J. E. (2007). *Economic mobility: Is the American dream alive and well?* Washington, DC: Economic Mobility Project, The Pew Charitable Trusts.

Reading 2.1

Sitting in the Dark With Max

Classical Sociological Theory Through Film

Michael Kimmel[1]

There's an old British joke that goes something like this:

Two Oxford professors, a physicist and a sociologist, were walking across a leafy college green. "I say, old chap," said the physicist, "What exactly do you teach in that sociology course of yours?"

"Well," replied the sociologist, "This week we're discussing the persistence of the class structure in America."

"I didn't even know they *had* a class structure in America," said the physicist.

The sociologist smiled. "How do you think it persists?"

While class—identity, structure, and inequality—is a core concept in sociology, few concepts prove more elusive. Ours is a doggedly psychological culture, in which structural problems are perceived as aggregated individual problems to be solved at the level of psychological motivation and individual good deeds. Thus poverty is not a social problem requiring structural solutions but a bunch of poor people who could get out of poverty if only they worked harder.

The entertainment industry typically is not much help in this matter. In fact, it is sort of the problem. The classic Hollywood movie, after all, proposes an individual heroic solution to virtually any problem. Under attack by aliens? Call Bruce Willis or Tom Cruise. Bad guys bent on destroying the world? Call Harrison Ford or Bruce Willis. Mean outlaws riding into town? Call Clint Eastwood (and probably Bruce Willis). Social problems, we learn, are caused by bad people; eliminate them and the problems go away, the hero gets the girl, and they ride into the sunset (except for John Wayne; he is so mythic a hero that he often eschews romantic entanglement with his "reward" and rides off alone, the last real man in America). We live in a culture founded not on the recognition of the "power of the social," in Durkheim's famous phrase, but rather on its denial.

But social scientists also have, for decades, drawn on cultural expression—art, music, literature, film—to illustrate the sociological. For example, in his

[1]The origins of this essay lie in graduate school conversations with Jeff Weintraub and Jerry Himmelstein about how to teach classical theory to undergraduates. Its birth, though, was stimulated entirely by conversations with Jean-Anne Sutherland. I'm grateful to her and Kathy Feltey for the opportunity to write it down, and for their judicious editing.

path-breaking look at art, *Ways of Seeing,* John Berger (1972) uses a well-known painting "Mr. and Mrs. Andrews" by Thomas Gainsborough to illustrate the transformation of class relations in Europe in the 18th century and to chart the rise of the bourgeoisie. First, he shows us a painting of a 16th-century landscape—bucolic, rustic, pastoral. Rolling meadows, a cow grazing placidly, perhaps a church in the background. It's the "before" image. By the 18th century, though, that landscape painting has changed; Gainsborough puts an English squire on a bench, his wife next to him, and a hunting rifle across his lap. "They are not a couple in Nature as Rousseau imagined nature," Berger writes (1972, p. 107). "They are landowners and their proprietary attitude towards what surrounds them is visible in their stance and their expressions." What Berger exposes is the transformation of the bucolic countryside when it changes into "property." "You see this land behind me," the squire seems to be saying. "I own it. It's mine." (And, given the gender politics of 18th-century England, he's also saying that about his "trophy wife.") Our view of the countryside changes as the countryside itself changes.

There is a long tradition of using other sorts of cultural texts—literature and music, as well as art—to illustrate these classical sociological themes. Examining the patronage system in the 18th-century Habsburg Empire sheds light not only on what subjects Mozart could, and could not, compose about, but also the sorts of musical techniques and instruments he could use. Ian Watt (1959) argued the rise of the novel is coincident with the rise of capitalism. The novel is the literary form of the rising bourgeoisie; with its disconnected individual hero seeking to make his fortune in a competitive and cynical world, the novel captures that decontextualized individual's struggle against the constraints of traditional society. For example, the early novels of Richardson, Defoe, or Fielding describe how upwardly mobile individuals like Tom Jones or Barry Lyndon calculate their next move up—only, nearly inevitably, to fall back where they came from, where they morally belong.

In the 19th century, the novel also became a weapon of the working class, as in Dickens's *Hard Times,* which imagined the noble, long-suffering workers graining under the (Grad) grinding weight of bourgeois regimentation and exploitation. Reading that work next to Engels's contemporaneous *Conditions of the Working Class in England* gives one a palpable sense of the concreteness of class relations in a way that theoretical texts alone cannot often convey.

Film can also convey the themes of classical sociological theory. In this essay, I examine four films that suggest different core themes addressed by classical theorists:

1. The rise of the market, the historical emergence of the individual, and consent of the governed as the contractual basis of the state (presociological contract theorists);
2. The centrality of class and class conflict as the dynamic motor that drives historical development (Marx);
3. The importance of culture, and especially religion, in shaping the social and personal experience of class dynamics (Weber); and
4. The influence of ethnicity and gender on class relations, and the effects of rationalization into the iron cage (Weber).

Burn and the Rise of Capitalism

One of the precursors to classical sociological theory was 17th-century British empiricism and contract theory. (The other was continental romanticism, as in Rousseau or Goethe.) Hobbes and Locke reasoned that society is not a social fact, as Durkheim said, *sui generis,* but rather it is formed through individuals coming together by rational contract to protect life (Hobbes) or property (Locke).

In his major work, *Leviathan* (1659), Hobbes provides a vigorous defense of absolute monarchy. In the state of nature, Hobbes argued, everyone is free and equal, which he saw as a terrifying prospect. Each of us is interested in furthering our own interests against everyone else. If we're all free and equal, and we all want what everyone else has, then life would be very unstable and insecure. Hobbes believed that in our natural state, life is a "war of each against all" and our lives would be "solitary, poore, nasty, brutish, and short" (in Kimmel, 2007, p. 7).

Fortunately, Hobbes believed, human beings also possess reason, which enables them to create society. Since the natural state of nature is that of war, the first move to build society must be to keep us safe and secure. That is the sole purpose of society—to ensure that war does not doom us all. No one can be trusted to refrain from harming others when it might benefit him or her to do so. We are thus happy to give up our individual power to a higher authority, the state (or *leviathan,* the biblical term for the whale that swallowed Jonah). As long as we all give up our individual freedom, we can be safe and secure. This is the social contract—each individual agrees to give up his or her individual power. Only when the state has all the legal and military power can individuals feel safe. Anything less makes government unstable and social life impossible.

Locke had a similar theory of the state of nature and the social contract, but came to rather different conclusions. Like Hobbes, Locke believed that human beings existed first in a state of nature. But his view was that nature was relatively harmless, not a war of each against all. Nature was abundant, but nothing could be guaranteed from generation to generation. Locke believed that we enter a social contract to secure property. In his version of the social contract, people remain free and just surrender the right to enforce laws. Government is limited; it exists only to resolve disagreements among individuals. If the government goes too far, Locke believed, and becomes the sort of omnipotent state that Hobbes advocated, the people have a right to revolution and to institute a new government. Locke's *Two Treatises on Government* served as the guiding document for Thomas Jefferson as he drafted the Declaration of Independence, which announced the American Revolution in 1776.

Adam Smith's sociology also deserves mention as a founding figure of the presociological enlightenment era. His book, *The Wealth of Nations* (1776), is the most influential economics book ever written, but its importance for sociology is profound. Smith set forth the doctrines of liberal capitalism and the ideology of laissez-faire (French for "leave alone") or minimal government. (Laissez-faire meant that the government should not interfere with the marketplace.) For Smith, competition in the marketplace of each industry against others was the source of prosperity. Smith's most original idea was that society as much as the individual is a system, or machine, whose workings are not the product of human intentions but of the unconscious "invisible hand." Contrary to those who today see Smith as an advocate of completely unrestrained capitalist greed, he saw individual appetites as naturally limited—why would people want more than they could possibly use, he asked? He also saw government as the active provider of public works, and he suggested, "It is not very unreasonable that the rich should contribute to the public expense, not only in proportion to their revenue, but something more than in that proportion."

For all British presociological contract theorists, the move from the state of nature to society is a rational move. It is a morally positive move as well, preserving individual freedom and property. By contrast, Rousseau's contract theory is quite a departure. In a state of nature, he argued in *Discourse on the Origin of Inequality* (1754), people are naturally good and innocent, distinguished from other animals only by capacities for self-improvement and for compassion or sympathy. But private property causes inequality, and with that, unhappiness and immorality. "Man was born free and everywhere he is in chains," he wrote in *The Social Contract.* Our goal, then, is to try and recover the freedom and happiness of the state of nature. Through what Rousseau called the "general will," a kind of moral ethics that lives inside each person as well as in society as a whole, people become a community.

These two themes—the British emphasis on individual liberty and the French idea that society enhanced freedom—collide in the 1970 film *Burn,* directed by Gillo Pontecorvo. Pontecorvo takes up these themes as they are emerging in the 18th century; that is, at the time that Locke's and Rousseau's ideas are gaining sway throughout Europe—though the obscurity of the film today is based largely on its transparency as an allegory critical of American involvement in Vietnam.

The film follows the successful journey of a professional mercenary, Sir William Walker, played by Marlon Brando, who is dispatched by English merchants to the Caribbean island of Queimada, a Portuguese colony. The Portuguese control the island, and its lucrative sugar plantations, in a traditionally colonial way, treating indigenous laborers as slaves and extracting all profits for repatriation to Portugal. The lives of local workers are, to use Hobbes's memorable phrase, "nasty, brutish, and short."

Brando is sent to the island by British capitalists to see if he can foment a slave rebellion and drive the Portuguese from the island, thus opening it up for a more rational sugar trade—namely, theirs. And in this he is quickly successful. True to the historical record of British trade, Walker enlists the help of a charismatic slave laborer, José Dolores (Evaristo Marquez), to lead the rebellion, and the Portuguese are quickly driven out.

A decade passes. We next see Walker, now a disillusioned alcoholic, suffering from bouts of conscience about what he had created. He is again approached by those same merchants, now firmly ensconced in Parliament, because the rebellion Walker helped ignite has never really died. Once the slaves tasted freedom from direct colonial oppression, they are finding capitalist wage slavery less to their liking than the sugar magnates had hoped. The rebellion has continued, with José Dolores now the leader of a popular, nationalist, and anticapitalist insurrection. Walker is dispatched to quell the very insurrection he helped begin because it now threatens those trade interests with ideas of national liberation and independence.

Dissolute and cynical, Walker is dispassionately ruthless in its suppression, ultimately worse than the Portuguese.

Burn both explores and critiques the theories of presociological British philosophical ideas, especially Locke's contract theory and Smith's optimistic notions that the marketplace will tend toward morality. In both theories, disaggregated individuals pursue rationally calculated self-interest, and the result is a society that is more efficient and more moral. Both proclaim freedom and autonomy as the goal, rational calculation as the means to achieve it.

Pontecorvo doesn't believe it for a moment. Both Locke and Smith believed that the acquisitive appetite would be sated by "enough," and so neither could have predicted the avaricious amassing of stupendous fortunes by these rationally calculating individuals. The relentless rational pursuit of profit is, in his mind, a mental illness—that is, rationality in all matters is, itself, irrational.

In its stead, Pontecorvo begins to outline a Rousseauian critique of capitalism, a critique that centers on the simply morality of the traditional community that had been upset, first by the direct oppression of the Portugese and later by the capitalist economic program. Whether you are a slave or a "wage slave," Pontecorvo suggests, makes little difference: market relations poison interpersonal relations. British contract theorists celebrated what capitalism could create; Rousseau (and Pontecorvo) also mourns what capitalism destroys in the process.

The film is also a thinly disguised allegory of American involvement in Vietnam. In this allegory, the United States seduced the Vietnamese into throwing off the French colonialists, only to return a few years after Diem Bien Phu to quell the nationalist insurrection U.S. corporate interests had begun. The film's allegorical argument is that American efforts to suppress the human desire for freedom end up being far more brutal than French colonialism. And, as it does to Walker in the film, the process of substituting one immoral system for another in Vietnam is argued as utterly corrosive to the American soul. (This Rousseauian argument explains why Marlon Brando, famous for his romanticized support of Native Americans, was so keen to take the part. And perhaps it also helps explain why a film starring Brando, at the height of his popularity—two years before *The Godfather*—was never released commercially in the United States. Perhaps the distributors thought American audiences wouldn't be able to relate to the bluntness of the allegory.)

To a sociologist, contract theory is incomplete; understanding society requires more than seeing it as the aggregation of autonomous individuals pursuing their self-interest in a marketplace, unfettered by government control. *Burn* makes a convincing case for the birth of sociological theory in the decades immediately following the years the film depicts. In a sense, William Walker's dispirited disillusionment with the rational marketplace is the emotional foundation of sociology.

1900 and the Transition From Feudalism to Capitalism

Burn helps us to understand the development of the capitalist world-economy at the moment of its birth—through unequal trade relations in the 18th century. It amply illustrates Wallerstein's (1974, 1975) notions that capitalist class relations flow from this transformation of local economies as they are incorporated into capitalist trade relations. But we also have to understand how the events of the industrial revolution transformed everything about social life.

For example, between 1789 and 1848, the following words were first used with the meaning they have today: *industry, factory, middle class, democracy, class, intellectual, masses, commercialism, bureaucracy, capitalism, socialism, liberal, conservative, nationality, engineer, scientist, journalism, ideology*, and, of course, *sociology* (see Hobsbawm, 1962, p. 17). Think for a moment about how different our world would be without those words! As language is a window into the idea of the time, you can easily get a sense, as a sociologist, that whether you think this was for the better or for the worse, it is undeniable that the new world was dramatically different from the old. How can we "see" this transition? How can we understand the consequences on our social and emotional lives?

I can think of few films better to illustrate this than *1900,* Bernardo Bertolucci's 1976 epic masterpiece. Running more than five hours, the film is panoramic in its sweep, tracing the transformation of Italian society from 1900 to the postwar era. At its center is the story of two boys, both born on New Year's Day, 1900, and their lives over the course of the first six decades of the century. Alfredo Berlinghieri (Robert De Niro) is the effete grandson of a wealthy landowner (Burt Lancaster), who has only family money to shield him from the consequences of his coddled fecklessness. His "other," his mirrored class foil, is Olmo Dalcò (Gerard Depardieu), the grandson of the central peasant family that works on the Berlinghieri estate.

The relationship between the two is a microcosm of the changing relationship between ruling class and working class in the transition from feudalism to capitalism. As children, they play together, competitively, but in a relationship both take to be as natural as the land itself. Olmo is stronger and more sensuous. Alfredo is reserved, barely daring to imagine himself having fun, wide-eyed in wonder at his playmate's playful abandon.

With a Marxist perspective, Bertolucci examines the human costs of capitalism's incursion on rural-class relations. Marx had argued that initially capitalism was a revolutionary system itself, destroying all the older, more traditional forms of social life, and replacing them with what he called the "cash nexus"—one's position depended only on wealth, property, and class. But eventually, capitalism suppresses all humanity. We are not born greedy or materialistic; we become so under capitalism. In a most memorable passage from *The Communist Manifesto,* Marx describes this process:

> The bourgeoisie, wherever it has got the upper hand, has put an end to all feudal, patriarchal idyllic relations. It has pitilessly torn asunder the motley feudal ties that bound man to his "natural superiors," and has left remaining no other nexus between man and man than naked self-interest, than callous "cash payment." It has drowned the most heavenly ecstasies of religious fervor, of chivalrous enthusiasm, of philistine sentimentalism, in the icy water of egotistical calculation. (in Kimmel, 2007, p. 175)

This passage informs one of the many turning points in the film. In the first decades of the century, the conditions of the peasants became worse after several economic crises. (These economic shock waves were felt all across Europe and propelled the continent toward the First World War.) The peasants on the manor began to agitate for better wages and working conditions, and they asked the padrone to meet with them. Traditionally, he would have listened, he would have seen the peasants as "his" peasants and recognized his obligations to them, as well as theirs to him. But capitalism has hardened his heart. Now he sees only potential profits. He turns his back, and one peasant makes a symbolic sacrifice that exposes that newfound indifference.

And so, on the eve of the First World War, the peasants embrace communism as the alternative to rural feudalism. Both Olmo and Alfredo go off to fight in the war: Alfredo saw no action but returns in full uniform filled with ribbons and bows; Olmo was, naturally, a war hero. But the estate to which they return in the 1920s is vastly different from the one they left. Machines with gas engines have replaced peasant labor, and the new enterprise is run by a young man from the city, up and coming and on the make. Neither a landowner nor a peasant, he represents something new in agrarian class relations—the urban lower middle class. Neither owner nor worker, he is a middleman, a manager. Attila Mellanchini (Donald Sutherland) is a cruel sadist, and he, of course, becomes a fascist Black Shirt. (Note that he is named after the famous marauding Hun; Bertolucci is never subtle.)

One of the film's greatest achievements, in my view, is that it so accurately portrays the historical origins of Italian fascism within a displaced urban lower middle class, a wedge between aristocratic frippery and an increasingly mobilized red peasantry. And the process is set in motion by the constant need for profits as the old agrarian system becomes increasingly commercialized. This alliance between the old aristocracy and the rising fascists thrust Italy into that earlier Axis of Evil in the Second World War—an insertion from which the nation has yet to fully recover.

Photo 2.1 Class struggle in the transition to capitalism in *1900*. Attila Mellanchini (Donald Sutherland), the lower-middle-class estate manager, confronts Olmo Dalcò (Gerard Depardieu), the leader of the peasantry, while the foppishly dressed son of the padrone, Alfredo Berlinghieri (Robert De Niro), looks on.

Though its vista is panoramic, the tone of the film is heavy and didactic, and its characterization often follows political ideology and not developmental psychology. Historical accuracy often slows the narrative to a crawl. Characters act because their class as a whole acted that way; form dictates content. And yet, the three characters, and their class experience, express in miniature the interlocking fates of landlords, peasants, and urban classes in the 20th century. And, at the film's end, one realizes that the dynamic tension, the camaraderie and the conflict between both these two men and the classes they represent remain the animating social forces in contemporary Italian society.

Babette and the Culture of Guilt

If Marx provided the broad outlines of the transition to modern class society, Weber described how it felt and what motivated social action. Capitalism required so much self-sacrifice, so much rational calculation, it's a wonder that it could succeed at all. Weber argued that capitalist economics wasn't enough; there had to be a social psychology of capitalism, a mind-set that established it, sustained it, animated it. After all, there had been other forms of capitalism than the one that Marx had described emerging in Western Europe in the 16th and 17th centuries. Ancient Jews, ancient China, and ancient India all had forms of capitalism; yet none became the self-sustained force that European capitalism did.

Weber's essay, *The Protestant Ethic and the Spirit of Capitalism,* explores that social psychology of capitalism. Weber begins with the observation that what distinguished European capitalism from other forms is its rationality (as Marx observed in that quote provided above). But in Europe, capitalism did not appear first in the most "advanced" countries—France, Spain, and Italy—which were also Catholic, but rather in Protestant Britain, the Protestant parts of Germany, and the Netherlands. Much of the book explored the way that Protestantism created a different mental landscape than Catholicism, one far more conducive to self-perpetuating capitalism. In that sense, the book charts *how* this religious ethic became the secular driving force of an entire economic system—how the Protestant Ethic *became* the spirit of capitalism—and then how it ended up eroding the very sensibility it was intended to create.

What was that spirit of capitalism? Rational calculation. In order to succeed, any capitalist enterprise must reinvest profits. If you are a baker, you cannot eat all your bread, or eventually you will go hungry. You have to deny yourself. You need to take what you earn and plow it back into the enterprise. You cannot get ahead if you take all the money you earn during the day, and go out partying all night and spend it all. You need to save for the future. But why would you do that? What could possibly lead someone to deny immediate pleasures for the sake of a rational enterprise? Why would people deny themselves?

One can actually feel this contrast in the film *Babette's Feast* (1988). It is, in one sense, a movie about the most sumptuous meal ever created—a movie about appetite. Directed by Gabriel Axel, based on a novel by Karen Blixen (Isak Dinesen), the film is a gastronomical companion to Weber's masterpiece.

In the early 19th century, two sisters lived in an isolated seaside village with their father, the pastor of a small Protestant church. The community of parishioners is an ascetic sect, plain clothed, plain spoken, and utterly abstemious and self-denying. One day, a French woman, a refugee of the Revolution and the Terror, arrives at their door, penniless, desperate, and pitiful, and begs them to take her in. She agrees to work for them as a maid, housekeeper, and cook. They pay her an utterly paltry wage, which she saves dutifully (especially since there is nothing at all in the village to spend it on).

Babette works for them for 14 years. Her sole connection to her past is a lottery ticket that a friend renews for her every year. One day, she receives news that she has won the lottery, and a sizeable fortune has come her way. Does she put it in the bank, buy stocks and shares, invest it in a restaurant? No! She begs the sisters to allow her to use all of it to prepare a dinner party to commemorate the centennial of their father's birth. It is to be a celebration of Babette's gratitude for her station, an act of epic sacrifice.

The sisters hedge: Babette is a Catholic, French, a foreigner. But they allow her to proceed. Massive amounts of food are delivered, including a live sea turtle, cases of wine, and quails. The meal preparation takes months. Indeed, Babette prepares what may be the most delicious meal in history. The sisters, meanwhile, come to believe that Babette is a witch and the meal is a trick to see if sensuous pleasure can break their iron-willed ascetic commitment. They determine to make no mention of the food during the dinner, and convince themselves that each sumptuous bite is really the blandest tasting gruel, the equivalent, I should think, of eating a cardboard box.

So they sit, chewing each bite with hardly a hint of pleasure. They talk of everything *except* the food. Until, that is, another guest arrives. A lieutenant at the Swedish court, whose aunt has invited him to the dinner, sits down at the table and takes one bite and realizes that the chef is Babette (whom he also loved as a young soldier) perhaps the most famous chef in prerevolutionary aristocratic France. (She had escaped the guillotine by coming to the Danish village.) His pleasure provides the catalyst for the enjoyment of the meal.

I cannot recall ever seeing a starker comparison between the two cultures—aristocratic and bourgeois, Catholic and Protestant, shame and guilt. Aristocratic abundance and sensuousness is seen by this Lutheran sect not with envy but with morally indignant horror. Shame-based cultures enable sensuous pleasures because they also promise some form of forgiveness. By contrast, guilt is relentless, unstoppable, and one simply can never do enough to unlock its grip. Thus, Weber focuses initially on those countries in which rational, ascetic capitalism first thrived (Protestant Europe in the 16th and 17th centuries). Only there, with guilt as its implacable taskmaster, could people deny themselves enough pleasures to reinvest profits sufficiently to enable capitalism to take off.

Having established this, Weber shifted to the place where capitalism reached its apex: the United States. First, he finds in the aphorisms of Benjamin Franklin the full secularization of the Protestant Ethic. Franklin's abstemiousness ("a penny saved is a penny earned") was matched only by his industrious inventiveness. A century later, though, Weber's mood turned sour as he considered what happens to people in this relentless pursuit of profit. The book ends with a haunting image of the "iron cage" of rationality, Weber's indictment of the dehumanizing effects of the modern world's overly controlled social order. Like Marx, Weber believed that the modern capitalist order brought out the worst in us. "In the field of its highest development, in the United States, the pursuit of wealth, stripped of its religious and ethical meaning, tends to become associated with purely mundane passions, which often actually give it the character of sport," he wrote (Weber, 1905/1967, p. 182).

The bourgeois critique of aristocratic luxury is also a critique of the culture of shame: it lets people off the hook too easily. It promises forgiveness—which enables them to just return to their sinning ways. I remember my first encounter with this. "When I go to confession and I do my penance," a Catholic friend explained to me when I was 10, "God forgives and forgets." The bourgeoisie can afford neither succor and is driven into a frenzy of self-denying accumulation. When my neighbor told me that, I immediately went home and announced my intention to convert to Catholicism. I knew that "my" God, the Jewish God, forgot nothing, and would likely hold all sins against me until the next Day of Atonement. I wanted some of that forgiveness! (Of course, this was a preteen cosmology, and bears little relation to actual Catholic or Jewish teachings. It does, however, expose the contrasting cultures.)

The film, like Weber, sets the contrast between those two cultures, based on nationality, religion, and temperament. Sensuous aristocratic culture may need discipline; ascetic Protestantism needs to loosen up. Marx, by contrast, saw the conflict as between distinct class cultures; the bourgeoisie is propelled more by anxiety than by guilt:

> The bourgeoisie cannot exist without constantly revolutionizing the instruments of production. . . . Constant revolutionizing of production, uninterrupted disturbance of all social conditions, everlasting uncertainty and agitation distinguish the bourgeois epoch from all earlier ones. All fixed, fast-frozen relations . . . are swept away, all new-formed ones become antiquated before they can ossify. All that is solid melts into air, all that is holy is profaned, and man is at last compelled to face with sober senses, his real conditions of life, and his relation with his kind. (in Kimmel, 2007, p. 175)

Both Marx and Weber foresaw the disenchantment and emptiness of such a culture, though they foresaw different political trajectories. *Babette's Feast* poses the question: can the bourgeoisie ever have any fun?[2]

---

[2]A gastronomical note: In the early 1990s, Jakob de Neergaard, the chef who actually created the dishes in the film, reproduced the meal at Sollerod Kro, his restaurant outside Copenhagen. Having loved the film, I ate at the restaurant. I found it disappointing, the emphasis more on splendor of presentation than the food itself. But as an apposite coda you can't beat this: the restaurant, still billed as one of Denmark's best, is now owned by Michael Jordan.

Capitalist Rationalization as Intergenerational Succession in *The Godfather*

All these films illustrate the transition from feudalism to capitalism, the birth of modern society. Francis Ford Coppola's masterpiece, *The Godfather* (1972), as well as *The Godfather: Part II* (1974), illustrates these structural themes of the birth of capitalism and the rational capitalist enterprise, as well as the social psychological themes of rationalization, disillusionment, and loss of meaning—all the while intertwining them in a discussion of ethnicity, immigration, and family dynamics. The canvas could not be grander: the two films together encompass the grand sweep of the American experience.

The Godfather illustrates many of the central themes of classical theory. For example, there is ethnicity (the contrast between Kay's aloof WASPishness and the Corleones' boisterous ethnicity); the experience of different groups of immigrants in the early years of the 20th century in America (especially gaining an economic foothold in an ethnic enclave); gender development and the masculinities (in the contrast between passionate and violent Sonny, cold and calculating Michael, and happy-go-lucky and ultimately venal Fredo). It's a meditation on the intersections of ethnicity and gender relationships, as well as on interracial dynamics and interethnic conflict in the mid 20th century. It's a story of ethnicity and family, and the "proper" roles for women as the men go about their legitimate and illegitimate business. (Many doors are closed in women's faces in these films; some things are simply for men.) And, of course, it's a saga about law enforcement's efforts to stop organized crime in New York, a gripping crime drama, in which the police are more corrupt than the criminals.

The first film begins at the 1945 wedding of the daughter of Vito Corleone (Marlon Brando), the head of a New York Mafia "family." The sequel flashes back to his childhood in Sicily and his beginnings as a young Italian immigrant on the make in turn-of-the-century New York (where Corleone is played by Robert De Niro). As the head of the family, Vito is traditional, patriarchal, and benevolent to those in his circle, steadfastly loyal and visibly loving to his old friends and family, but ruthless and merciless to those outside. Vito has three sons: the tempestuous hothead, Sonny (James Caan), sexually predatory and emotionally volatile; Fredo (John Casale), a ne'er-do-well, slightly retarded from an infant malady; and Michael (Al Pacino), a war hero who is the youngest and smartest, and the son Vito has shielded from the family business.

Approached by another family to provide protection for the drug trade, Vito refuses, because although he knows there are great profits to be made, he considers drugs an ugly business, not dignified for his family operations. (His traditional authority joins effortlessly with his racism to suggest that the drug business be left to the black community.) This refusal sets the plot in motion: an assassination attempt on Vito, Sonny's murder, and Michael's eventual ascension to the role of Godfather.

Michael is both untroubled by traditional constraints and driven by a vision of taking the family business entirely legitimate in Las Vegas gambling. He is both more rational and calculating than his father and more ruthless and murderous. He leaves all old loyalties at the door; he sacrifices everything for the family business. He is the ultimate expression of the rational, calculating capitalist of which Weber warned.

Perhaps the central leitmotif of the film is the contrast between traditional and rational authority as discussed by Weber. The film's core theoretical insight is expressed in just three words spoken simply (and often) by Michael Corleone: "it's just business." In that phrase, he captures the transition from family-based benevolent patriarchy to a rational business enterprise, from traditional to legal rational authority.

Vito's way is the old way, eventually supplanted as Michael takes over. Michael takes those reigns reluctantly; he doesn't want power, and his approach is simply rational. Michael is cold, calculating, bureaucratic, and successful; he drowns all "heavenly ecstasies in the icy water of egotistical calculation." "Just business" is "only business" to Michael. It is he who has all of the means but knows not the ends. He avenges every betrayal with cold-blooded murder, taking out his own brother, his brother-in-law (immediately after Michael stands in as his baby's godfather), and his father's best friend Tessio. Deliberate and decisive, he doesn't peacefully coexist with the other crime families in New York, as did his father; he methodically eliminates them and consolidates power entirely at the top.

Even though it is abstract, formal, and bureaucratic—indeed, as Weber reminds us, precisely *because* it manifests those qualities—legal rational authority is far more pervasive and far more powerful than traditional authority ever could be. By embedding its power in the formal rules, power is massively enhanced, and can be far more ruthless as it is accountable to no one but itself. What Michael calls "just business" can be very unjust.

In the end, Michael has more power than his father could have ever imagined—and more than he ever would have felt he needed. And for what? The film ends with an elegiac Weberian tone. Michael, desperate to take the rational business model legitimate, has rationally calculated himself into the iron cage. He sits alone, having abandoned all he loved, recalling the happy moments of that earlier family life—a man of enormous power, but empty inside, trapped by the machine of his own making.

Classical Theory and Pleasure

In one sense, the core of classical sociology has been to chart the transition from feudalism to capitalism, the birth of modern class society, the triumph of bureaucratic rationality over traditional authority relations based on kinship or religion. The grand narrative of capitalism is usually a triumphalist celebration of progress. And indeed, capitalism unleashed creativity and innovation beyond the wildest imaginations of those luxury-infected aristocrats.

But sociology's task has always been to peek underneath that triumphalist narrative, to tear back the curtain on the fraudulent wizard, to see the human costs of capitalist rationality, the deadening of the soul that invariably accompanies the accumulation of wealth and power. Sociology's unique contribution has always been to chart that grand irony: the way that efficiency becomes dull routine, that rational organization leads to a loss of meaning.

Each of the films I have discussed charts the passing of one historical era and the birth of the new, modern era. And one of the main characters in each of the films I have discussed—Olmo, Babette, José Dolores, and even Vito Corleone—represents the roads not taken in that move toward modernity: the communitarian collectivism of Olmo's communist peasantry; the embodied sensuousness of Babette's feast; José Dolores's dream of national self-determination for all colonized peoples; and the warm solidarity of familial ethics in Vito Corleone's world. They remind us that what is lost in the drive for profit may be the capacity for pleasure.

References

Berger, J. (1972). *Ways of seeing.* New York: Penguin.

Hobsbawm, E. J. (1962). *The age of capital.* New York: Scribner's.

Kimmel, M. S. (Ed.). (2007). *Classical sociological theory.* New York: Oxford University Press.

Wallerstein, I. (1974). The rise and future demise of the world capitalist system: Concepts for comparative analysis. *Comparative Studies in Society and History, 16*(4), 387–415.

Wallerstein, I. (1975). *The modern world system. Vol. 1.* New York: Academic Press.

Watt, I. (1959). *The rise of the novel.* Berkeley: University of California Press.

Weber, M. (1967). *The Protestant ethic and the spirit of capitalism.* New York: Scribner's. (Original work published 1905)

Reading 2.2

Understanding Social Mobility Through the Movies

James J. Dowd

Movies are a particularly important vehicle for the transmission of cultural norms and understandings. As audience members, we view movies in a relaxed mode, not fully appreciative of the ways that film narrative is structured to be consistent with the ideals, norms, and expectations of the surrounding culture. Movies in this way may be said to support the dominant culture and to serve as a means for its reproduction over time. But the question may be asked why audiences would find such movies enjoyable if all that they do is impart cultural directives and prescriptions for proper living. Most of us likely would grow tired of such didactic movies and probably would come to see them as propaganda, similar in a way to the cultural artwork that was common in the Soviet Union and other autocratic societies.

The simple answer to this question is that movies do more than present two-hour civics lessons or editorials on responsible behavior. They also tell stories that, in the end, we find satisfying. The bad guys are usually punished; the romantic couple almost always finds each other despite the obstacles and difficulties they encounter on the path to true love; and, in general, the way we wish the world to be is how, in the movies, it more often than not winds up being. In fact, it is this utopian aspect of movies that no doubt accounts for why we enjoy them so much. The movies provide us with the happy endings and the just solutions that we cherish in our hearts, even as we understand in our heads that they are not always found in the real world (Jameson, 1990). Movies, then, offer both the happy, utopian ending that we love *and* the more conservative support of the dominant culture that guides behavior in the "real world."

Cultural ideas are transmitted to audiences without our discursive awareness and contribute to the social reproduction of society (Giddens, 1979).[1] Though we may conduct ourselves effectively in routine social interactions, we do so without having to explain why we are engaged in this particular line of behavior. While we know what we are doing, we may be unable to explain how it is that we came to know what is usually done in these types of situations. This is where the movies and

[1]Social actors possess three types of consciousness: *Discursive consciousness* refers to those motivations we can analyze, articulate, and put into words. *Practical consciousness* lies behind our capacity to perform the routinized or habituated practices of everyday life, which we generally do not dwell on and which, therefore, are not available to the actor's discursive comprehension. And the third is the *unconscious.*

television shows that we watch, the music that we listen to, and the novels and magazines that we read come into play. They are all vehicles that transmit cultural information from cultural producers to audiences.

A core belief in American culture is that of social mobility. Social mobility can be understood as the "movement of individuals and groups between different class positions as a result of changes in occupation, wealth, or income" (Giddens, Duneier, & Applebaum, 2007, p. 234). Change in class position across the life of an individual is referred to as *intragenerational mobility.* Generally, individuals change class positions over the course of their lives as a result of marriage, inheritance, illness, acquisition of human capital (such as education), promotion at work, and/or becoming a business owner. The belief in the possibility of upward social mobility and the dream of upward mobility in one's own life are ideas that continually are reinforced both in Hollywood movies and in the wider American culture more generally. In the next section, we examine three Hollywood films in which upward social mobility is central to the plot and the outcome of the story.

The Social Mobility of Deserving Individuals

We all are cognizant of the fact that certain individuals are economically more successful than others, and we are also aware that economic success tends to favor certain social groups more than others. Yet the reality of social inequality has only occasionally been the source of social conflict within American society. More often, we accept the existence of social classes and economic inequality, believing in one way or another that inequality is inevitable, normal, and perhaps even beneficial for society. Even if we recognize the ideological basis of the social Darwinist views of social theorists from previous eras, such as Herbert Spencer and William Graham Sumner, we still may consider social inequality to constitute an inevitable part of human existence. As long as the system remains at least somewhat fluid, with the possibility of upward mobility existing for at least some if not all of us, we generally turn our attention to other matters that might be more within our capacity to affect. Generally, we accept the reality of social classes and social inequality as legitimate, normal, and reflecting the underlying differences in ability and ambition among individuals. It is this last point that finds its way into the stories told by Hollywood movies.

Movies that in some way purport to depict society in a naturalistic, if not perfectly realistic, way often will present characters whose natural abilities or positive qualities allow them to achieve a level of success that characters in similar positions but without the necessary abilities or qualities are highly unlikely to experience. When we view such movies, we often find the story to be compelling, uplifting, and enjoyable. We want the protagonist to succeed and are pleased that the movie's conclusion almost always allows this mobility to occur. Though we may spend little time analyzing our reactions to the film, we know at some level that such stories are moving because they confirm our beliefs that hard work and meritorious efforts can, at least sometimes, be recognized and rewarded.

The films in this reading, in one way or another, deal with social mobility. These films include the romantic comedy *Maid in Manhattan,* the animated children's film *Lady and the Tramp,* and the coming-of-age film *Good Will Hunting.*[2] In each of these films, the central character, although possessing definite virtues and talents, lives a life deeply rooted within the working class. All of these films end with the central characters, though not others who associate with them, becoming socially mobile. The protagonists of these films demonstrate through their behavior, talents, and other qualities that they have earned their mobility. Using these films we can explore cultural understandings about social mobility in American society, particularly the belief that the possibility of upward mobility is open to everyone.

Maid in Manhattan (Wayne Wang, 2002)

Maid in Manhattan is a romantic comedy about the unlikely relationship between Christopher "Chris" Marshall (Ralph Fiennes), a New York assemblyman who is running for a seat in the U.S. Senate, and Marisa Ventura (Jennifer Lopez), a maid in a fancy New York City hotel and a single mother of a gifted child, Ty (Tyler Posey). They meet when Chris takes a room at the hotel where Marisa works. Chris mistakes Marisa for a wealthy socialite when he sees her in an expensive dress that actually belonged to one of the hotel's rich guests. The two characters begin a romantic relationship, though Marisa's hidden identity as a maid constitutes the main plot device that moves the story along to its conclusion. When Chris discovers Marisa's deceit, he breaks off the relationship, only to realize his mistake later when Ty confronts him at a press conference about the importance of giving people second chances. Chris and Marisa are reunited and, in the final few scenes of the film, the audience learns that both of them are successfully pursuing their chosen career goals.

The main story line, then, concerns the romance between Chris and Marisa. But the secondary story line, without which this film would not succeed, concerns the class differences between the two characters and Marisa's efforts to move up in the world. Although Marisa works as a maid, the audience is quickly made to see that she is both capable and deserving of far more prestigious and lucrative employment. Marisa defies any stereotype that we might hold of working-class single

---

[2]It is fair to say that the three films discussed here have all been widely viewed by American audiences. They have also all been quite profitable movies. *Good Will Hunting* cost approximately $10 million to produce. During its first weekend of wide release in early 1998, it recouped the $10 million and then some. By mid-1998, this film had grossed more than $138 million in the United States alone, and it turned out to be one of the top-grossing films released in 1997. *Maid in Manhattan* didn't do quite so well, though the movie clearly turned a profit for its investors. The movie cost an estimated $55 million to make, but a few months after its initial release in late 2002, it had taken in almost $100 million in gross receipts in the United States. And the children's animated film, *Lady and the Tramp,* which cost in the neighborhood of $4 million to make in 1955 (hand animation being an extraordinarily difficult and time-consuming process and, therefore, a very expensive one), also turned a tidy profit for the Disney company. This film grossed almost $100 million by 1987, with almost half of that amount in VHS and DVD rentals.

mothers. She is energetic, dependable, and intelligent, and she possesses a pleasant disposition and vibrant personality that draws others to her and would be difficult not to notice. But the most telling evidence that Marisa deserves a better place in the world is her son. Ty is a remarkable child, a true gem. He is a dedicated student, an obedient son, and—like his mother—a charming and sweetly genuine personality. His tastes in music run not to hip-hop or other genres popular among youth but to the poetic lyricism of early Simon and Garfunkel. Ty also knows more about Richard Nixon than most adults who may actually have lived through his presidency, further testifying to his individuality and warrant for more education and middle-class status. Such a child does not develop by accident but requires the loving guidance of a caring and dedicated parent.

To add one last element to Marisa's quest for upward mobility, the film interjects into the story additional information concerning her difficult background. It is clear that Marisa could easily adopt a defeatist attitude about life, considering her low pay, long hours, demanding family responsibilities, and the few opportunities that exist for someone like her to move up. This "realistic" understanding of Marisa's life is given voice by her mother, another hard-working but somewhat embittered woman whose presence in the film serves as a model for the type of life that Marisa could easily anticipate as her own future. When her mother learns of her daughter's relationship with the rich politician, and learns as well of the means by which her daughter began this relationship, she advises Marisa to get back to work and think seriously and realistically about rent payments and similar issues of adult life. Marisa recognizes that her mother is well-meaning, but she refuses to accept the inferior status that her mother has settled for. When her mother offers to help Marisa find employment as a private household domestic,[3] Marisa responds:

> You're right, Ma. I'm a good cleaning lady. I'll start over. But not with Mrs. Rodriguez. I'm gonna find a job as a maid in some hotel. After some time passes, I'm gonna apply for the management program. And when I get the chance to be a manager—and I will, Ma. I know I will. I'm going to take that chance without any fear. Without your voice in my head telling me that I can't.

Marisa's happy ending, then, demonstrates that mobility and success are possible even for a working-class daughter of Hispanic immigrants. With Marisa's mobility, the American dream of success is further embellished and reinforced in the minds of the audience. We know this is "only a story" but it is a story that is central to the larger cultural narrative of the American dream. Generations of immigrants have come to this country, worked hard, and found success. Further, the

[3]In urban centers such as New York City, where the film takes place, women from Mexico, the Caribbean, and Central America now predominate as nannies, housekeepers, and housecleaners. Unlike European immigrant women of the early 20th century, these women find themselves, generation after generation, stuck in the occupational ghetto of domestic work (Hondagneu-Sotelo, 2007).

gendered story of women's ability to "marry up" in social class is reinforced in the film. While cross-national research indicates that, in general, marriage improves women's chances of upward mobility (Li & Singelmann, 1998), women tend to marry men in the same social class as their fathers (Kearney, 2006). Marisa's story is one of victory on both fronts: she works hard, has ability and talent, and can attract a man from a higher social class. Her story is an instance of the proven formula that ability, merit, and attractiveness eventually yield success.

Lady and the Tramp (Clyde Geronimi, Wilfred Jackson, and Hamilton Luske, 1955)

The classic Disney animated feature *Lady and the Tramp,* although with similar elements, tells a very different story of a romance involving two mismatched lovers. One is a respected and protected member of the upper-middle class; the other is an uncollared, disreputable, free-wheeling rascal. Unlike *Maid in Manhattan,* however, this time it is the male who is the outcast. Released by Disney in 1955, *Lady and the Tramp* follows Lady, the cocker spaniel who lives a cosseted, insular existence in the home of Jim Dear and Darling, and who, following the birth of her owners' first child, feels somewhat ignored and unloved. Leaving the protected environs of her suburban home, she finds herself in the unfamiliar slums of the city. Lady is pursued by a pack of menacing-looking city dogs, who eventually corner her in a back alley. Coming to her rescue, however, is Tramp, who—like a true "knight in shining armor"—defeats the pack and triumphantly watches as they scamper away. In this act of chivalry and heroism lie the beginnings of the romance between Lady and Tramp. The movie reaches its climax when Tramp is taken away to the pound after being falsely blamed for knocking over the baby's bed, presumably attempting to injure the infant in some way. As every viewer of this film will remember, however, Tramp was actually trying to protect the baby, who was in danger of being bitten by a large and menacing rat.

Demonstrating their decency, perceptiveness, and sense of noblesse oblige, Jock and Trusty—two respected members of the neighborhood's canine community—jump into action to save Tramp. The attempt is successful and a happy ending ensues. Lady's human family adopts Tramp, who is awarded with the preeminent symbol of respectability, an official collar. Tramp also wins the most cherished prize of all as he and Lady become a couple. The movie concludes with a Christmas scene with Lady and Tramp, along with Trusty and Jock, playing with Lady and Tramp's litter of four energetic puppies.

Much of the film's plot revolves around the obstacle to romance and general social acceptance posed by social class and associated understandings of status hierarchy. When placed in the context of its own era, these themes are particularly resonant. When the film was released in 1955, the McCarthy hearings were fresh on the minds of Americans. The fear of being labeled a communist prompted middle- and upper-middle-class Americans to hew closely to conservative principles and disavow any deviation from convention and propriety. Tramp, in one sense, is the rebel who rejects the restrictions on his freedom demanded by the middle-class way of life. Like Johnny

Strabler, the character played by Marlon Brando in the 1953 film *The Wild One*, who when asked what he was rebelling against famously answered, "What'd ya got?," or Sal Paradise, Jack Kerouac's alter ego from his beat novel *On the Road* (1957/2007), Tramp's life is not one of quiet desperation and neither does he seem a likely candidate "to go to the grave with the song still in . . . [him]" (Thoreau, 1854/1995).

Tramp tells Lady to open her eyes to what a dog's life can really be. When looking down at the suburbs with Lady, Tramp points out that the middle-class, picket-fence livelihood is the "world with a leash." This statement might be seen as a critique of the American Dream and of the contentedness of America during the Eisenhower era. However, echoing the sentiments of conservative middle America, Jock the Scottie warns Tramp that "we have no need for mongrels and their radical ideas." Not surprisingly, the voices of both Jock and Trusty sound like old men, while Tramp speaks with a youthful exuberance that is consistent with the age of those who later brought the so-called youth counterculture into public notice.

In the end, however, the ideological theme of the American Dream and the status quo is upheld rather than being changed, or even slightly altered by Tramp. Despite all of Tramp's philosophies against "life on a leash," he ends up living with a middle-class human family and even becomes the typical middle-class husband/father. Only after he has acquired his collar and license, the badge of a dog's respectability, is he totally accepted by Jock and Trusty. His collar and license parallel the suburban home and nice car that mark one's status as middle class in today's America. Tramp's contentedness with family and friends around the Christmas tree at the end of the movie suggests that even active members of the counterculture, if fortune smiles on them, can become upwardly mobile and, with age and maturity, even accept middle-class values. The outspoken, streetwise Tramp has little to say in the film's final scene, merely observing his puppies at play.

Certainly this ending could also be viewed from a utopian standpoint if one focuses on the romance between Lady and Tramp. This, however, would be a mistake if it caused one to overlook this film's continual references to the ideal of American middle-class family life. Though Lady and Tramp presumably live "happily ever after," one wonders if Tramp has settled in his own mind whether his incorporation into the home of Jim Dear and Darling, however comfortable his new existence might be, is worth the loss of his freedom. The film seems to couch this story in terms of what in the 1950s might have been understood as the traditional story of the rambunctious male who settles down following a youthful period of sowing wild oats. Yet this traditional view of gender roles is not the only lens through which the story of Tramp can be analyzed. There is also the mobility lens.

Viewed as a children's story that is filled with images of social class and ideas about the possibility of moving among the various classes, we can see Tramp as a character who lends credence to our culture's cherished myths of upward mobility. Tramp—like Marisa Ventura and many other characters from well-known Hollywood films (Vivian Ward, the beautiful prostitute in *Pretty Woman* is another instance)—may love his freedom but he also has the mettle and true grit of a successful entrepreneur, the verbal dexterity of a high-priced trial attorney, and the heart and spirit of an officer and gentleman. That it is Tramp, and not Boris, Pedro, or one of the other dogs that Lady meets during her stay at the pound, who eventually moves up in the world and wins his own collar is hardly surprising since it is Tramp who the film shows to be truly meritorious. His place in the bosom of Lady's human family is not an anomaly or a happenstance but, rather, a rectification of an earlier misplacement. Tramp demonstrated his value to society and, in return, society welcomed Tramp to its upper echelons. Mobility is possible but requires a demonstration of merit, talent, and virtue.

Good Will Hunting (Gus van Sant, 1997)

Good Will Hunting, written by its two costars, Ben Affleck and Matt Damon, was a very popular coming-of-age drama set in the working-class areas of South Boston. At the center of the story is Will Hunting (Damon), a troubled but brilliant youth blessed with a singular intelligence and a love of learning who works as a janitor at one of America's most prestigious engineering schools, MIT. Will hangs out with a regular group of guys, foremost among them is Chuckie (Affleck), a construction worker, nonpareil raconteur, and contented regular at the L Street Bar & Grille in Southie. The film sets out, then, with this puzzle: why does someone so brilliant content himself with a dead-end job and a social life spent with a group of childhood friends who, however loyal and funny, show no apparent signs of sharing any of Will's intellectual interests, let alone his mathematical brilliance? As the story unfolds, Will's internal rage and self-loathing continually gets him into trouble until a judge finally determines that this bright but difficult lad needs to see a psychiatrist.

This film deftly develops four entangled story lines: (1) Will's mathematical genius, which brings him to the attention of an MIT professor, Gerald Lambeau (Stellan Skarsgård); (2) Will's relationship with Sean Maguire (Robin Williams), his widowed psychiatrist who also was once years ago a classmate of Professor Lambeau; (3) Will's budding romantic relationship with Skylar (Minnie Driver), a Harvard student he meets one evening in a bar near the college; and (4) the relationship of Will and his pals, primarily Chuckie. It is the fourth thread in this tangle that is of particular interest here since it is the relationship of Will and Chuckie that reflects most clearly the movie's point-of-view concerning intragenerational social mobility.

Good Will Hunting, it is important to note, gets much that is right about social class. Will, Chuckie, and the lads who are drawn into their orbit are basically decent human beings. They are witty, verbally agile, and enjoy a good laugh, albeit often at the expense of one or another of them, most usually Morgan. They probably drink too much, however, and have little if any ambition other than to meet some girls, go to a ballgame, or just hang out. Their futures are not difficult to imagine. As they grow older, they almost certainly will continue in a life of generalized and low-paying, physically demanding labor, looking forward to little but the possibility of early retirement when their aching bodies might be able to rest. We know little if anything about their relationships with women, other than the fact that none of them is currently in what might be termed a steady relationship. They entered adulthood as working-class youth and likely will eventually exit this vale of tears as working-class older men. The odds of their intragenerational upward mobility are quite long.

In this context, Will emerges as the one who is different (as in night from day). Although he shares in the jokes and verbal banter with his friends, he seems more comfortable off to the side, so to speak, rather than—like Chuckie—a person who enjoys being the center of attention. Will is also the only one in his crew whose abilities would almost certainly secure him a steady place among the upper-middle class, the home to those who Max Weber (1968) described as the "propertyless intelligentsia." He is also the one among the lads who demonstrates an interest in developing a romantic friendship with a woman. As it turns out, it is Skylar's move to California that finally spurs Will to leave the comfortable but hopeless surrounds of South Boston in order to take a chance on love. But what is this film saying about social mobility?

Good Will Hunting's theory of social mobility can be discerned through a juxtaposition of the lives of Chuckie and Will. Chuckie, as the sun around which his friends predictably move, is clearly a young man with charisma, ability, and intelligence. Though we in the audience can only know about Chuckie what the film allows us to know, he has a quality about him that suggests the possibility of upper mobility. Yet the trajectory of the story clearly indicates that Chuckie has resigned himself to a peripheral existence solidly entrenched among his working-class friends he knows and trusts. Out of this environment, as when the lads visit the Bow & Arrow, a bar in the Harvard neighborhood frequented mostly by the gifted and affluent students who attend schools like MIT and Harvard, Chuckie's working-class habitus is clearly a liability. Will, in contrast, possesses the cultural capital that allows him to navigate the social scene at the Bow & Arrow just as easily as he does back home at the L Street Bar & Grill. Will comes to Chuckie's rescue as he deflates the pretensions of the Harvard graduate student who was attempting to humiliate Chuckie in order to impress Skylar and her girlfriend. As indicated in the following dialogue between Chuckie and Will, Chuckie is not blind to the differences between him and his closest friend, Will. At the same time, Will feels great ambivalence about leaving his social class and community of origin. Research on working-class individuals who experience upward mobility reveals that they often feel like they are letting down their own group by selling out and becoming part of the dominant, privileged class (Granfield, 1991).

Chuckie: Are they hookin' you up with a job?

Will: Yeah, sit in a room and do long division for the next 50 years.

Chuckie: Yah, but it's better than this shit. At least you'd make some nice bank.

Will: Yeah, be a fuckin' lab rat.

Chuckie: It's a way outta here.

Will: What do I want a way outta here for? I want to live here the rest of my life. I want to be your next door neighbor. I want to take our kids to little league together up Foley Field.

Chuckie: Look, you're my best friend, so don't take this the wrong way, but in 20 years, if you're livin' next door to me, comin' over watchin' the fuckin' Patriots' games and still workin' construction, I'll fuckin' kill you. And that's not a threat, that's a fact. I'll fuckin' kill you.

Will: Chuckie, what are you talkin' . . .

Chuckie: Listen, you got somethin' that none of us have.

Will: Why is it always this? I owe it to myself? What if I don't want to?

Chuckie: Fuck you. You owe it to me. Tomorrow I'm gonna wake up and I'll be 50 and I'll still be doin' this. And that's all right 'cause I'm gonna make a run at it. But you, you're sittin' on a winning lottery ticket and you're too much of a pussy to cash it in. And that's bullshit 'cause I'd do anything to have what you got! And so would any of these guys. It'd be a fuckin' insult to us if you're still here in 20 years.

Will: You don't know that.

Chuckie: Let me tell you what I do know. Every day I come by to pick you up, and we go out drinkin' or whatever and we have a few laughs. But you know what the best part of my day is? The ten seconds before I knock on the door 'cause I let myself think I might get there, and you'd be gone. I'd knock on the door and you wouldn't be there. You just left. . . . Now, I don't know much. But I know that.

Finally, Will's ability to solve the math problems that Professor Lambeau posted on a hallway bulletin board demonstrates, if additional proof was needed, that Will Hunting was meant for better things than pushing a broom along with the other night cleanup crew at MIT. The fact that, once his talent became known to Lambeau and others, offers of employment from both private corporations and secret government agencies came rushing in seemed both inevitable, right, and satisfying, thus attesting both to our human longing for distributive justice and happy endings *and* to our collective belief in the flexibility and porousness of the American class structure.

Like Tramp and Marisa Ventura, Will Hunting is granted the opportunity to move up in the class structure. The main characters in all three films, but not their friends and acquaintances, create the opportunity for upward mobility through their talents, merit, and good character traits. They succeed because they deserve to. Implicit in this understanding of social mobility is the tacit belief that those who do not move up are those who lack the requisite talent and skill.

Conclusion

Pierre Bourdieu (1984), sociologist and leading French intellectual, once was asked how sociologists explain the constant flux experienced by individuals in the postmodern world. In his answer, Bourdieu acknowledged the reality of change as experienced by the individual but pointed out that we often overlook the extent to which social life remains much the same, reproducing itself from generation to generation. Inequality persists in part because we are socialized to different ways of thinking and perceiving that are connected to our class position. We grow up immersed in a particular *habitus,* which provides us with ways to understand our immediate social worlds that we eventually become so accustomed to they become a sort of habit. We develop a "feel for the game," so to speak, that pushes into the background of human consciousness all of the information and behaviors that are necessary to know in order to negotiate a particular piece of social interaction.

Culture operates in ways that we can consciously consider and discuss, but also in ways of which we are far less cognizant. When we have to offer an account of our actions, we consciously understand which excuses might prove acceptable given the particular circumstances in which we find ourselves. In such situations, we use cultural ideas as we would a particular tool: certain jobs call for a Phillips-head while others require an Allen wrench. Whichever idea we insert into the conversation to justify our actions, the point is that our motives are discursively available to us. They are not hidden. In some cases, however, we are far less aware of why we believe a certain claim to be true or how we are to explain why certain social realities exist. Ideas about the social world become part of our worldview without our necessarily being aware of the source of the particular idea or that we even hold the idea at all. Beliefs about social mobility, I would argue, are like this.

We may never consciously give the social class structure of our society much thought, but this is not to say that we do not have particular beliefs about the class structure, as well as the reasons why some people are successful while others are not. In American society, it is commonly believed that individual success (or failure) is an accomplishment for which the individual is responsible. Indeed some of our most cherished cultural beliefs have to do with individuals who struggle against the conventions of the day and despite obstacles of all kinds persevere and succeed. Many are satisfied merely to achieve only a modicum of success, not willing to exert themselves to the extent necessary to reach the higher echelons. Others simply are incapable of a strong performance in a demanding situation because they lack the skills, talents, or aptitude required.

While no one could argue persuasively that individual talent plays only a minor role in success, the more widely held view is that individual characteristics are the single most important explanation of why individuals succeed or fail. The class structure is the eventual outcome, it is believed, of countless instances of individuals of varying levels of ability encountering situations requiring certain levels of skill. My intent in this section is to demonstrate how our beliefs about the class structure and social mobility more generally (1) serve to reinforce existing levels of social inequality as inevitable and normal, and (2) are replicated and supported through the countless popular films we have enjoyed from early childhood well into adulthood.

Hollywood movies are often considered by theatergoers to be vehicles for entertainment, opportunities for mindless enjoyment and the pleasure of viewing our favorite actors engaged in romantic, heroic, or otherwise interesting behaviors. Film producers also like to tout the educational aspects of film, pointing particularly to ways in which movies will increase our understanding of particular historical events or the lives of people in other places. Films do this, but the educational function of film is not restricted to such manifest lessons. Films also serve as vehicles or arenas for cultural learning, much of which is the learning of the basic, presumably commonsensical, ideas that are part of the dominant culture.

References

Bourdieu, P. (1984). *Distinctions: A social critique of the judgment of taste.* Cambridge, MA: Harvard University Press.

Giddens, A. (1979). *Central problems in social theory: Action, structure, and contradiction in social analysis.* Berkeley: University of California Press.

Giddens, A., Duneier, M., & Applebaum, S. (2007). *Introduction to sociology* (6th ed.). New York: W. W. Norton.

Granfield, R. (1991). Making it by faking it: Working class students in an elite academic environment. *Journal of Contemporary Ethnography, 20*(3), 331–351.

Hondagneu-Sotelo, P. (2007). *Doméstica: Immigrant workers cleaning and caring in the shadows of affluence.* Berkeley: University of California Press.

Jameson, F. (1990). *Signatures of the visible.* New York: Routledge.

Kearney, M. S. (2006). Intergenerational mobility for women and minorities in the United States. *The Future of Children, 16*(2), 37–53.

Kerouac, J. (2007). *On the road.* New York: Viking Adult. (Original work published 1957)

Li, J. H., & Singelmann, J. (1998). Gender differences in class mobility: A comparative study of the United States, Sweden, and West Germany. *Acta Sociologica, 41*(4), 315–333.

Thoreau, H. D. (1995). *Walden.* New York: Houghton Mifflin. (Original work published 1854)

Weber, M. (1968). *Economy and society: An outline of interpretive sociology.* New York: Bedminster Press.

Reading 2.3

Class in the Classroom

Hollywood's Distorted View of Inequality[1]

Robert C. Bulman

To the casual viewer, a film about high school may be nothing more than simple entertainment. When the films are viewed collectively, however, the high school film genre reveals patterns that transcend entertainment and teach deeper lessons about American culture. Motion pictures do not necessarily reflect the high school experience accurately. Hollywood routinely twists and shapes reality to maximize dramatic or comic effects. Films must also frame complicated social relationships within two hours and on a two-dimensional canvas. Nevertheless, even if they are not precise social documents of real high schools and real adolescents, these high school films are still culturally meaningful. That is, they have something to teach us about how Americans make sense of education, adolescence, and class inequality.

As I argue in *Hollywood Goes to High School: Cinema, Schools, and American Culture* (2005a), there are significant differences in how Hollywood represents poor students in urban schools, middle-class students in suburban schools, and wealthy students in private schools. This reading focuses primarily on the urban school films and the way in which Hollywood depicts social class.

By analyzing the representation of schools in film, we find that Hollywood has a double standard in how students from different social classes are depicted. Specifically, poor students are rarely allowed to be the heroes in Hollywood films (adults always must "save" them) while middle-class students are nearly always the heroes of the films in which they are featured (they are always wiser than the adults in the film). This double standard, I argue, reflects the middle-class bias of American culture. The middle-class perspective is always the predominant and heroic perspective in film. While it would be nice to simply blame Hollywood for this double standard, the truth is that these distorted images of poor and middle-class teens are a reflection of the distorted image that Americans in general have of different social classes. These representations are a fantasy of the middle class (that poor students are troubled and middle-class students are wise), but it is fantasy that transcends the middle class because the middle-class perspective in American society is hegemonic.

The middle class occupies a special place in the American economy and in American culture. In very rough terms, the middle class can be defined as that class of people who have a college degree, are employed in professional white-collar jobs,

[1]Significant portions of this chapter were previously published as "Teachers in the 'Hood: Hollywood's Middle Class Fantasy." *The Urban Review*. Vol. 34, 3 (2002), 251–276.

exercise autonomy on the job, work for a salary rather than a wage, and own their own homes. After World War II, the American economy expanded and with it expanded a new middle class to manage the industries and bureaucracies of the growing society. With the shift from an industrial-based economy to an economy based in the information and service sectors, the educated and professional middle class has become even more important to the American economy. The middle class is not necessarily the most powerful class politically or economically in the United States. However, it is the most powerful class culturally. The image of the middle class has become the image of America. The definition of what it means to be an American in the popular imagination is closely linked to the stereotypical cultural images of middle class life—suburban home ownership, heterosexual marriage and family life, educational and occupational achievement, and financial security (but not opulence).

I do not disagree with analyses of film that argue male or white perspectives tend to be privileged in Hollywood films. Indeed, Hollywood films are often framed in such a way as to highlight the perspective of male or white characters. Hollywood films are also, by and large, told from the perspective of the middle class and tend to privilege middle-class characters, middle-class values, and middle-class assumptions about the social world. The middle-class perspective of films is even harder to recognize by film audiences than racial or gender perspectives because class itself as a social category is difficult for Americans to wrap their minds around. Americans do not like to think in terms of class because we like to think that people regardless of their background, can achieve anything in life provided they work hard enough. We often fail to recognize that there can be significant class barriers preventing individuals from achieving upward social mobility.

The High School Film Genre

The theme that high school movies have in common is an ethic of individualism. Adolescents in these films are expected to transcend the limitations of their communities, the narrow-mindedness of their families, the expectations of their parents, the conformity of their peers, the ineffectiveness of their schools, their social class status, and the insidious effects of racism in order to express themselves as individuals apart from social constraints. The source of their academic success and/or personal fulfillment is to be found within the heart and mind of each individual regardless of social context. There are dramatic differences, however, in the ways the theme of individualism plays out in films based in urban, suburban, and elite private high schools.

In the urban high school film, a classroom filled with socially troubled and low-achieving students is dramatically transformed by the singular efforts of a new teacher or principal. All of this is accomplished to the consternation of the inept administrative staff and other teachers, who never believed that these students had such potential. This lone "teacher-hero" is always an outsider, one who has a troubled and mysterious past, little teaching experience, a good heart, and an unorthodox approach to teaching (Ayers, 1996; Burbach & Figgins, 1993; Considine, 1985; Heilman, 1991; Thomsen, 1993). Invariably, the outsider succeeds where veteran professional teachers and administrators have repeatedly failed. The outsider can defeat the culture of poverty that had previously inhibited academic achievement. In these films, the poor and mostly nonwhite students must change their behavior and accept middle-class values and cultural capital in order to achieve academic success.

In the films based in suburban high schools, however, academic success is not a central focus of the plot. The suburban school films depict schools less as actual places of learning and more as social spaces where middle-class teenagers search for their identities and struggle with each other for the rewards of social status and popularity. In these suburban school films, schoolwork is secondary to the real drama of teen angst. Students must reject the conformity of their peers, the culture of popularity, and the constraints of adults in order to express their true selves. The hero is almost never an adult as in the urban school films, but always a student who can overcome the conformity of teen society or the authoritarianism of adult society.

In the films based in elite private high schools, academics are once again featured as an element of the story. However, whereas in the urban school films academic achievement is valued as the answer to the culture of poverty plaguing inner-city students, in the elite private school films the narrow focus on academic achievement is portrayed as an oppressive burden on students. The students in these films must conform to the wishes of their parents and the school in order to protect their social class status. The hero of these films is usually an outsider who challenges the culture of privilege that pervades the upper-class institution. This working- or middle-class hero works to expand the horizons of the upper-class students away from narrow academic achievement.

The upper-class students are challenged to risk their taken-for-granted position in the class hierarchy by finding and expressing their true selves independent of the expectations that elite culture has of them.

It is important to note that these different representations do not arise because of the actual differences between real urban, suburban, and private high schools. To be sure, there are real differences between these types of schools. However, the subgenres differ in significant ways because American culture makes sense of poor, middle-class, and wealthy youth in very different ways. The differences depicted on screen reflect the fantasies that Americans have about social class and inequality more than they reflect the realities of social class and inequality in the United States.

Because these high school films are made by and largely consumed by members of the middle class, and because middle-class culture is the hegemonic culture in the United States, these high school films tend to reflect middle-class worldviews and assumptions. Suburban school films represent middle-class frustration with the conformity and status hierarchy of suburban middle-class life and express fantasies of self-expression and individual rebellion against such a system. The elite private school films reflect middle-class resentment of the rich and a fantasy that to be truly happy it is not necessary to be rich, but it is necessary to be true to oneself as an individual.

The urban high school film genre represents the fantasies that suburban middle-class Americans have about life in urban high schools and the ease with which the problems in urban high schools could be rectified—if only the right type of person (a middle-class outsider) would apply the right methods (an unconventional pedagogy with a curriculum of middle-class norms and values). This teacher-hero represents middle-class hopes that poor students in urban schools can be rescued from their troubled lives not through significant social change, but by the individual application of common sense, good behavior, a positive outlook, and better choices.[2]

A Cinematic Culture of Poverty

In most of the urban high school films, the plot revolves almost exclusively around the activities of one particular classroom of rowdy students and their heroic teacher in a troubled and violent school. The students in this class are depicted homogeneously: they all have similar social class characteristics and similar problems. We are rarely offered a glimpse into the complexities of their individual characters, their histories, their identities, or their families, as is the case in many of the suburban high school films. The urban high school students, for the most part, are from lower- and working-class homes, are often nonwhite (but not exclusively so), come from broken families that neither understand nor care much about their child's education, have low educational aspirations and expectations, behave poorly in the classroom, and express a great deal of frustration with the formal structure of the school.

The students in these films represent the working- and lower-class populations as they are stereotypically imagined by suburban middle-class Americans. These students represent what middle-class people fear most about the poor urban youth: they are out of control, loud, disobedient, violent, and addicted to drugs; have no family values; and reject the dominant social institutions. The rejection of the school is particularly offensive to members of the middle class since they depend on educational credentials and because schools have served them quite well (Eckert, 1989).

One argument is that such stereotypical notions are the result of psychological projections—that the suburban middle class projects these images onto the residents of inner cities to relieve the burden of carrying such negative characteristics themselves. In other words, the identity of the middle-class suburban resident is formed in opposition to that of the inner-city resident, who is imagined to be impoverished both economically and morally. The growing social distance between suburban and urban America is reflected in the exaggerated representations of inner-city residents in the popular media. In response to the anxiety the middle class feels about life as it imagines it in the inner city, the suburban middle class seeks to impose its particular values and strategies for success on the residents of

[2]While the students in these urban school films are very often African American and Latino, the social class differences between the students and their teacher-heroes are more significant than the racial differences between them. The middle-class protagonists of *To Sir, With Love, Lean on Me, 187, Only the Strong,* and *Stand and Deliver* are all African American or Latino. Also, white working-class students are in need of salvation from a middle-class hero in *Blackboard Jungle; To Sir, With Love; Class of 1984; Teachers; Summer School;* and *Cheaters.*

the inner city (McCarthy et al., 2004). What prevents inner-city residents from achieving educational and occupational attainment is believed by many in the middle class not to be a political or economic problem, but a moral one. Hollywood reinforces these middle-class fantasies about how best to address the problems of the inner city.

Hollywood's depiction of urban life and urban schools generally reflects the culture-of-poverty thesis. This view holds that residents of poor inner-city neighborhoods are poor not because they face racial and/or class discrimination or because they lack access to stable employment opportunities. Rather, it is argued that the urban poor are impoverished because they have the wrong values and the wrong attitudes about school, work, and family. In contrast to what is considered the normative cultural values of the middle class (material goals, rational calculation, and a belief in the efficacy of individual effort), the culture-of-poverty thesis implies an impoverished culture—a culture that is lacking in the requisite values to achieve individual success. The urban poor remain poor due to their failure to adopt middle-class values and to fully integrate into the dominant culture of the United States (Banfield, 1968).

Much social science research, however, has discredited the idea that cultural values are responsible for either success or failure in life (see, for instance, Bourdieu, 1977; Gibson & Ogbu, 1991; Lareau, 1987; MacLeod, 1995; Swidler, 1986; Willis, 1981). This research has shown that while cultural values and attitudes vary, they do so primarily as they adapt to larger historical, social, political, and economic conditions. As sociologists have studied the inner city, they have found that many of the social problems found there are less the result of cultural values and more the result of low levels of public investment in infrastructure, poor public housing, inadequate health care, poor schools, and a disappearing employment base (Wilson, 1996).

Nevertheless, the culture-of-poverty framework has found its way into the popular imagination, and it is difficult to dislodge. Rather than focusing on the social, political, and economic sources of the problems in the inner city, Americans prefer to place the blame on the moral failings and bad decision making of the residents of the inner city. National surveys have found that a majority of white Americans believe that a lack of personal motivation is the primary reason African Americans, on average, have lower socioeconomic status than white Americans (Schuman, Steeh, Bobo, & Krysan, 1998). It is generally assumed that a solution to the problems in the inner city must be applied individually rather than structurally. As President George W. Bush remarked early in his first term, "Much of today's poverty has more to do with troubled lives than a troubled economy. And often when a life is broken, it can only be restored by another caring, concerned human being" (Hutcheson, 2001). Explaining poverty as the result of individual failure helps to relieve the suburban middle class of its share of responsibility for having politically and economically neglected the inner city. The frame that Hollywood uses to make sense of problems in urban high schools vividly reinforces the culture-of-poverty thesis and assists the middle class in its displacement of responsibility from troubled social structures to troubled lives.

Welcome to the Jungle: The Urban School in Hollywood Films

Many of the urban school films do acknowledge that inner-city students face the challenges of poverty, racial discrimination, and poor schools. However, the films portray the individual attitudes of the students as the primary obstacle to their academic achievement. These students don't have the right manners, the right behavior, or the right values to succeed in school. They have low aspirations and a low self-image, and they believe the odds are stacked against them. The schools, therefore, cannot effectively educate these students. The reproduction of their low social status seems inevitable.

In the classic *Blackboard Jungle* (1955) a class of working-class New York boys is depicted at first as nothing but a street gang who spend their days causing havoc in their vocational high school. A female teacher is nearly raped, a baseball is heaved at a teacher's head, a teacher's wife is harassed, and a newspaper truck is stolen. The metaphor in the film's title is all too literal; these students are seen as working-class animals. These are "beasts" that even music won't soothe; in one scene the students destroy a teacher's priceless collection of jazz records. In *The Principal* (1987), one teacher compares the students to animals only to have another claim she would rather teach animals because at least animals do not carry knives. In *Teachers* (1984), the song "In the Jungle" plays while police search student lockers for drugs. In *Lean on Me* (1989), the high school is depicted explicitly as an untamed jungle. In the opening moments, we see students selling drugs, assaulting teachers, harassing women, and generally running amok. All of this takes place as the movie soundtrack plays Guns N' Roses' loud and angry "Welcome to the Jungle" in the background.[3]

The jungle metaphor conveniently summarizes the imagined difference between middle-class suburban Americans and the poor urban students portrayed in these films. These are not students as middle-class Americans expect students to act. Their depiction as "animals" suggests that the problems in these schools are rooted in student behavior and, furthermore, that their behavior is rooted in an inferior culture.

In the opening scenes of *Teachers,* a student is stabbed, a student bites a teacher, the school psychologist has a nervous breakdown, and we see a teacher pack a gun in her briefcase. The assistant principal of the school casually explains these events as typical problems for a Monday. In *The Principal,* Rick Latimer (James Belushi) single-handedly breaks up a gang fight on his first day on the job as the principal of an inner-city school. In *The Substitute* (1996), gang members have such firm control over an urban public high school that they attack a teacher with impunity. In *Dangerous Minds* (1995), the white, middle-class, and somewhat naive

[3]In contrast to the songs that refer to the urban high school students as jungle animals, several of the suburban school film soundtracks feature Pink Floyd's anthem of adolescent resistance, "Another Brick in the Wall," with the lyrics "Teachers, leave those kids alone!"

Ms. Johnson (Michelle Pfeiffer) walks into her class for the first time only to walk right back out after encountering nothing but abusive and hostile students, who first ignore and then ridicule her. These are the same students who, by the end of the movie, Ms. Johnson (and we the audience) will embrace warmly.

Or is it that by the end of the movie "they" (the at-risk, poor, and inner-city students) will have learned to embrace "us" (the educated, middle-class, and suburban audience as represented by Michelle Pfeiffer's portrayal of Ms. Johnson)? This distinction is an important one. Will the audience learn that these students are not animals after all? Have the students simply been misunderstood? Will the audience be the ones who learn a lesson? Or will the students radically change their behavior as they come under the civilizing influence of the middle-class teacher who will socialize them in the culture of middle-class life? With few exceptions, it is the students who must learn and change, not the audience.

The School Staff: Inept Bureaucrats and Incompetent Teachers

If the students are portrayed in a negative light, the school administrators and teaching staff are not depicted much more generously. The teachers and staff are generally shown as uncaring, cynical, incompetent, and ineffective educators. In short, the administrative and teaching staffs in these movies represent the worst fears that suburban residents have of urban public schools. These characters represent what many Americans believe to be typical of the urban public school "crisis"—a selfish, inept, wasteful, and uncaring bureaucracy.

These are schools with no soul—just troubled students, failed educational methods, burned-out personnel, too many arcane rules, and too much paperwork.

If the harshest critics of public education (such as Chubb & Moe, 1990; Lieberman, 1993) were to make a movie about the public schools, their fictional schools would look much like the schools in these films.

In *Blackboard Jungle,* the stern principal is offended by the suggestion that there are discipline problems in his school. He seems unaware of the disobedience that surrounds him. In *Dangerous Minds,* the soft-spoken principal is so narrowly focused on teaching the students to follow the most minor of rules that he is blind to their real life-and-death problems. Similarly, the administrators in *Up the Down Staircase* (1967) are more concerned that teachers follow the strict rules, obey the proper procedures, and fill out the right forms than they are with the welfare and education of their students. The principal in *Teachers* is blissfully ignorant of all the chaotic events in his school. Most of the administrative energy in the school is spent fighting a lawsuit filed by the family of a student who graduated without knowing how to read. The school authorities in *Stand and Deliver* (1988) have little faith in their students and do not believe that they could possibly do well in an advanced math class. In *Coach Carter* (2005), the principal of the school is furious that the coach is focused more on helping the players academically than he is focused on winning basketball games. In *Freedom Writers* (2007), the chair of the English Department doesn't believe the low-income students in her school have the capacity to learn. She refuses to allow a young idealistic teacher to distribute *Romeo and Juliet* or *The Diary of Anne Frank* to her students, books that are otherwise collecting dust in the warehouse. In *Lean on Me,* the dramatic deterioration of the high school over the years is blamed on the actions of the selfish teachers' union and the corrupt politicians in City Hall. In *The Principal,* the teachers complain bitterly when the principal insists that the "thugs" of the school actually attend their classes. In *The Substitute,* the principal is actually one of the thugs. He has established an alliance with the dominant gang in the school to distribute drugs throughout the school district.

The vast majority of the teachers in these films have cynical attitudes about their jobs, and they seem to believe that most of the students are beyond hope. As one teacher from *Up the Down Staircase* summarizes her pedagogical philosophy, "You keep them off the streets and you give them a bit of fun and you've earned your keep." These veteran teachers are burned out and have failed to do what was assumed to be their professional obligation—to reform these students into respectable, educated, and well-behaved citizens.

The Outsider as the Teacher-Hero

While all of the students, all of the administrators, and most of the teachers are depicted as impediments to education, there is one bright light of hope in these films: the teacher-hero (or, in the case of *Lean on Me* and *The Principal,* the principal-hero). This lone figure can ignore the cynicism of veteran teachers, escape the iron cage of the school bureaucracy, and speak directly to the hearts and minds of these troubled youth who are, by the end of the film, transformed from apathetic working-class and poor students into studious and sincere students with middle-class aspirations.

The heroes of these films do not need teacher training, smaller class sizes, a supportive staff, strong leadership, parental participation, technological tools, corporate partnership, school restructuring, a higher salary, a longer school day, vouchers, or more financial resources. All they need to bring to the classroom is discipline, tough love, high expectations, and a little good old-fashioned middle-class common sense about individual achievement and personal responsibility.

In each of these movies, the hero is someone new to the school, and often new to teaching entirely. The teacher-hero is a mysterious figure who literally becomes the savior of these students (Ayers, 1996). All hope would be lost if not for the intervention of this unconventional new teacher who breaks from the failed methods of the school and effectively reaches the students with a unique approach. The teacher-hero represents a likely fantasy of the suburban middle-class audience: a character they can identify with goes into a troubled urban high school and single-handedly rectifies its problems. The teacher- or principal-hero can clearly see through the confusion that has bewildered many educators and policymakers for years. She or he can identify the faults in these students and the problems in these schools and knows just what it takes to correct them. The teacher-heroes teach the students to escape the depressing and limiting world of their parents, to appreciate art and poetry, to learn manners and cultural skills, to develop new study habits, to set high goals, to have an optimistic attitude, and to believe that hard work pays off. In short, the teachers show the students how to overcome their culture of poverty. It is through this figure of the heroic outsider that the audience feels some sense of control over an otherwise chaotic situation.

In *Blackboard Jungle,* Mr. Dadier (Glenn Ford), a white man with plenty of upper-middle-class cultural capital (he recites Shakespeare in his job interview with the principal), enters the "jungle" (the "garbage can of the educational system," as one teacher puts it) and attempts to reform unruly thugs who don't even seem to care about an education. Mr. Dadier's wife wishes he would retreat to a middle-class school with well-behaved students. Mr. Dadier, however, is determined to reach the students in his "jungle." He wants them to care about an education, to learn "to think for themselves," and to make something positive of their lives. He takes a special interest in Gregory Miller (Sidney Poitier), the charismatic black leader of the class, and tells him that he should not settle for being an auto mechanic, that in 1955 racial discrimination and poverty are no longer excuses for blacks not to make something of their lives in the United States. Through his persistence and dedication, Mr. Dadier manages to convince Miller to stay in school. They create a pact: Mr. Dadier will not quit his job if Miller doesn't drop out of school. In addition to Miller, Mr. Dadier eventually wins the respect and admiration of most of the other inhabitants of his classroom "jungle."

In *Dangerous Minds,* Ms. Johnson finds herself teaching some of the most difficult students in the school ("rejects from hell"). Her primary message to these students is that they can achieve anything they want, provided they put their minds to it. With only a superficial nod to their community, their poverty, their race, or their families, Ms. Johnson declares that their lives are defined by their individual choices, nothing more. As she tells her students, "If you want to pass, all you have to do is try." To give them the confidence that they can achieve anything they choose, she breaks from the traditional curriculum and uses "college-level" poetry to teach her students. Her class engages in intellectual debate about the similarities between the poetry of Dylan Thomas and Bob Dylan. The upper-middle-class cultural capital she imparts to them is in stark contrast to the poor and working-class family lives they lead. The grandmother of two brothers in her class doesn't see the point of all this book learning and withdraws the boys from Ms. Johnson's class. Nevertheless, most of her students begin to care about schooling and begin to believe that education, including poetry, can make a difference in their lives. Ms. Johnson develops a particular interest in one student, Raul (Renoly Santiago), and develops a pact with him: she loans him $200 but will allow him to pay back the money only on the day he graduates from high school. Ms. Johnson's love (and the candy bars she uses as bribery) inspires her students to believe in themselves

Photo 2.2 The teacher-hero in *Dangerous Minds*. Ms. Johnson (Michelle Pfeiffer) tries to reach her "difficult" class through alternative teaching strategies, such as adopting supposed "urban" language in a grammar lesson and tossing candy bars to students who provide correct answers.

and in the power of an education in spite of the hardships they face in the world outside the school.[4]

In *Coach Carter,* Ken Carter (Samuel L. Jackson) is not an educator. He is not even a basketball coach. He is a successful owner of a sporting goods store. His only qualification to coach or to teach, it seems, is that he was a basketball star at his high school in the 1970s. During a visit to his old school he notices with disappointment the lack of discipline among the players and their constant bickering. He is shocked to see the impoverished state of the school and the destructive attitude of the students. "That school was tough when I went there," he says, "now it's off the charts. . . . Those boys—so angry and undisciplined." It's up to him to set things straight. He sacrifices most of his free time to coach (and to teach life lessons to) the boys on the basketball team.[5]

---

[4]In his critique of *Dangerous Minds,* Henry Giroux (1996, p. 46) argues that the movie represents "whiteness" as the "archetype of rationality, authority, and cultural standards." While I agree generally with Giroux's critique of *Dangerous Minds,* I believe that these urban school films as a whole represent middle-class values, not whiteness, as the archetype of rationality, authority, and cultural standards. Americans generally lack the cultural language to make sense of social class. To the extent that they recognize social class, they often name it in racial terms. Similarly, racial differences in these films are very often conflated with and often stand in for social class differences. I agree with Barbara Ehrenreich (1989, p. 94) when she notes in her review of *The Wild One, Rebel Without a Cause,* and *Blackboard Jungle* that these films deliver "impeccable middle-class messages: crime doesn't pay; authority figures are usually right; you can get ahead by studying."

[5] Some of my analysis of Coach Carter was originally published as "Coach Carter: The Urban Cowboy Rides Again." *Contexts: Understanding People in Their Social Worlds,* Summer (2005), 73–75.

In *Freedom Writers*, Erin Gruwell (Hilary Swank) is a young, naïve, and idealistic teacher who has never taught before. She declined opportunities to study law in order to become a teacher, so that she can reach young people before they find themselves in the criminal justice system. At first she faces stiff resistance in the classroom by hostile students. Her students are embroiled in gang wars and distrust anyone not of their own race. Ms. Gruwell eventually "reaches" her students by telling them about the greatest gang there ever was—the Nazis. She takes her students on a fieldtrip to the Museum of Tolerance and invites Holocaust survivors to speak to her class. By teaching about the Jewish Holocaust she breaks down the irrational prejudice her students have of each other. The students begin to exercise compassion toward each other and begin to care about schooling. The film reduces the problems in urban schools to gang violence and reduces the problems of gang violence to individual prejudice. Once prejudice is overcome, anything becomes possible. Nothing else seems to be a barrier to individual success.

In *Stand and Deliver*, Mr. Escalante (Edward James Olmos) leaves a lucrative engineering job to teach high school math to Latino students in an East Los Angeles high school. Mr. Escalante insists on teaching calculus to students who normally would take regular or remedial math. His unconventional methods and his high expectations succeed. He gets his students to believe in themselves in spite of the doubts that their parents and the school authorities continue to have. His students pass the advanced-placement exam in calculus, and he inspires many of them to aspire to college. They begin to believe, as he tells one student who is covered in grease from working on his car, that it is better to design automobiles than to fix them. The only thing preventing them from designing cars, apparently, is a belief in themselves and the application of their abilities.

This same message about the ethic of hard work is repeated in film after film. As Mr. Thackeray (Sidney Poitier) tells his students in *To Sir, With Love*, "You can do anything you want with hard work." Similarly, in *Cooley High* (1975), Mr. Mason (Garrett Morris) tells a student whose future may be working on an assembly line, "What is it you want? With your brains you can have it. Knowledge will get it for you." The simple power of knowledge to open up opportunities and to transform troubled lives is echoed by Ms. Johnson in *Dangerous Minds* as well: "The mind is like a muscle. If you want it to be really powerful you have to work it out. Each new fact gives you another choice." Coach Carter tells his players, "Go home tonight and look at your lives. Look at your parents' lives and ask yourself, 'Do I want better?'" As I have written elsewhere (Bulman, 2005b), it is not entirely clear what these poor and mostly African American adolescent boys in a depressed inner-city neighborhood and underresourced school are supposed to do to get a better life. The coach advises his students to show up to class and to study. It is hard to argue that this is bad advice. But the underlying message of the film is that the path to a better life is *entirely* in the hands of these low-income boys. It is a message of rugged individualism that denies the structural obstacles that these boys face on a daily basis. The message of the film is captured well by a poster hung prominently in the coach's office: "Life Is Full of Choices: Choose Carefully." Life *is* full of choices. The film, however, doesn't even begin to suggest that some lives are offered more choices than others and that it takes much more than individual determination to succeed in life.

Coach Carter and the other films about poor urban youth suggest to audiences that for those who dream big, live disciplined lives, and work hard enough, anything is possible. The assumption in these movies (and too often in actual schools) is that aspiring to fix cars, to work on an assembly line, or to live a life similar to that of your parents is a sign of personal failure. This attitude serves to condemn those students who, for whatever reason, do not have college in their future. Furthermore, it is disingenuous to assume that the only obstacle standing in the way of middle-class occupational attainment for these students is their individual attitudes and their failure to exercise their brains.

Hollywood's Politically Conservative Worldview

What messages do these urban high school movies send to audiences about urban education? Do these films implicitly endorse any particular political solution to the problems in urban schools? In each of these movies the answer to the students' problems is revealed to be primarily an individual one: to reform the individual student, not the educational system or the wider society. The teacher-hero in all of these films must teach students the right values and manners, to convince them they have the power to improve their lives, and to insist they make better choices and take responsibility for those choices. As Principal Joe Clark (Morgan Freeman) in *Lean on Me* tells the students in his high school, "If you fail I want you to blame yourself. The responsibility is yours."

While there is certainly nothing wrong with encouraging personal responsibility among students, these movies dramatize only a portion of the story when they portray a lack of individual effort as the only reason the future of poor students is often limited. The serious business of school reform or revitalization of the inner-city economy takes a distant back seat to the individual reformation of these poor and working-class students. Success is a choice that each individual student must make. In the absence of a portrayal of the social, political, or economic context in which these individual choices are made, an implicit (and sometimes explicit) conservative political message is conveyed in each of these films.

Near the end of each movie, the teacher- or principal-hero faces a crisis that almost causes her or him to give up the mission. In each case, however, the crisis is heroically dealt with, and the teacher or principal stays on the job, having found her or his true calling in life. The dilemma facing the hero in the climax of each of these movies tells a significant political story about urban policy choices in late 20th-century America. Should the state play an active role in the structural reform of urban schools and urban economies? Or should the state retreat and let market forces work the magic of the invisible hand? Or is there a "compassionately conservative" third way, in which public policy addresses inequality, but only at the level of the individual?

In *Stand and Deliver,* Mr. Escalante's students are accused by the testing authorities of having cheated on their advanced-placement calculus exams. Mr. Escalante begins to doubt himself and wonders if he placed excessively high expectations on his students. In the end, however, the students retake the test, and they all pass. The students are redeemed. Even more important, however, Mr. Escalante is redeemed.

In the face of bureaucratic resistance, he, as a newcomer to the profession of teaching, is able to apply new pedagogical methods and his students succeed beyond any level the school has experienced before.

Several crises face Ms. Johnson in *Dangerous Minds.* One of her students, the charismatic leader of the class, is shot and killed by a crack addict. Another of her students, Callie, gets pregnant and is pressured by the school administration to attend an alternative school. Still other students drop out of school altogether. Ms. Johnson begins to lose hope and announces she will not return to teach the following year. Her students, however, protest vigorously. They feel angry and, ironically, victimized by Ms. Johnson's apparent betrayal. Callie returns to school on Ms. Johnson's last day to ask her to stay. Callie refers to one of the poems they have studied in class to make her point:

> I thought you'd always be here for me. . . . I decided, we decided, we aren't going to let you leave like that . . . you have to rage against the dying of the light . . . we see you as being our light. You are our teacher. You got what we need.

Moved by her students' testimonies, Ms. Johnson decides to continue teaching at the school.

In *Up the Down Staircase,* Sylvia Barrett (Sandy Dennis), a new teacher in a rough New York City high school, decides to resign after less than one semester on the job. Frustrated and angry with the school bureaucracy, saddened by a student's attempted suicide, and disheartened because several of her students plan to drop out of school, Ms. Barrett decides she is not up to the challenge of teaching at a "problem-area school." "A teacher should be able to get through to her students, even here," she complains. Near the end of the term, Jose (Jose Rodriguez), a quiet, shy, and apparently depressed student, comes out of his shell and presides as the judge in a mock trial Ms. Barrett has organized for her English class. "I'm sorry you are leaving us," says Jose. "English was the greatest course I ever took." Thrilled that she has "gotten through" to Jose, Ms. Barrett changes her mind about quitting.

In *Blackboard Jungle,* several of Mr. Dadier's students assault him and harass his wife. Mr. Dadier loses the hope he had for all of his students and nearly takes a teaching job at an elite high school. In a fit of frustration toward the end of the film, Mr. Dadier asks, "What's the point of teaching if kids don't care about an education? And make no mistake about it. They don't!" However, Mr. Dadier soon regains faith in his students when they team up to defeat the most incorrigible troublemakers in class after they threaten Mr. Dadier with a knife.

In *Lean on Me,* the corrupt fire chief and mayor arrest Principal Clark for putting chains on the high school doors to keep the drug dealers out. The students, however, rally to his defense. They surround the jail and demand his release with testaments to how much he has helped them. The news that 75% of the students have received a passing grade on the state's minimum skills test redeems him. He is released from jail and returns to lead his flock.

In *Freedom Writers,* Erin Gruwell's freshman and sophomore students learn that school policy prevents Ms. Gruwell from being their teacher in their junior year.

The students, having finally learned to care for each other, to respect Ms. Gruwell, and to apply themselves in school, are incredulous that the school would make them take an English class with another teacher. In an attempt to allow an exception to the school policy, Erin Gruwell goes over the heads of the school administrators. After her efforts fail, the students contemplate engaging in civil disobedience to keep their teacher. At the end of the film the superintendent of schools unexpectedly intervenes and allows Gruwell to continue to teach her students in their junior and senior year. The students rejoice. Apparently, in spite of the progress they have made academically, the students still need Ms. Gruwell.

Coach Carter, angry that his students have not been performing well in school, locks the team out of the gym and forfeits several basketball games. The students, the parents, and the community at large are furious with the coach. The school board holds a meeting to discuss the lockout. Carter announces at the board meeting that he will quit if the school board votes to end the lockout. Nevertheless, the board votes to end the lockout. The basketball season resumes, apparently without Coach Carter. The dejected coach clears out his locker and walks into the gym one last time. There, on the basketball court, he finds his players. They are not playing basketball. They are studying. The student he earlier threw against a wall emotionally recites a poem for his coach. He has apparently been transformed from a gangster to a scholar during the week of the basketball lockout. He says emotionally to the coach, "Sir, I just want to say thank you. You saved my life." With such powerful testimony of the way in which his players *need* him, Carter has a change of heart—he will not quit.

In each of these moments of crisis, the teachers are at the end of their rope. They are disappointed that they have failed as teachers, angry that the students have not responded to their lessons, or frustrated that the administration has tied their hands. This moment epitomizes the anxiety and frustration with urban schools expressed by politicians and many middle-class suburbanites; it is a representation of the neoconservative impulse to retreat from state efforts to solve social problems. It is as if the teacher-hero says, "Well, I've done my best to help these people but it failed. Let's cut school funding, eliminate affirmative action, end welfare, and insist on personal accountability. Their failure is no longer my responsibility."

In Hollywood, the well-intentioned middle-class reformer ultimately succeeds just when failure seems imminent. Success, however, is measured not by any institutional or social changes, but by the adoration of the students for the teacher-hero. With such admiration from the students, the "compassionately conservative" teacher-hero continues to work with the students. This is the moment of truth in these movies—proof to the teacher-hero that the students have been successfully reformed. They have progressed from lower-class animals to respectable middle-class students who finally understand and appreciate the efforts of their middle-class hero. Their troubled lives have been compassionately transformed by a caring and concerned human being.

However, in spite of an emphasis on the value of individual transformation and self-reliance, the students in these films continue to express a need for a relationship with their teacher. This is a need that the teacher-hero, in all good conscience, cannot ignore. Who else will save these students? In *Blackboard Jungle,* Miller agrees to stay in school provided that Mr. Dadier does not quit. In *Dangerous Minds,* Raul agrees to graduate only if Ms. Johnson does not leave the school. In *Up the Down Staircase,* Jose's transformation as a student is due entirely to the efforts of Ms. Barrett (who decides not to quit because of Jose's transformation). In *Teachers,* one student's decision to care about school is implicitly predicated on Mr. Jurel's (Nick Nolte) decision to care about teaching. In *Lean on Me,* the crowd of students who gather to demand that Principal Clark be released from jail proclaim, "We don't want a good principal. We want Mr. Clark!" In *Coach Carter* the coach stays on the job only after a student credits him with saving his life.

An implicit assumption in most of these movies is that if the teacher- or principal-hero does not agree to remain at the school, the students would quickly jettison the lessons they have learned and return to their apathetic underperformance and violent behavior. No other teacher (and certainly no school reform) can reach these students. There is a dependence on the middle-class teacher by these lower-class students, which points to an inherent contradiction in these movies.

The teachers encourage their students to transcend their dysfunctional families, their rotten peers, their lousy schools, and their culture of poverty. The teachers encourage their students to use their power as individuals to compete successfully and to attain a higher social status. Yet, to reach this goal, the students are necessarily placed in a position of dependence on the teacher- or principal-hero. For all of the rhetoric about independence and individual achievement, we never see the students in these urban high school films fully express their autonomy. Rather, their individualism is firmly embedded in their relationship with the teacher-hero.

The lessons that these urban school films teach about autonomy, competition, and individual achievement ironically require a relationship of interdependence, cooperation, and shared goals. However, the lesson about interdependence, cooperation, and shared goals is left implicit. Independence and achievement, on the other hand, are heralded explicitly. They are heralded, however, without an awareness of the social connections and social institutions required to sustain them. This reflects the American culture's unwillingness to acknowledge our reliance on community. It also contributes to the sense of emptiness and loneliness that some sociologists have argued are part of the dark side of middle-class American life (Bellah, Madsen, Sullivan, Swidler, & Tipton, 1985).

Conclusion: The Urban School Frontier

With few exceptions, these urban high school films are a celebration of the middle-class values of rational calculation and individual achievement. There is no suggestion that a longer-term solution to the problems in urban public high schools must address employment in the inner city, equitable school funding, sensitivity to racial and class differences, or the restructuring of urban schools. In true Hollywood fashion, these teachers and principals have saved the day as solitary heroes. These educators—mysterious, troubled, well-intentioned, alone, selfless, and heroic—are the cowboys of the dangerous and untamed urban-high-school frontier. They represent the essence of American individualism—they stand outside society in order to save it. The students, meanwhile, are explicitly grateful for their salvation. However, the salvation the teacher-heroes offer is inevitably tangled up with the contradictions of American individualism: the independence they demand of students requires a relationship of dependence to achieve.

Certainly, a high score on a test, an emotional tribute to a beloved teacher, and happy and optimistic students make for a good dramatic conclusion. But what do these endings imply for the public's image of urban schools? The audience is left feeling triumphant and optimistic about the potential for improvement in urban public schools. However, by simplifying the many problems of urban public education and turning inner-city students and public-school teachers into caricatures of their respective social classes, Hollywood is reflecting middle-class anxiety about the problems of inner-city schools and the naive hope that such problems need not be a sustained political commitment from all members of society, but merely the individual moral conversion of poor students.

References

Ayers, W. (1996). A teacher ain't nothin' but a hero: Teachers and teaching in film. In W. Ayers & P. Ford (Eds.), *City kids, city teachers: Reports from the front row.* New York: New Press.

Banfield, E. (1968). *The unheavenly city.* Boston: Little, Brown.

Bellah, R. N., Madsen, R., Sullivan, W. M., Swidler, A., & Tipton, S. M. (1985). *Habits of the heart: Individualism and commitment in American life.* New York: Harper & Row.

Bourdieu, P. (1977). Cultural reproduction and social reproduction. In J. Karabel & A. H. Halsey (Eds.), *Power and ideology in education* (pp. 487–511). New York: Oxford University Press.

Bulman, R. (2005a). *Hollywood goes to high school: Cinema, schools, and American culture.* New York: Worth.

Bulman, R. (2005b). Coach Carter: The urban cowboy rides again. *Contexts: Understanding People in Their Social Worlds, 4,* 73–75.

Burbach, H. J., & Figgins, M. A. (1993). A thematic profile of the images of teachers in film. *Teacher Education Quarterly,* Spring, 65–75.

Chubb, J. E., & Moe, T. M. (1990). *Politics, markets, and America's schools.* Washington, DC: Brookings Institution.

Considine, D. M. (1985). *The cinema of adolescence.* Jefferson, NC: McFarland.

Eckert, P. (1989). *Jocks and burnouts: Social categories and identity in high school.* New York: Teachers College Press.

Ehrenreich, B. (1989). *Fear of falling: The inner life of the middle class.* New York: Harper Press.

Gibson, M. A., & Ogbu, J. U. (Eds.). (1991). *Minority status and schooling: A comparative study of immigrants and involuntary minorities.* New York: Garland.

Giroux, H. (1996). Race, pedagogy and whiteness in "Dangerous Minds." *Cineaste, 22*(4), 46–50.

Heilman, R. B. (1991). The great-teacher myth. *American Scholar, 60*(3), 417–423.

Hutcheson, R. (2001, May 21). Bush calls for faith-based "assault on poverty." *Sacramento Bee,* p. A15.

Lareau, A. (1987). Social class differences in family-school relationships: The importance of cultural capital. *Sociology of Education, 60,* 73–85.

Lieberman, M. (1993). *Public education: An autopsy.* Cambridge, MA: Harvard University Press.

MacLeod, J. (1995). *Ain't no makin' it: Aspirations and attainment in a low-income neighborhood.* Boulder, CO: Westview.

McCarthy, C., Rodriguez, A., Meecham, S., David, S., Wilson-Brown, C., Godina, H., et al. (2004). Race, suburban resentment, and the representation of the inner city in contemporary film and television. In M. Fine (Ed.), *Off white: Readings on power, privilege, and resistance* (pp. 163–174). New York: Routledge.

Schuman, H., Steeh, C., Bobo, L., & Krysan, M. (1998). *Racial attitudes in America: Trends and interpretations.* Cambridge, MA: Harvard University Press.

Swidler, A. (1986). Culture in action: Symbols and strategies. *American Sociological Review, 51,* 273–286.

Thomsen, S. R. (1993, April). *A worm in the apple: Hollywood's influence on the public's perception of teachers.* Paper presented at the Southern States Communication Association and Central States Communication Association joint annual meeting, Lexington, KY.

Willis, P. (1981). *Learning to labor.* New York: Columbia University Press.

Wilson, W. J. (1996). *When work disappears.* New York: Knopf.

CHAPTER 3

Race and Ethnicity

Victor Joseph:	You gotta look mean or people won't respect you. White people will run all over you if you don't look mean. You gotta look like a warrior! You gotta look like you just came back from killing a buffalo!
Thomas Builds-the-Fire:	But our tribe never hunted buffalo—we were fishermen.
Victor Joseph:	What! You want to look like you just came back from catching a fish? This ain't "Dances With Salmon" you know!

Smoke Signals (1998)

Furious Styles:	Why is it that there is a gun shop on almost every corner in this community?
The Old Man:	Why?
Furious Styles:	I'll tell you why. For the same reason that there is a liquor store on almost every corner in the black community. Why? They want us to kill ourselves.

Boyz n the Hood (1991)

The history of race and ethnicity in Hollywood films reflects the racial and ethnic inequalities of American society. In the early 20th century, movies were created and exhibited to Eastern European immigrants to encourage assimilation. Whiteness was a status to be achieved, a way of life to emulate, with homogeneity as the goal, at least for immigrants from Europe. However, the portrayal of most nondominant groups in American film was the story of the "objectified Other" (Collins, 2000), created and controlled through practices of exclusion, expulsion, and violence.

In 1915, D. W. Griffith made history with the first blockbuster film, *The Birth of a Nation.* Historians have called the film a major landmark for American cinema, and a "sacrifice of black humanity to the cause of racism" (Cripps, 1977, p. 40). Demonstrating the power of film to influence, and even shape, social events, this film sparked a political response from African Americans and supportive whites across the country. The release of the film, with its demonizing of African Americans (played by white actors in blackface) and sympathetic portrayal of the Ku Klux Klan, led to a nationwide protest and educational campaign organized by the National Association for the Advancement of Colored People (NAACP).

The NAACP published a 47-page pamphlet titled "Fighting a Vicious Film: Protest Against *The Birth of a Nation,*" in which they referred to the film as "three miles of filth" (Lavender, 2001). Sociologists were among those who protested the film. W. E. B. Du Bois published scathing reviews of the film in *The Crisis,* the official publication of the NAACP, and Jane Addams, in an interview for the *New York Evening Post* about *The Birth of a Nation* called it "unjust and untrue," noting that "one of the most unfortunate things about this film is that it appeals to race prejudice" (American Social History Project., n.d.). The most important, and perhaps defining, outcome of the film was the nationwide heightened awareness and collective identity of African Americans, reinforced by success in both censoring the most objectionable scenes from the film and banning the film from theaters in major cities (Cripps, 1977).

Griffith's work, based on exploitation of nativist and miscegenation themes, became central to the cinematic construction of the American racial order (Denzin, 2005). Movies throughout the 20th century reflected not just race prejudice, but the subjugation of people through their absence or, when included, in narrowly defined, stereotypical roles. This was true for all groups defined as Other, with some variation in the specific stereotyped roles between broadly constructed categories, such as Native American, Asian American, Hispanic American, and African American. As Victor Joseph tries to explain to his hapless friend in the movie *Smoke Signals,* there is a stereotype associated with being Indian that is important for him to learn and use to survive in the white man's world. While many would agree that there has been significant progress in addressing racial inequality over the past century, there have also been major roadblocks.

In recent years, there has been a reversal of the achievements of the mid-20th-century movements for racial justice, as evidenced by "the dramatic increase in black prisoners, and the growth of the prison-industrial complex, crumbling city infrastructures, segregated housing, soaring black and Latino unemployment . . . and deepening inequalities of incomes and wealth between blacks and whites" (Giroux, 2002, p. 237). Additionally, the "new racism" (Collins, 2004) operates in the guise of a color-blind society where the successes of the civil rights movement have been achieved and racial equality has been realized.

Racist portrayals of African Americans in contemporary films are plentiful, with persistent linking of violence with men and "sexual promiscuity to the nature and identity of African American women" (Littlefield, 2008, p. 677). Such negative portrayals are not limited to African Americans, but are also seen in images of Latinos/as (as criminals), Native Americans (as clichéd savages), and Asian Americans (as immigrant shopkeepers or martial artist experts). Oftentimes, a (white) hero in a film is established through the creation of a minority enemy, or "second-fiddle" such as Tonto to the Lone Ranger or Kato to the Green Hornet.

The readings in this chapter focus on race and ethnicity in film. In the first reading, Ed Guerrero examines contemporary black screen violence, tracing its origins from Hollywood's blaxploitation period, to the 'hood-homeboy films of the 1990s to "historical agonies" in black/African history. A central theme is the possibility of collective social change in the tradition of the black liberation movement versus the "grim, violent struggle for individual survival." Even when the latter is presented as a heroic action-adventure story, any political significance is undercut by what Guerrero calls "popcorn" violence. Ultimately, he asks the question, "How does one make a feel-good Hollywood movie, with big box office expectations, about some of history's most wicked crimes: racism, genocide, and slavery?"

In the next reading, Susan Searls Giroux and Henry Giroux analyze the popular, award-winning film *Crash,* pointing out that the film begins with a paradox about the loss of community in modern society and ends in an exchange of racist slurs. In keeping with the "new racism" noted above, Searls Giroux and Giroux claim that "most Americans believe that racism is an unfortunate and bygone episode in American history, part of a past that has been more than adequately redressed and is now best forgotten." They are critical of *Crash* for presenting racism as a function of private discrimination, rather than a "systemic political force with often dire material consequences." Perhaps most tragic of all is that the conditions that produce structural racism also destroy the possibility for an engaged democracy.

In the last reading, Carleen Basler addresses the image of Latino/as in film. Beginning with a brief historical overview of Latino/as, Basler examines the stereotypes that have been cemented in the American mind, noting that throughout history, there has existed an "us" (white Americans) and "them" (Latinos) attitude that is evident in the stereotyping of Latinos in film. She concludes by exploring the ways Latinos have found to subvert and resist negative stereotypes, typecasting, and the status quo way of making films.

The readings in this chapter draw our attention to the ways that race and ethnicity are socially constructed and how systems of inequality are structured and maintained. At the same time that film projects racial and ethnic inequality, it can also be a vehicle for challenging racist ideologies. As we engage with film as viewing audiences, we can apply the sociological imagination to understand the historical contexts, individual experiences, and social consequences of racism—old and new.

References

American Social History Project. (n.d.). *Reformer Jane Addams critiques* The Birth of *a* Nation. Retrieved April 22, 2009, from http://historymatters.gmu.edu/d/4994/

Collins, P. H. (2000). *Black feminist thought: Knowledge, consciousness, and the politics of empowerment.* New York: Routledge.

Collins, P. H. (2004). *Black sexual politics: African Americans, gender, and the new racism.* New York: Routledge.

Cripps, T. (1977). *Slow fade to black: The negro in American film.* Oxford: Oxford University Press.

Denzin, N. (2005). Selling images of inequality: Hollywood cinema and the reproduction of racial and gender stereotypes. In M. Romero & E. Margolis (Eds.), *The Blackwell companion to social inequalities* (pp. 469–501). Malden, MA: Blackwell.

Giroux, H. A. (2002). *Breaking in to the movies: Film and the culture of politics.* Malden, MA: Blackwell.

Lavender, C. (2001). *D. W. Griffith: The Birth of a Nation (1915).* New York: Department of History, The College of Staten Island of The City University of New York. Retrieved April 22, 2009, from http://www.library.csi.cuny.edu/dept/history/lavender/birth.html

Littlefield, M. B. (2008). The media as a system of racialization: Exploring images of African American women and the new racism. *American Behavioral Scientist, 51,* 675–685.

Reading 3.1

The Spectacle of Black Violence as Cinema[1]

Ed Guerrero

The raw synergy of comedy and history is a revealing place to start an exploration of black violence in American commercial cinema. African Americans living today, more than 40 years after the end of the civil rights movement, are confronted with many daunting ironies and unsettling paradigm shifts. As Chris Rock, one of the incisive, funky heirs to the politicized early comedic style of Richard Pryor, acidly implores in his stage monologue, wherever you find a boulevard named after this nation's great apostle of nonviolence, Dr. Martin Luther King, Jr., "run . . . because there's some violence going down." One must now read the post–civil rights era in the broader, less optimistic twilight of "postslavery."

With declining resources, and the hegemony of race and class privilege reasserted in the neoconservative language of a post-9/11, post-Katrina, world, the complexities of American society are sharp, and often very obvious. The 2008 election of the first black U.S. president, Barack Obama, is cast into ironic relief by the fact that far too many black people (along with much of the nonindustrial world) are slipping deeper into poverty and (globalized) ghettoization. For the restless many at the bottom, the only social "progress," it now seems, is that the main boulevards and public schools of these urban zones of disenfranchisement are named after the previous generation's martyrs. Worse still, the great collective aspirations of the black freedom struggle, across the political spectrum from nonviolent action to "by any means necessary" militancy, have eroded into an ambivalent, self-focused consumerism of brand-name jeans and sneakers, and intracommunal annihilation through gun-drug-gang violence.

In neoconservative, "running on emptiness—posteverything" America, where prisons surpass public education as lines in state budgets, black people have little say over their structurally determined condition at the bottom of the economic and social heap.[2] Despite the gains of a thinning, black middle class and the ornamental rhetoric of "black progress," for too many African Americans, stunted horizons and mean ghetto streets are the lived reality, now refracted in the 'hood-homeboy action flicks that have become such an influential, though fading, staple of what

---

[1]An earlier version of this chapter appeared as "Black Violence as Cinema: From Cheap Thrills to Historical Agonies" in *Violence and American Cinema,* edited by J. David Slocum (New York: Routledge, 2001).

[2]The racial income gap reveals the relative position of black and white earners. In 1975, blacks made $605 for every $1,000 that whites earned; by 1995 that figure had dropped to a black $577 for every white $1,000 (Hacker, 1997). The "wealth gap" is even more telling; white households have a median net worth that is 14 times that of black households ("Study finds," 2004).

critics have dubbed the "new black film wave." While 'hood-homeboy violence is big box office, it's only one of the varied expressions of black violence on contemporary screens.

In this reading I would like to interrogate, historicize, and critically comment on some of the varied social, political, and psychic conditions factoring into black violence on the commercial cinema screen. In this process, perhaps I can address a few salient questions raised about black violence in the movies. For instance, what is the general framework within which dominant cinema violence, black and white, expresses itself? What are the origins of black violence in contemporary commercial cinema, and are there variations on the theme? And is black violence held to prefigured historical codes and a double standard by the dominant movie industry and its mainstream audience?

Violence in American Film

First and importantly, one must note that most black-focused films or black characters in mainstream films are not unique in portraying violence, and that blacks were latecomers to the screen violence game. With few exceptions, the stylistic range of black violence follows the overall configuration of mainstream white cinematic violence, which escalated in 1966 with the collapse of the Production Code and the advent of such technology-driven, stunningly violent, box office successes as *Bonnie and Clyde* (1967), *Bullitt* (1968), and *The Wild Bunch* (1969) (Amis, 1996).

Since then, we as a nation have become increasingly entertained by, and addicted to,[3] ever more graphic representations of violence expressed across a broad field of commercial movies with two loosely defined categories at either end of the violence continuum. At the most popular end is action-adventure or "popcorn" violence, with an emphasis on shootouts, car chases, pyrotechnics, quadraphonic noise, and ever increasing body counts. Other than shouted threats, curses, and screams, these films are light on dialogue, character development, and intellectual or psychological complexity. In industry terms, they are "sensation driven" and are made for Hollywood's biggest and most influential demographic, the young (Gabler, 1997). Consequently, the violent, action-adventure blockbuster delivers a jolting cinematic experience that's more akin to the thrill of a hyperkinetic amusement park ride or the action-packed computer games that, in terms of profit, outperform the average movie.

At the less profitable end are the social, psychological, or political dramas, and historical epics, made for older, baby boomer audiences seeking an aesthetic or intellectual experience at the movies. These films are considered by the industry as "plot and character driven." When violence is depicted, it is aesthetically, socially, or morally edifying, erupting in such contexts as dramatic, character-focused conflicts or the broader sweep of history as in genocidal cataclysms, mass movements, political struggles, and social upheavals.

---

[3]According to Monk-Turner et al. (2004), "A taste for violence, like any addiction, needs more of a fix over time. We believe that it will take more violence, and more graphic violence, to capture the attention of modern audiences" (p. 3).

Responding to Hollywood's biggest audience (adolescent males 16–24) violence driven, crime cinema, horror, and action-adventure flicks like the *Terminator* films (1984–2009), the *Die Hard* films (1988–2007), *True Romance* (1993), *Independence Day,* (1996), *The Replacement Killers* (1998), *Kill Bill: Vol. 1* and *Vol. 2* (2003–4), *Saw I–V* (2004–8), and *No Country for Old Men* (2007) abound. As a result, popcorn hyperviolence is one of Hollywood's biggest moneymaking ingredients. At the other end of the spectrum, films that explore the consequences of violence, like *The Pawnbroker* (1964), *Dr. Zhivago* (1965), *Gandhi* (1982), *Schindler's List* (1993), *Michael Collins* (1996), *The Passion of Christ* (2004), or *Apocalypto* (2006), are less common in the industry.

However, since some sort of violence is a necessary plot ingredient for box office success these days, these polarities are not discrete, and we find a great many feature films alloyed with violent moments regardless of the overall theme. Moreover, in the great mix of the middle resides films that sample graphic violence in an action-adventure style yet still strive to make an aesthetic, ideological, or psychological point, such as *Falling Down* (1993), *Boogie Nights* (1997), and *The Game* (1997). Even the industry's biggest blockbuster ever (at this writing), *Titanic* (1997), which is a romantic adventure staged in a historical context, cannot escape deploying liberal doses of the popcorn disaster violence first seen in its action-adventure-disaster ancestor *The Poseidon Adventure* (1972). There are also films that are graphically violent but claim to take an ironic view of violence, or to be statements against it, such as *Natural Born Killers* (1994) or the classic Vietnam War films such as the symbolic *The Wild Bunch* (1969) and the literal *Platoon* (1986).

Violence in Black-Focused Commercial Films

In black-focused commercial films, we find a similar spectrum between the polar ends of cheap thrills and historical agonies. Features like *New Jack City* (1991), *Juice* (1992), *Trespass* (1992), *Menace II Society* (1993), *Posse* (1993), *Bad Boys I* and *II* (1995, 2003), and *Set It Off* (1996) are all centered on violence as action-adventure entertainment and make up the popular end of the field. Then there is the hyperviolent romanticization of "the life" in *Belly* (1998), with its poster bright colors, slick music video surface. At the other end, examples of black-focused films that attempt to depict violence in a historic, philosophic, or epic context would include *Malcolm X* (1992), *Amistad* (1997), *Rosewood* (1997), and *Beloved* (1998). Films that are a mix of action-adventure violence and politicized or historicized themes would include films like *Boyz n the Hood* (1991), *Dead Presidents* (1995), and *Panther* (1995), along with the more sanguine and cautionary tale of "the game," the *noir* 'hood tour de force, *Paid in Full* (2002). Moreover, perhaps ironically, both *Boyz n the Hood* and *Menace II Society,* and more successfully *Clockers* (1995), all make varied claims that their violent tales are meant to help stem black urban violence.

However we map this tangled field, one can convincingly argue that the origins of contemporary black screen violence are located in Hollywood's blaxploitation period, which consists of about 60 cheaply made black-focused action-adventure flicks released between 1969 and 1974. A scene in *The Dirty Dozen* (1967) marked a shift in Hollywood; star athlete turned actor Jim Brown sprinted down a line of ventilators, dropping grenades into them and blowing up a gathering of the elite Nazi German command (to the cheers of blacks and the approval of the mainstream white audience). The industry's long unspoken but strictly observed rules regarding the expression of black violence toward whites, and the sanctity of the white body, were beginning to erode under the political pressures of the civil rights movement and the surging "black power" aspirations of urban blacks.

Before this defining cinematic moment, with rare exceptions, like Paul Robeson killing a white prison guard (a scene which was cut by censor boards in the United States after the film's release) in *The Emperor Jones* (1933), or the obligatory threat of savage (Hollywood) tribesmen in the Tarzan flicks, or the Mau-Maus in *Something of Value* (1957), or the great siege battle in *Zulu* (1964), black lives were expendable and, in many instances, spectacularly devalued. Except for the functional purposes of staging threats and challenges to European supremacy to which whites could heroically respond, nonwhites were prohibited from inflicting violence upon whites, for compared to black life, white life was sacrosanct on the cinema screen (Shohat & Stam, 1994).

Blaxploitation Film

It was not too long after Jim Brown's grenade attack that a black superhero outlaw with revolutionary pretensions emerged in *Sweet Sweetback's Baadasssss Song* (1971) to maim or kill several white policemen, enjoy various dubious sexual escapades, and then escape a citywide dragnet to brag about it all. Thus the blaxploitation formula, which generally consisted of a black hero out of the ghetto underworld, violently challenging "the Man" and triumphing over a corrupt, racist system, was born. What followed was a succession of detectives, gangsters, ex-cons, cowboys, dope dealers, pimps, insurgent slaves, and women vigilantes in flicks like *Shaft* (1971), *Across 110th Street* (1972), *Super Fly* (1972), *Black Caesar* (1973), *Black Mama, White Mama* (1973), *Boss Nigger* (1975), and *Drum* (1976), in which the protagonists shoot, punch, stab, and karate chop their way through a series of low-budget features that garnered megaprofits for Hollywood (Guerrero, 1993).

A couple of things are notable about the construction of violence in most blaxploitation period pieces. For one, because the technological advances of today's cinematic apparatus allow the industry to ever more convincingly represent or simulate anything that can be written or imagined, blaxploitation violence in many instances now appears crudely rendered and visually "camp" or naive in comparison to graphic blood- and brain-splashing shoot-outs in contemporary 'hood-homeboy action flicks. Also, the blaxploitation genre had a place for macho women in its pantheon of fierce action stars, women who echoed and upheld the cultural moment's call for reclamation of black manhood in the most violent, masculinist terms. Pam Grier in *Coffy* (1973) and *Foxy Brown* (1974) and Tamara Dobson in *Cleopatra Jones* (1973) play sexy black women adventurers configured to the social message of the times and black male adolescent fantasies that largely determined the success of blaxploitation films at the box office.

Blaxploitation films are full of fantastic moments of popcorn violence, like Foxy Brown (Pam Grier) triumphantly displaying the genitals of her archenemy in a jar, or vampire Blacula (William Marshall) (*Blacula,* 1972) gleefully dispatching several white L. A. cops. However, in most cases, these films referenced black social reality, or transcoded,[4] however fancifully, black political struggles and aspirations of the times. The historical context in the urban north during this time involved increasing disenchantment with the limited gains of the civil rights movement among African Americans and a rise in black militancy with insurrections in hundreds of U.S. cities; these social energies found barely containable expression on the blaxploitation screen.

Super Fly (Gordon Parks, Jr., 1972)

The box office hit *Super Fly* exemplifies the social grounding of blaxploitation violence in a variety of expressions, even though the film was considered regressive by many black critics because of its blatant celebration of cocaine use and the hero's self-indulgent, drug pushing, hustling lifestyle. In the film's opening, the protagonist dope dealer, Super Fly (Ron O'Neal), chases down and then brutally stomps a junkie mugger. However, the *mise-en-scène*[5] is insistently socially contextualized as the foot chase winds its way through the grimy alleys and dilapidated tenements of Harlem and culminates with this hapless derelict getting the vomit kicked out of him in front of his impoverished wife and children, and all this to the overdub refrain of Curtis Mayfield singing the hit "Little Child Runnin' Wild." No matter what we think of Super Fly, or his nameless junkie victim, the social setting of this vignette forces us into a disturbing awareness of urban poverty, drugs, and the wicked symbiotic power relation between the junkie and the pusherman.

In contrast to the gritty realism of the black underworld informing *Super Fly,* where, interestingly, no guns are ever fired and most of the violence wears the cool mask of macho gesture, threat, and intimidation, the movie concludes with a rather fantastic athletic explosion of fisticuffs. In a cartoonish allusion to Popeye's love of spinach, Super Fly toots up on cocaine and then single-handedly whips three police detectives (in slow motion, no less), all while the corrupt police commissioner, literally known as "the Man," holds him at gunpoint. The social issues of this violent denouement emerge when Super Fly informs "the Man" that he's quitting the dope business, with a million dollars in drug profits. However, this retirement speech is also meant to recuperate the film's reactionary attitude and align it more closely

---

[4]Transcoding can be understood as the way that the social world is represented in film.

[5]*Mise-en-scène* is a French term and originates in the theater. It means, literally, "put in the scene." For film, it has a broader meaning, and refers to almost everything that goes into the composition of the shot, including the composition itself: framing, movement of the camera and characters, lighting, set design and general visual environment, even sound as it helps elaborate the composition (Kolker, 2001).

Photo 3.1 Political consciousness and macho women in the blaxploitation genre. Foxy Brown (Pam Grier) makes a call for black unity.

with the political energies of the times. For Super Fly dramatically tells "the Man" off in the collective voice and terms of the black social insurgence of the late 1960s, thus framing what would be a totally implausible scene in the social yearnings of the historical moment. This type of politically conscious speech is fairly standard throughout the genre, from Ji-Tu Cumbuka's rousing gallows speech after an aborted slave revolt in *Mandingo* (1975) to Pam Grier's call to black unity and arms, against the backdrop of a wall-size George Jackson poster, in *Foxy Brown.*

'Hood-Homeboy Film

One can argue that society's racial power relation hasn't changed all that significantly in the past 20 years (Bell, 1995), but it's clear that the psychic and social influences impelling the construction of black cinematic violence most certainly have changed. In comparison to the blaxploitation era, the 'hood-homeboy films of the 1990s are less obviously politically focused, and in their violent nihilism (and sometimes self-contempt) they hardly suggest the possibility of social change. Violence in the blaxploitation genre, no matter how crude or formulaic, captures the black liberation impulse of the 1960s. By contrast, the depiction of violence in the new black film wave, and especially the homeboy-action flick, rather than mediating black social and political yearnings, is concerned more with depicting the grim, violent struggle for individual survival left in the wake of the faded collective dreams of the 1960s. *The Toy* (1982) and *Trading Places* (1983) certainly signal this shift for black-focused mainstream cinema as surely as the rise of their respective stars, Richard Pryor and Eddie Murphy, followed the end of blaxploitation. The concentration of Hollywood's attention and production budgets on the rise of Pryor and Murphy coincides with the political and cultural shift from the "we" to "me" generation in mainstream culture and in representations of African Americans in commercial cinema. For no matter how imperfectly rendered its narratives, violence in much of blaxploitation either depicted or implied shaking off "the Man's" oppression, and most important, moving toward the dream of a liberated future.

By contrast, violence in the new black film wave for the most part transcodes the collapse of those very hopes under the assaults of the Reagan and Bush years and their rollback policies on affirmative action, black social progress, multiculturalism, and the social progress of the welfare state in general. This is the ghettocentric 'hood-homeboy flick's most salient political point. With black inner-city neighborhoods ringed and contained by police departments, totally deindustrialized, chemically controlled with abundant drugs, fortified with malt liquor, and flooded with cheap guns, ghettos have become free-fire zones where the most self-destructive impulses are encouraged by every social and economic condition in the environment. As a cynical white cop in Spike Lee's Clockers (1995)coldly analogizes the situation, the urban ghetto has become "a self-cleaning oven." At best, then, turn of the century, cinematic 'hood-homeboy violence is socially diagnostic, an aesthetic attempt to raise consciousness by depicting the symptoms of a failed social and racial caste system. These depictions amount to endless variations on a theme: grotesque scenes of black and other nonwhite people trapped in ghettos and killing each other. Note Doughboy's (Ice Cube) "either they don't know, or they don't care" appeal for an awakening of conscience (and consciousness) at the end of *Boyz n the Hood.* The Hughes brothers, the directors of *Menace II Society,* rather disingenuously, say it another way, declaring that they are not here to give people hope but rather to depict the realities of those trapped in the urban 'hood (Giles, 1993). Thus, one can discern the great distance between the political consciousness underpinning even the cheapest of blaxploitation, ghetto thrillers, and the "keepin' it real" nihilism of contemporary 'hood-homeboy flicks.

Yet, as with blaxploitation, the success of any commercial black film today is still eminently configured by the tastes of its (and Hollywood's) biggest aggregate, the youth audience, which mainly consumes action-adventure violence and generalized comedy, to the avoidance of drama- and character-focused films (Fabrikant, 1996). Consequently, the popular end of the black-focused film spectrum is crowded with productions deploying violence very much in the style of dominant cinema, as a necessary action-adventure ingredient for box office success. Moreover, one can perceive a definite escalation or acceleration of violence from film to film. John Singleton's *Boyz n the Hood,* while punctuated with explicitly violent scenes, structured its narrative and ideology around the fates of good and bad brothers, Tre (Cuba Gooding) and Doughboy (Ice Cube). In this way the film at least holds out the possibility of escape from the 'hood through the time honored, black race-building route of education. Conversely, the fates of Caine (Tyrin Turner) and O-Dog (Larenz Tate) are sealed when O-Dog brutally murders two Korean grocers in the opening moments of *Menace II Society.* From this gruesome beginning, the trajectory is straight down. And compared to *Boyz* the body count increases exponentially; the action in *Menace* is best described as sort of a gruesome hyperviolence or autogenocide.

Correspondingly, many of the issues and debates regarding the depiction of extreme graphic violence as entertainment or realism were brought into focus with the release and box office success of *Menace II Society.* The Hughes brothers have claimed that their depiction of violence in *Menace* was a means to promote antiviolence efforts in the urban 'hood, telling the *New York Times,* "we wanted to show the realities of violence, we wanted to make a movie with a strong antiviolent theme and not like one of those Hollywood movies where hundreds of people die and everybody laughs and cheers" (Weintraub, 1993).

However, the film's violence and its impact on its audience would seem to annul this claim. As the brothers say, throughout most of *Menace,* violence is handled in an ugly, brutal fashion. The narrative is saturated with so many intensely violent scenes that violent action becomes the main structuring, captivating—and thus cumulatively entertaining—device in the film. What is more, in *Menace*'s final moments, when Caine is cut down in a hail of bullets, the scene is filmed in close-frame slow motion, thus fetishizing violence and evoking a style in the grammar of cinematic violence paying homage to Hollywood's foundational *Bonnie and Clyde* (1967) (Massood, 1993). But perhaps more disturbing was *Menace*'s social impact, with film and media critics noting that youth audiences cheered the violent scenes, particularly the one in which the Korean grocers were killed. Because film is an intensely visual medium, audiences, whether impressionable or sophisticated, will always look past what a director says about a film's lofty intent to the visible evidence of what a film actually depicts on the big screen.

It is also important to note that ghettocentric violence is not always revealed through the lens of the 'hood-homeboy action formula and that in various ways some black artists have tried to counter its exploitation on the screen. When he stands up to the neighborhood bully, Ice Cube's character, Craig Jones, comes to a critical moment of extreme provocation in Ice Cube's popular 'hood-homeboy comedy *Friday* (1995). Flush with anger, Craig Jones pulls a gun, but instead of following the protocols of homeboy realism and "bustin' a cap" on the neighborhood bully, he pauses and reflects on the dire outcomes of such an act. Spike Lee's psychologically complex, character-focused drama *Clockers,* about a low-level drug dealer, covers the same issues, and on the same deadly turf. Like Ice Cube, Lee turns his eye more to the destructive consequences and grief that violence brings to the community, people's lives, and families. The violence in *Clockers* is minimal, awkward, low-key though realistic, and decidedly not of the action-entertainment variety. Marking the contrast between antiviolence strategies in *Clockers* and *Menace,* critic Leonard Quart (1996) comments that "Lee truly wants to turn his adolescent audience away from violence, rather than ostensibly moralize against it like *Menace,* which simultaneously makes the gory spectacle of people being slaughtered so exciting that the audience could howl with joy while watching."

Perhaps one of the most artful, sanguine tales about "the life" or "game," and its violent, nihilistic consequences, is *Paid in Full.* Set in 1980s Harlem, at the dawn of hip-hop, pagers, and the rise of sophisticated gang-drug organizations, violence here is a business matter. Nothing personal, violence is disciplinary. Show up late on the corner, or short on the count, or talk too much, and "you subject to get popped." Or in *Paid*'s narrative, as Rico (the rapper Cam'ron) rather unsympathetically tells Ace (Wood

Harris) who has just survived a violent stick up, "Niggas get shot every day B." This cautionary tale articulated in the 'hood-home boy film *noir* flashback, opens with Ace, a gaping head wound, soaked in blood, being asked by first-aid attendants who did this to him. In full, gory close-up, understanding where his actions have led him, Ace whispers "I did." Violence here is socially and psychically mimetic. We traverse Ace's flashback year through the black world, and "the game" without benefit of guide or interpretation, to come full circle to the tale's reckoning and bloody conclusion.

Nowhere are the contrasts between sensation-driven and character/plot driven films more evident than between the filmmaking practices of black men and black women, and how these gender differences are perceived by the dominant film industry and played out on the screen. Since Julie Dash's film *Daughters of the Dust* (1991), which had a reflective, dreamlike surface and no action-adventure violence, Hollywood has released into mainstream distribution a meager handful of features directed or written by black women, films like *Just Another Girl on the I.R.T.* (1992) directed by Leslie Harris, *I Like It Like That* (1994) directed by Darnell Martin, and *Eve's Bayou* (1997), *The Caveman's Valentine* (2001), and *Talk to Me* (2007), all directed by Kasi Lemmons. When it occurs, violence in these films is understated. It causes the protagonists a great deal of anguish and is used to dramatize the complexities of broader social or psychological situations.

Like other realms of cultural production, exclusion from positions of influence in the film industry intensifies with the number of minority statuses an individual holds. Hollywood's executive offices tend to view women's narratives as "soft," centered on drama and character, and outside of their most reliable moneymaking formulas, which, of course, means liberal doses of action and violence. Black women experience "multiple jeopardy" based on sex and race (Collins, 2000), such that black male filmmakers have greater opportunities and are able to reach a wider (male) audience.

Differences in industry perception and audience consumption of black men and women's products, especially concerning the uses of graphic violence, in part explain the box office success of *Menace II Society* and the comparative flop of *Just Another Girl on the I.R.T.*, which were both modest-budget black films released at the same box office moment.

It is also interesting to note that the gender hybrid originating in blaxploitation, the woman-focused action-adventure flick, has risen again in a series of variations and specific moments in films. *Set It Off*, featuring Queen Latifah, Jada Pinkett, and Vivica A. Fox, is about an all-girl gang that tries to escape the ghetto by pulling a series of increasingly violent bank jobs. In Quentin Tarantino's blaxploitation reprise *Jackie Brown* (1997), the undisputed queen of black, campy violence, Pam Grier, returns to play a somewhat more subdued, middle-aged airline stewardess who rooks the streetwise gunrunner Ordell (Samuel L. Jackson) out of $500,000. One of Grier's big Freudian moments of reversal and castrating violence comes when she presses a gun to Ordell's dick and talks bad to him. Recalling blaxploitation's sexual ideology, in Ice Cube's *The Player's Club* (1998), homosexuality, the great threat to cultural nationalism, is punished as the beautiful stripper protagonist brutally defeats the "wicked" lesbian antiheroine in the film's culminating woman-on-woman fistfight.

Historical Agony: Racism and Violence

If violence is the principal and profitable cheap, popcorn thrill in the 'hood-homeboy action flick, it also finds expression at the other end of the black cinema spectrum as historical agony in such films as *Amistad, Beloved, Malcolm X, Rosewood,* and to a lesser degree in *Panther.* As one of Spike Lee's most ambitious and publicized projects to date, *Malcolm X* grapples with a routine industry contradiction: it is a Hollywood biopic with blockbuster, crossover moneymaking pretensions, but at the same time, it aims to portray the life of a black revolutionary hero with some historical veracity. In pursuit of that broad audience in the middle, Lee mixes his renderings of violence by first entertaining us with the adventures of Malcolm as a hoodlum with his sidekick Shorty as they pursue the transgressive adventures of the hustling life. Lee even throws in a dance-musical number in zoot suits. But in the latter half of *Malcolm X,* the mood shifts and culminates with the brutally drawn out and explicit assassination of Malcolm X before a public assembly; the moment is both stunning and ambivalent in its effects. This scene has complex crosscurrents of political meaning, first working as historiographic realism to psychically shock us into fully recognizing the sacrifice that Malcolm X (and his family) made for black freedom and social progress. Yet the graphically violent surplus of the scene also turns it, in the Foucaldian sense, into spectacle, a public execution by firing squad rendered in brutal detail, thus punishing Malcolm for his beliefs. Accordingly, his last speech can also be viewed as a gallows speech (Foucault, 1979).

Like Lee's *Malcolm X,* and in contrast to his own "Panther" film, *The Huey P. Newton Story* (2001), Mario and Melvin Van Peebles's *Panther* aspires to historical distinction, this time by sympathetically depicting the rise of the Black Panther Party, and the grievances of the black community that brought the party into being: police brutality, ghettoization, economic marginalization, and political disenfranchisement. However, while grappling with the same recurring contradictions between commerce and politics that appeared in *Malcolm X, Panther,* in pursuit of profits at the box office, relies mostly on the action-adventure violence of Hollywood's heroic individual. Thus, beyond the political insights about its historical moment the film articulates, like in the shoot-out in which Bobby Hutton is killed, the film's deployment of popcorn violence in other scenes tends to undermine *Panther*'s claims to a historically realist style. Moreover, this tangle of issues involving varied styles of violence is certainly relevant to John Singleton's *Rosewood* and Steven Spielberg's *Amistad.* And perhaps these issues can be best explored by considering a question salient to both films: How does one make a feel-good Hollywood movie, with big box office expectations, about some of history's most wicked crimes: racism, genocide, and slavery?

Rosewood (John Singleton, 1997)

In *Rosewood,* John Singleton formulates the story about the real-life 1923 destruction of the all-black Florida town of Rosewood by a white mob as a revisionist western, replete with an opening scene of the loner hero Mann (Ving Rhames) riding into town on his black "hoss," packing two .45 automatics, and "lookin' for a nice place to settle down." As Lee did in *Malcolm X*, Singleton opts for a mix of action-adventure moments of popcorn violence, enveloped in a more shocking overall rendering of historicized violence. In one scene depicting the former, Mann hurries out of town to escape a deputized lynch mob, but he is set on by a gang of whites and chased deep into the woods. Finally, Mann turns, stands his ground, and opens up with both of his .45s. Cut to the whites hauling ass out of the woods, with the punch line coming when they later excitedly exclaim that they were ambushed by a gang of "ten or fifteen niggers." The audience explodes with laughter. Singleton's timing and editorial touch with this classic scene from the archives of the cinematic Old West proves just right.

Overall, however, what happens to Rosewood on that gruesome night, as rendered by Singleton, is not at all funny or entertaining. At the height of the film's action, the disturbing sight of black men and women hanging from trees and telephone poles illuminated by the flames of their burning community seamlessly merges with those old *Life*, *Jet*, and archival photographs of real lynchings in America's gallery of historic horrors. Consequently, *Rosewood*'s panorama of violence is decidedly not escapist entertainment in the Hollywood sense. Violence here provokes the return of barely repressed collective nightmares and guilty complicities, as well as a painful examination of the national conscience that we as a national audience do not like to address, especially as entertainment, even in the darkness and anonymity of the movie theater. The historic agony of genocidal violence is brought into sharp focus in one of *Rosewood*'s culminating scenes, when one of the mob's prime instigators, proud of his crimes, forces his young son to look at a pile of black bodies awaiting disposal. Here, all of humanity's body counts are evoked, from Auschwitz to Wounded Knee to My Lai to Rwanda and beyond. Singleton's obvious point is that hope resides in the next generation, as the child rejects his father and runs away from home.

Rosewood is a mainstream commercial vehicle, and as such, its approach to violence is necessarily a mixed bag. If the film aspires to social conscience by shocking us with the historically repressed and oppressed, the lynch mob and its victims, it unfortunately lapses into the delusion of Hollywood formula in its portrayal of violence against women, black and white. Here we confront Hollywood's trickle-down theory of punishment, with the most powerless individuals in any hierarchy taking the rap for privileged elites hidden from critical interrogation. So, one must ask, why is it that the darkest black woman in the cast (Akosua Busia as Jewel), who opens the film with her legs spread, squealing from the pain of rape, then in the film's closure is gruesomely displayed as a murdered corpse, face up, eyes open in close-up? The argument here is that the narrative chain of significations, and the visual framing of her corpse, links the spectacle of Jewel's punishment to miscegenation, but also, implicitly, to her color. The whole issue of a devaluing, color-caste hierarchy, in this instance focused on those whom Alice Walker (1983) has referred to as "*black*, black" women, continues to be a troubling reality in mainstream commercial cinema.

Rosewood ends with the camera looking down in a long shot of a shack as we hear the screams of the white woman Fannie (Catherine Kellner), who initially yelled "nigger" to set things off, being brutally beaten by her husband, mixed with an overdub of lush, poignant cinematic music of the type used to signal narrative and ideological resolution. In Singleton's defense, one can speculate that a society that could burn an entire black town on drunken impulse would have no trouble thrashing one defenseless, lower-class white woman of loose reputation, especially one who has been set up by the narrative. Fannie does bear the historical burden of the oft-deployed false rape charge against black men. Yet with this beating the film reverts to a final cheap thrill, amounting to another act of symbolic punishment that displays the sacrificial offender/victim as spectacle, while hiding the intrigues of the much more guilty and powerful. Coming in the film's closing moment, then, this beating concentrates blame for the genocide of a racial minority on yet another Hollywood out-group, disenfranchised white women (Girard, 1972).

In dominant cinema, moreover, the representation of black violence in the service of white narratives still circulates powerfully, animating some of Hollywood's most popular feature films. This is in part because the sign of blackness has become indispensable as the implicit, negative standard in neoconservative political rhetoric and moral panics about family, education, welfare, and crime. Simultaneously, the stylistic inventions and expressions of black culture powerfully influence every aspect of mainstream culture, especially urban youth styles, language, music, and dance. In many ways the films of writer-directors David Lynch and Quentin Tarantino epitomize the utility and profitability of black violence, as well as the deeply rooted psychological fantasies about the sign of blackness in the white popular imaginary.

David Lynch's crime-action-romance *Wild at Heart* (1990) opens with the gratuitous and brutally graphic murder of a black man who gets his brains publicly stomped out for the entertainment, and perhaps wish fulfillment, of the action-adventure audience. In this scene, in fact, David Lynch treats us to a "lynching," which also happens to be a play on his name and his style as invoked by the popular press (Willis, 1997). More broadly, Quentin Tarantino, in films like *Reservoir Dogs* (1992), *True Romance* (1993), *Pulp Fiction* (1994), and *Jackie Brown* (1997), appears to be deeply disturbed by barely repressed, ambivalent feelings about race in general, black masculinity in particular, as well as the issues of violence, miscegenation, and sex. Black male delinquents, while hip and alluring in Tarantino screen plays, wind up eliminated, raped, or murdered, with black male–white female miscegenation always punished. Conversely, black women are the exotic trophies of white male desire.

Perhaps the most troubling and historically predictable of Tarantino's constructions of black violence occurs in *True Romance,* with the figure of the vicious pimp, drug-dealer Drexl (Gary Oldman). With this character, Tarantino's construction of the violent black male criminal becomes so grotesque and caricatured that it serves as an updated version of its explicitly racist, historical referent, the "renegade Negro" Gus of *The Birth of a Nation* (1915). The similarities between the two are instructive. Gus, like Tarantino's drug-dealing pimp in dreadlocks, is a blackface caricature portrayed by a white man. Most significantly, both Gus and Drexl represent sexual threats to white women and are ultimately punished with violent deaths. Gus's pursuit of white women becomes Griffith's rationalization for the organization and glorification of the Ku Klux Klan, which lynches Gus. Asserting the same basic threat, Tarantino evokes the rape-rescue paradigm in the actions of Clarence (Christian Slater), who, in a spectacularly violent scene, kills the black criminal Drexl to redeem his white girlfriend from sexual slavery. Considering this scene, and Tarantino's obsession with (for pleasure and profit, no less) the word *nigger,* which he says he wants to shout from the rooftops (Groth, 1997), one can speculate as to how little power relations and systems of representation have changed in the dominant film industry.

Amistad (Steven Spielberg, 1997)

The mid-to-big-budget, mainstream feature at the end of the scale of black violence as historical agony and realism is Steven Spielberg's *Amistad,* weighing in with production costs totaling $75 million. *Amistad* caps a trajectory well established by a number of films made during the blaxploitation period addressing the issue of slavery, like the action-sexploitation-driven *Mandingo* and *Drum,* and sustained in the 1990s with *Sankofa* (1993) and *Beloved,* all of which sharply reverse or debunk Hollywood's genteel sentimental depiction of slavery over the 80-year span of its "plantation genre" (Guerrero, 1993). Because *Amistad* recounts the actual events of a historic slave revolt on the high seas, and the successful repatriation of the rebellious Africans after an extensive court case, the film's narrative struggles with two issues pertaining to black violence that are not given much exposure in dominant commercial cinema. The first concerns the right, the necessity even, of the slave to rebel against the tyranny of his or her oppressors. The second issue has to do with the graphic revelation of the violence and oppression routinely inflicted on blacks by the daily operations of the slave system, and how that revelation recasts the perception of slavery in the American psyche, white and black.

Of course it's a long way from historical actuality to the big-budget Hollywood canvas, with its ultimate imperative that everybody's story be measured by its box office potential—that is, reduced to the compromises and revisions of its commodity status. Given the pressures on any blockbuster to return a profit at the ratio of three to one on an already hefty investment, it seems that in search of the broadest audience (read: white patronage), Spielberg approaches the issue of black revolt against European systems of tyranny, or even the frank depiction of those systems, with narrative restraint, to say the least. Yet, due to the inherent violent and irrepressible surplus of the subject matter, *Amistad,* unavoidably perhaps, unmasks the extreme cruelty of the slave system in historical realist terms. Consequently, as in Spielberg's *Schindler's List,* as well as the work of Lee and Singleton, here again we see the containing power of delimiting form and formula aimed at ensuring the biggest audience, in direct conflict with the insurgent emancipation theme of *Amistad.*

Amistad opens with a spirited, furiously successful shipboard rebellion that soon goes adrift, with the black rebels being recaptured, imprisoned, and put on trial in New England. From here on, the Africans are enslaved by the representational chains of Hollywood neoliberalism as they are portrayed as confused, exotic creatures, denied all agency, voice, and centrality in what one would expect to be their story. The narrative drags, turning into a two-hour civics lesson about noble, well-intentioned whites defining and securing black freedom (Sale, 1998).[6]

When it comes to Hollywood's standard depiction of black freedom struggles, black characters tend to have no agency in the production of their own history and are reduced to passive victims emancipated by courageous whites (like the FBI in 1988's *Mississippi Burning* or the white lawyers in 1996's *Ghosts of Mississippi*). In fact, this stratagem is so cliché that Matthew McConaughey was recruited from *A Time to Kill* (1996) to play yet another dedicated white lawyer in defense of black liberation in *Amistad.* This convention is also present in *Schindler's List,* in which the narrative focus is on factory owner Oskar Schindler and concentration camp commander Amon Goeth, with the Jews mostly reduced to the passivity of victims, a grim historical backdrop. Whatever happened to the Warsaw Ghetto's resistance and rebellion?

However, any commercial film as visual commodity bears complex, contradictory forces and meanings that are not entirely containable, and are contested by many, often opposing, social perspectives. Regarding the depiction of violence in *Amistad,* this is no less the case. The slave revolt in *Amistad*'s opening, a revolt of oppressed against oppressor, has considerable action-adventure impact and historical appeal. Yet its placement is a subtle form of co-optation. Staging the insurrection in the opening moments of the narrative denies it any conclusive, cathartic force, thus displacing the insight that resistance against oppression in defense of one's freedom is a form of social justice, one afforded every white hero from John Wayne, Gary Cooper, and Henry Fonda onward. As we come to see, *Amistad*'s version of justice is a weary court case that essentially defines and celebrates constitutional definitions of *white* freedom. It is important to note that while the *Amistad* court case freed 38 Africans, it did nothing to alter the fate of millions of black people enslaved in the United States, or the efficient workings of the slave system itself. If anything, the slave system was further legally entrenched and refined by such sanctions as the Fugitive Slave Act of 1850 or the *Dred Scott* decision of 1857. In contrast, note who has agency and the narrative positioning of the violent slave insurrection in Haile Gerima's *Sankofa.* Coming at the film's conclusion, this slave rebellion works as a resolution, underlining the brutality inflicted on *Sankofa*'s blacks, broadcasting that they have agency, history, and justice in their own hands. What is at issue here is how the big-budget mainstream blockbuster, regardless of what its producers claim its subject is, almost always winds up talking about whiteness, guided by the persistent refrain of Eurocentrism that subtends its narrative (Dyer, 1997).

However, *Amistad* is punctuated with scenes that cannot be entirely repressed, scenes that are quite disturbing and challenging to comfortable, dominant notions of slavery, cinematic, psychic, or otherwise. As noted, Hollywood's depiction of slavery has experienced a sharp reversal of meaning since the resistant blaxploitation flicks of the 1970s. By now, the plantation more closely resembles the blood-soaked ground of

[6]Besides an excellent historical survey *of* the *Amistad* affair, Sale here gives a good account of how the *Amistad* rebels were constructed as passive "happy-go-lucky children in need of protection" at the bottom of the abolitionist defense council's hierarchy of concerns.

the concentration camp than it does the majestic site of aristocratic Southern culture and gentility. So the irrepressible horror of the slave trade comes into disturbing focus with *Amistad*'s depictions of the infamous Middle Passage, and with one particularly stunning scene when commerce and mass murder converge with brutal clarity. This happens when the Spanish slavers discover that they don't have enough rations to keep their entire human cargo alive for the Atlantic crossing. Echoing the death camp scene in *Schindler's List,* in which the weak are culled from the strong, in *Amistad* the cargo's weak and sickly are stripped naked, chained together, and very efficiently thrown overboard. Besides the naked, chained bodies, what is more disturbing about the scene's violence is the banal utility seen in the practices of maintaining human beings as slaves for profit. This insight is reinforced by other scenes of maintenance and discipline, in which slaves are fed, flogged, and washed to enhance their commodity value on the auction block in the New World. Very much like *Schindler's List, Amistad*'s true force resides in its visual shock value, and in those resistant images, currents, and arguments that escape the policing of Hollywood's neoliberal, paternalist discourse. Relevant to the fundamental problem with both *Amistad* and *Schindler's List,* one critic puts it succinctly when he asks why *Schindler's List* "is so complicit with the Hollywood convention of showing catastrophe primarily from the point of view of the perpetrators" (Bernstein, 1994, p. 429).

No discussion of black violence as cinema would be complete without the mention of dominant cinema's depiction of contemporary, black Africa on the big screen. As Africa resides in the Western imaginary, *at best,* as a vast zone of war, poverty, disease and underdevelopment (without really understanding how it got, or is persistently kept, that way), Hollywood's recent mid-budget productions have depicted "the dark continent" mostly through the lens of violence as an assortment of historical agonies. Certainly recent examples, *Hotel Rwanda* (2004), *Blood Diamond* (2006), and *The Last King of Scotland* (2006), depicting genocide, dictatorship, and neoslavery confirm this trope. Don Cheadle and Sophie Okonedo of *Hotel Rwanda* do an excellent job of dramatizing the Hollywood formula of presenting historical violence through the lens of the heroic individual, which flattens out and backgrounds the historical and social circumstances of this genocide.[7]

Besides their depiction of contemporary violence in the service of historical epic, *The Last King of Scotland* and *Blood Diamond* share variations on "the black/white buddy" theme. In *The Last King,* the paranoiac tyrant, Idi Amin (Forest Whitaker), is coupled with his Scotsman, physician confidant (James McAvoy); in *Blood Diamond,* work slave (Djimon Hounsou) is coupled with white redeemer (Leonardo DiCaprio). Considering the historical process of producing diamonds in Africa, from colonial conquest to apartheid to bush wars, there resides a nagging, ironic question implicit in the title *Blood Diamond:* is there only one? But also, if there are more, how can you tell which ones are drenched, or not, in blood?

[7]The HBO feature *Sometimes in April* (2005) provides a politically complex counterpoint to *Hotel Rwanda.*

Conclusion

I have no doubts that the cinematic depiction of black violence expressed in a variety of mixtures between entertainment and edification will continue to evolve as an inseparable part of Hollywood's accelerating commitment to all forms of cinematic violence as a technology-enabled, profit-driven industry strategy. Predictably, even though the 'hood-homeboy genre has been played out (Guerrero, 1998), the film industry will continue to produce a certain number of popcorn violence-saturated, empty entertainments, like *Belly* or *State Property* (2002), or the more restrained, complex, and socially grounded *Slam* (1998). Hope resides in the more innovative ways in which black filmmakers deploy violence in the service of their takes on realism, or the revision of social or historical issues from a black point of view. One can see suggestions of new directions in several recent black-inspired, black-cast commercial productions, including the Spike Lee–produced *Tales From the Hood* (1995) and the Oprah Winfrey–inspired and –produced *Beloved.*

A horror flick, *Tales From the Hood* makes it clear that for blacks, the horrific repressed fears returning in the form of the monstrous are markedly political and have to do with the great violent horrors of African American life: police brutality, lynching, racism, domestic abuse, and the catastrophic effects of overall social inequality. In its mix of popcorn and historical violence, the monsters that arise in *Tales* are more literal than metaphorical: corrupt, brutal cops framing black citizens, criminally violent 'hood-homeboys bent on autogenocide, and spouses who turn into violent monsters in front of their wives and children.

Similarly, *Beloved* struggles to innovate and frankly depict the reality and consequences of slavery's violence through the metaphor of scarring, both physical and psychic. The violent horror of slavery is revealed in "rememory," in a series of flame-lit, nightmarish flashbacks of hangings, whippings, and bizarre mutilations that continue to haunt and scar the psyches, the narrative present, and the bodies of the film's black cast. Ultimately, then, what will continue to subtly influence the trajectory of black screen violence and suggest new creative directions for its expression and critical understanding will be the persistent and conscientious visions of wave after wave of new black filmmakers, in all of their racial and heterogeneous incarnations as gays, women, men, subalterns, artists, intellectuals, and rebels, and their ability to bring fresh narrative, social, and representational possibilities to the big screen.

References

Amis, M. (1996). Blown away. In K. French (Ed.), *Screen violence* (pp. 12–15). London: Bloomsbury.

Bell, D. (1995). Racial realism—After we're gone: Prudent speculations on America in a post-racial epoch. In R. Delgado (Ed.), *Critical race theory: The cutting edge* (pp. 2–8). Philadelphia: Temple University Press.

Bernstein, M. A. (1994). The *Schindler's List* effect. *The American Scholar, 13,* 429–432.

Collins, P. H. (2000). *Black feminist thought: Knowledge, consciousness, and the politics of empowerment.* New York: Routledge.
Dyer, R. (1997). *White.* London: Routledge.
Fabrikant, G. (1996, November 11). Harder struggle to make and market black films. *New York Times,* p. D1.
Foucault, M. (1979). *Discipline and punish: The birth of the prison.* New York: Vintage Books.
Gabler, N. (1997, November 16). The end of the middle. *New York Times Magazine,* pp. 76–78.
Giles, J. (1993, July 19). A "Menace" has Hollywood seeing double. *Newsweek,* p. 52.
Girard, R. (1972). *Violence and the sacred.* Baltimore: Johns Hopkins University Press.
Groth, G. (1997). A dream of perfect reception: The movies of Quentin Tarantino. In T. Frank & M. Weiland (Eds.), *Commodify your dissent* (p. 186). New York: W. W. Norton.
Guerrero, E. (1993). *Framing blackness: The African American image in film.* Philadelphia: Temple University Press.
Guerrero, E. (1998). A circus of dreams and lies: The black film wave reaches middle age. In J. Lewis (Ed.), *The new American cinema* (pp. 328–352). Durham, NC: Duke University Press.
Hacker, A. (1997). *Money: Who has how much and why.* New York: Scribner.
Kolker, R. (2001). *Film form and culture.* McGraw-Hill.
Massood, P. (1993). Menace II Society. *Cineaste, 20*(2), 44–45.
Monk-Turner, E., Ciba, P., Cunningham, M., McIntire, P. G., Pollard, M., & Turner, R. (2004). A content analysis of violence in American war movies. *Analyses of Social Issues and Public Policy, 4,* 1–11.
Quart, L. (1996). Spike Lee's *Clockers:* A lament for the urban ghetto. *Cineaste, 22*(1), 9–11.
Sale, M. M. (1998). *The slumbering volcano: American slave ship revolts and the production of rebellious masculinity.* Durham, NC: Duke University Press.
Shohat, E., & Stam, R. (1994). *Unthinking Eurocentrism.* New York: Routledge.
Study finds white families' wealth advantage has grown. (2004, October 18). *New York Times,* p. A13.
Walker, A. (1983). *In search of our mothers' gardens.* New York: Harcourt.
Weintraub, B. (1993, June 10). Twins' movie-making vision: Fighting violence with violence. *New York Times,* p. C13.
Willis, S. (1997). *High contrast: Race and gender in contemporary Hollywood film.* Durham, NC: Duke University Press.

Reading 3.2

Don't Worry, We Are All Racists!

Crash and the Politics of Privatization

Susan Searls Giroux and Henry A. Giroux

Public reaction to the summer 2005 blockbuster *Crash* is a study in paradox. Well before it received the Oscar for Best Picture of the Year in 2006, Paul Haggis's directorial debut generated a great deal of discussion and no small amount of controversy in the media. Reviewers of the film tended to fall into two camps, describing it as gritty realism or sentimental manipulative melodrama; characters as believable or utterly stock; the central theme as committed to unmasking racism or to explaining it away in pious nostrums about the tragically flawed propensity of all humans to fear and distrust "the Other." The heated debates over *Crash* show little sign of abatement—especially given its new life on the college film circuit. Indeed, on that most combustible of public issues—race—some critics even went so far as to publicly challenge colleagues who penned unfavorable reviews of the film, reducing intellectual integrity to a political litmus test in which disparagement of the film translated into an unqualified endorsement of racist expression and exclusion.

Crash narrates the interconnected lives of some 15 characters—black, white, Latina, Asian—within a 36-hour time frame on a cold Christmas day in postriot, post-9/11 Los Angeles. The film begins with a paradox that captures the dominant mood about the politics and representations of race in America. Rear-ended on their way to a murder scene, Detective Graham Waters (Don Cheadle) and his partner and girlfriend Ria (Jennifer Esposito) respond to the accident in ways that communicate beyond the literal collision of metal and glass parts on foggy Mulholland Drive in car-obsessed Los Angeles. The scene reveals what happens when strangers are forced to rub against and engage each other across the divisive fault lines of race, class, ethnicity, and fear. For Waters, the collision gives rise to a doleful rumination about the loss of contact, if not of humanity, in a city where people appear atomized and isolated in their cars, homes, neighborhoods, workplaces, and daily lives. Desperate for a sense of community, if not for feeling itself, melancholic police detective Waters ponders, "In any real city, you walk, you know? You brush past people, people bump into you. In L.A., nobody touches you. We're always behind this metal and glass. I think we miss that touch so much that we crash into each other, just so we can feel something." Ria is less philosophical about the collision and jumps out of the car to confront the Asian woman who hit them. Any pretense to tolerance and human decency soon disappears as the crash becomes a literal excuse for both women to hurl racial epithets at one another.

A scene that begins with a rumination about the loss of community and meaningful contact ends in an exchange of racist slurs inaugurating the polarities and paradoxes that are central to the structural, ideological, and political organization of the remainder of the film. Like *Magnolia* (1999), *Amores perros* (2000), *21 Grams* (2003), *Syriana* (2005), and *Babel* (2006), among other celebrated films, *Crash* maps a series of interlocking stories with random characters linked by the gravity of racism and the diverse ways in which they inhabit, mediate, reproduce, and modify its toxic values, practices, and effects. Episodic encounters reveal not just a wellspring of seething resentment and universal prejudice, but also a vision of humanity marked by internal contradictions as characters exhibit values and behaviors at odds with the vile racism that more often than not offers them an outlet for their pent-up fear and hatred. Put on full display, racism is complicated by and pitted against the possibility of a polity enriched by its diversity, a possibility that increasingly appears as a utopian fantasy as the film comes to a conclusion.

The unexpected confrontation, the nerve-racking shocks, and the random conflicts throughout the film often require us to take notice of others, sometimes forcing us to recognize what is often not so hidden beneath the psychic and material relations that envelop our lives. But not always. Crashing into each other in an unregulated Hobbesian world, where fear replaces any vestige of solidarity, can also force us to retreat further into a privatized world far removed from the space of either civic life or common good. *Crash*'s director and screenwriter, Paul Haggis, plays on this double trope as a structuring principle that makes people uneasy, yet also in a more profound and troubling sense, more comfortable, especially if they are white. When racist acts are exposed in a progressive context, it is assumed that they will be revealing, furthering our understanding of the history, conditions, and agents that produce such acts. Equally important is the need to undo the stereotyping that gives meaning and legitimacy to racist ideologies. Haggis moves beyond the opening scene of his film to explore in detailed fashion the power of racial stereotyping and how it complicates the lives of the perpetrators and the victims. But ultimately, *Crash* does very little to explore the historical, political, and economic conditions that produce racist practices and exclusions and how they work outside of the visibility of serendipitous interpersonal collisions in ways that might reveal the continuous, structural dimensions of everyday racism. *Crash* foregrounds racism, mimicking the logic of color-blindness currently embraced by both conservatives and liberals, but Haggis takes the truth out of such criticism by an appeal to the liberal logic of "flawed humanity" in an attempt to complicate the individual characters who allow deeply felt prejudices to govern their lives. *Crash* unfolds another paradox in which social life becomes more racially inflected and more exclusionary just as dominant discourses and representations of racism become increasingly privatized. In short, for all its pyrotechnics, *Crash* gives off more heat than light. It renders overt racist expression visible, only to banish its more subtle articulations to invisibility, along with its deeply structural and institutional dimensions. Theorizing racism as a function of private discrimination—a matter of individual attitude or psychology—denies its role as a systemic political force with often dire material consequences.

Yet the public conversation about *Crash* in all its vivid contradiction has something to teach us, revealing the widespread confusion over the meaning and political significance of race in an allegedly color-blind society. That the same film can be read as a searing indictment of America's deep-seated racism and also as its very denial, in the form of grand and forgiving assessments of the all-too-human tendency toward misunderstanding and fear, reflects the kind of schizophrenia that marks the politics of race in the post–civil rights era. That such a film can win an Oscar at a time marked by rabid racial backlash only underscores this tension. With the formal dismantling of institutional, legal segregation, most Americans believe that racism is an unfortunate and bygone episode in American history, part of a past that has been more than adequately redressed and is now best forgotten, even as informal, market-based resegregation proliferates in the private sector. In spite of ample research on race and inequality in the United States, many white Americans believe that white equality, integration, and racial justice are not just established ideals, but accomplished realities.[1] A recent report by the Pew Research Center found that 68% of whites did not believe that the government's response to Hurricane Katrina "revealed the persistence of racial inequality," whereas 71% of blacks viewed the response as an expression of discrimination on the part of the racial state (Allen, 2005). If inequality persists today between blacks and whites, so mainstream opinion goes, it is a function not of structural disadvantage (e.g., dilapidated, dysfunctional schools; rampant unemployment or underemployment; unequal access to loans and mortgages; police harassment, profiling, or mass incarceration), but rather of poor character. Reinforced by the neoliberal mantra that negotiating life's problems is a singularly individual challenge, forms of racist expression and exclusion are denied their social origin and reduced to matters of personal psychology—a function of individual discrimination (or discretion, as arch conservatives like Dinesh D'Souza would insist) (Goldberg, 2002). Not only are racial logics personalized, they are also radically depoliticized. As a complex set of historical and contemporary injustices, racisms—the plural underscoring their multiple and shifting nature—are analytically banished from the realm of the political. The role of the state, political economy, segregation, colonialism, capital, class exploitation, and imperialism are excised from public memory and from accounts of political conflict. As Wendy Brown (2006) observes, when emotional and personal vocabularies are substituted for political ones, when historically conditioned suffering and humiliation are reduced to basic forms of a presumably universal "difference," calls for "tolerance" or "respect for others" are substituted for real political transformation. Social justice and action devolves into sensitivity training, and the possibilities for political redress dissolve into self-help therapy. The oddly despairing and redemptive ethos of *Crash* marks the transition from an earlier generation's commitment to civil rights to the compromised contemporary insistence on civility, for which Rodney King's plaintive inquiry "Can't we all just get along?" provides the dominant refrain.

---

[1]Richard Morin, "Misperceptions Cloud Whites' View of Blacks," *Washington Post* (July 11, 2001), p. A1. See, for example, David R. Williams, "Race, Socioeconomic Status, and Health: The Added Effects of Racism and Discrimination," *Annals of the New York Academy of Sciences*, Vol. 896 (December 1999), pp. 173–188; Eduardo Bonilla-Silva, *Racism Without Racists: Color-Blind Racism and the Persistence of Racial Inequality in the United States* (Boulder, CO: Rowman & Littlefield, 2006).

The intersection of racism and what we call its "humanization" comes into full view in a number of scenes in *Crash* that weave a diverse tapestry of actions held together by the intersections of difference, fear, and violence. Haggis begins the task of making racism visible—while implicating its victims and perpetrators—by structuring one of the most important scenes in the film around a carjacking. Haggis and his wife were actual victims of a carjacking in Los Angeles, an experience that provided the backdrop for writing the script.

Circling back to the afternoon before the first depicted crash on Mulholland Drive, two young black men in their twenties emerge from a bistro in an upscale white neighborhood. Anthony (Chris "Ludacris" Bridges) complains to Peter (Larenz Tate) about the daily stereotyping and humiliations visited upon poor people of color from both blacks and whites. He launches into the beginning of what will be an ongoing treatise on antiblack racism in America, ranging from the bad service he received in the restaurant by a black waitress to corporate hip-hop as a way of perpetuating black-on-black violence. Given a city wracked by two major racial uprisings and infamous for the racist violence of its police force, Anthony's assessment of internalized white supremacist beliefs would appear on target. His tirade is interrupted when he spots a wealthy white woman react in fear upon seeing him with his friend. Anthony is outraged:

> Man, look around you, man! You couldn't find a whiter, safer, or better-lit part of the city right now, but yet this white woman sees two black guys who look like UCLA students strolling down the sidewalk, and her reaction is blind fear? I mean, look at us, dog! Are we dressed like gangbangers? Huh? No. Do we look threatening? No. Fact: If anybody should be scared around here, it's us. We're the only two black faces surrounded by a sea of overcaffeinated white people patrolled by the trigger-happy LAPD. So you tell me, why aren't we scared?

Anthony responds with irony to this classic Fanonian moment of being "caught in the gaze," making it clear that if anyone should be afraid of racial violence it should be he and Peter, not white people, and especially not rich whites. Rather than allow his audience to ponder this insight, Haggis performs a cheap reversal. Peter responds to Anthony's rhetorical question in wry tones, "Because we got guns." In a startling turn of events, the two young men then force the wealthy couple out of their black Lincoln Navigator, and speed away from the crime scene. An encounter that at first seems to underscore the indignity and injustice of the racist gaze is dramatically cancelled out when the white woman's fear proves legitimate. The stereotype of the dangerous black man is suddenly made all too real, an empirically justified fact. In an effort to underscore the basic contradictions that mark us all as human, Haggis undercuts whatever insight the film offers about the contemporary racist imagination, which often equates the culture of blackness with the culture of criminality.

Haggis plays on and extends the dominant presumption of black criminality in the following sequence. Waters and his partner soon find themselves at the scene of another crime: a white undercover detective named Conklin (Martin Norseman) has shot and killed a black man driving in a Mercedes. Unfortunately for Conklin, the black man turns out to be a police officer in the Hollywood division. Conklin claims he shot in self-defense, but this does not ring true to Waters (Don Cheadle), who quickly discovers that this is the third black man that Conklin has killed in the line of duty. Haggis toys with a liberal sensibility that would assume this to be another tragic example of the violence of racial profiling, and Conklin as a pernicious profiler. But he soon shatters this liberal presumption as the detectives discover that the black cop is dirty, having cut himself into the L.A. drug trade. In a series of intertwined scenes, black people are characterized as complicit with racist practices or as flawed individuals whose sorry plight has less to do with racial subjection than with their own lack of character, individual fortitude, or personal responsibility. And the stereotypes cut across gender and class lines.

For all of the pious defense of Haggis's rigorous antiracism among film critics, it is curious to note that none reflect on the utterly offensive portrayal of black women in the film, of whom we are introduced to three. The most damaging portrayal is of Waters's mother (Beverly Todd), who is represented as cruel and dysfunctional, a crack addict whose lack of discrimination and judgment results in her doting on her wayward son, Peter, who is a carjacker, to the exclusion of the more responsible and hard-working Waters, the only functional member of the family. Waters's mother is a grown-up version of the infamous "welfare queen," a slander coined by the presidential campaign of Ronald Reagan and kept in circulation by conservative and liberal ideologues who pushed for dismantling the welfare state. Similarly, the portrayal of an African American women named Shaniqua (Loretta Devine) as the heartless, bureaucratic insurance supervisor who works for the L.A. public health service is equally vicious. In an instance of alleged "reverse racism," she denies an elderly and probable prostate cancer victim in severe pain a visit to a specialist because his son is a white, racist cop. And if Shaniqua's bigotry is not clear enough in this instance, we meet her again at the end of the film when she is rear-ended at a traffic light. Emerging from her car, she angrily shouts at the Asian driver who hit her car: "Don't talk to me unless you speak American." Shaniqua stands in as the bad affirmative action hire willing to punish poor whites because of her own racial hostility. Why bother with a critique of the social state or the crumbling and discriminatory practices of an ineffectual welfare system when heartless, hapless, and cruel black women like Shaniqua can be conjured up from the deepest fears of the white racist unconscious. The third portrayal is of wealthy, light-skinned Christine (Thandie Newton), first encountered when she performs oral sex on husband Cameron (Terrence Howard) while he drives home from an awards dinner. She quickly becomes the overly sexed and out of control black woman with a big mouth who has no idea how to negotiate racial boundaries. After being pulled over by a white racist cop to whom she mouths off, she is subsequently subjected to a humiliating body search. As humiliating as is the violence she experiences from the racist and sadistic cop, Christine appears irresponsible and unsympathetic, if not quite deserving of the racial and sexual violence she has to endure.

Even Waters, who provides a sense of complexity and integrity that holds together the different tangents of the story, is eventually portrayed as corrupt cop willing to corroborate in a lie about the shooting of a black officer; he will end up framing a not quite innocent white cop in order to keep his brother out of jail, get a promotion, and provide political advantage with the black community for the L.A. District Attorney (Brendan Fraser). Such scenes participate in a "color-blinding racism" that tends to focus on white victimization, while denying blacks grounds for protestation by rendering them complicitous with their own degradation.

But Haggis tempers his own confusion about the constitutive elements of the new racism, who perpetuates it, and under what conditions by powerfully organizing *Crash* around the central motif that everyone indulges in some form of prejudice. As Haggis explains in an interview for *LA Weekly*: "We are each such bundles of contradictions. . . . You can conduct your life with decency most of your days, only to be amazed by what will come out of your mouth in the wrong situation. Are you a racist? No—but you sure were in that situation! . . . Our contradictions define us" (Feeney, 2005). In other words, some racists can be decent, caring human beings and some decent caring human beings can also be racists. In this equal opportunity scenario, racism assumes the public face of a deeper rage, fear, and frustration that appears universally shared and enacted by all of L.A.'s urban residents. This free-floating rage and fear, in Haggis's worldview, is the driving force behind racist expression and exclusion. Thus racism is reduced to individual prejudice, a kind of psychological mechanism for negotiating interpersonal conflict and situational difficulties made manifest in emotional outbursts and irrational fears. Without denying its psychological dimensions, what such a definition of racism cannot account for is precisely racism's collusion with rationality, the very "logical" use to which racism has *historically* been put to legitimate the consolidation of economic and political power in favor of white interests.

Perhaps more important, once questions of history and power are excised from public consciousness, racial inequality can be "transformed from its historical manifestations and effects perpetuated for the most part by whites against those who are not white into 'reverse discrimination' against whites who now suffer allegedly from preferential treatment" (Goldberg, 2002, p. 230). We offer two elaborated examples, perhaps the two most remarked on scenes in *Crash*—the first sequence involving two LAPD cops, the utterly venomous, racist veteran cop, Ryan (Matt Dillon), and his sympathetically drawn rookie partner, Hanson (Ryan Phillipe), who pull over the upper-middle-class black couple to whom we have already alluded. The other scene portrays a domestic dispute between the utterly opportunistic District Attorney, Rick Cabot (Brendan Fraser), and his spoiled, overtly racist wife Jean (Sandra Bullock).

The first sequence involves Officer Ryan on the phone with the aforementioned HMO representative, Shaniqua—a conversation that ends on a very sour note. This prelude is important because it is intended to provide a context for the even more disturbing event about to unfold. When Ryan returns to the squad car, he and his partner witness a black SUV glide by, and they decide to follow. It is not the car involved in the earlier carjacking, but Ryan spies something amiss. In the afterglow of a little sexual foreplay, the couple is amused, but things turn more serious when Ryan asks a very sober husband, Cameron (Terrance Howard), to step out of the vehicle to see if he has been drinking. Christine, who has been drinking, takes offense and proceeds to verbally lambaste the officer as, unbidden, she too steps out of the SUV. Ryan calls the rookie for backup and insists the two put their hands against the vehicle to be patted down, a request that further infuriates Christine, whose assault has turned utterly profane and inflammatory, against the protests of her husband who demands that she stop talking. But it is too late, she has crossed the line and Ryan feels like the couple needs to be taught a lesson. He proceeds to sexually molest Christine in the guise of police procedure while demanding from her husband, who helplessly looks on, an apology for their illicit behavior on the road. The producer and his wife return to their vehicle silenced, humiliated, and broken.

While the scene renders Ryan entirely unsympathetic and hateful, our judgment is presumably to be tempered by taking account of the context in which the abuse has unfolded. It also sets us up for Ryan's later redemption when he happens upon a car wreck and saves Christine from a blaze that is about to consume her and her SUV. In keeping with Haggis's understanding of how racist outbursts occur—racism is repeatedly represented in a series of isolated incidents, rather than as a systemic and institutional phenomenon that informs every aspect of daily life. From Ryan's perspective, he has endured the assaults of two black women one after the other. He is the white victim of an allegedly incompetent affirmative action hire, as he later reveals, and of "reverse racism," since Christine called him "cracker," among other names, in a litany of personal assaults. Apparently, Ryan has regained some equilibrium when, in a rather incredible coincidence, he saves Christine's life, in spite of her aggressive efforts to resist his initial attempt. Having staged a profound disidentification between the audience and Ryan, Haggis now repositions us to admire the officer's bravery and selflessness. Curiously, critics have read this scene as indicative of Ryan's contriteness, even moral growth, his coming to terms with the consequences of his earlier actions and his efforts to transcend a racist attitude. But is this really what we witness? Does remorse drive his "heroic actions" in saving Christine, or is he simply doing his job, which is to serve and protect the public—a commitment he utterly violated scenes ago? Is there anything truly productive or uplifting in the pendulum swing of his character from racial victim to racial savior (with rapid downward momentum generating a brief sadistic lapse quickly forgotten if not forgiven in the upward arc of his transcendence)—or are these precisely the subject positions open to whites who refuse to engage self-reflectively and self-critically on deeply historical and power-infused social relations?

Photo 3.2 From racist to savior in *Crash*. Ryan (Matt Dillon) saves Christine (Thandie Newton) from a near-fatal car accident.

We are similarly set up for a "surprise" reversal (though by the end of the film the gesture has become well-nigh predictable) in the sequence involving the carjacking. As we've already argued, Haggis has flipped the script on two young African American men, who initially transcend the "thug" stereotype and appear educated, even critically conscious of such racist positioning, only to morph back into mainstream America's worst fears—a pair of gun-toting nightmares materialized in the flesh.

Not only does the scene insert a kind of empirical validity to the presumption of black criminality, it manages to pathologize critical thought in the translation, a double demonization entirely in keeping with the conservativism of the color-blind commitments of the post–civil rights era. But Haggis's intentions become more curious still. Shortly, we find the victimized DA and his wife at home. The district attorney consults his staff, preoccupied with how to spin the situation so as not to lose either the "law-and-order vote" or that of the black community—a tongue-in-cheek moment satirizing the pretence to governmental color-blindness while the language of race is invoked in private policy decisions to strategize favorable outcomes. His wife Jean, meanwhile, bristles in the kitchen as she watches a young Latino male, who she quickly surmises is a threatening gangbanger, change the locks. As with Officer Ryan, the stress of the evening has gotten the better of her and she explodes in an emotional fit before her husband. She demands to have the locks changed again, referring to the evening's earlier incident when she had a gun pointed at her face:

> It was my fault because I knew it was going to happen! But if a white person sees two black men walking towards her and she turns and walks in the other direction, she is a racist, right. Well, I got scared and I didn't say anything and ten seconds later I had a gun in my face. And now I'm telling you that your amigo in there is going to sell our key to one of his homies and this time it would be really fucking great if you acted like you actually gave a shit!

Jean's comments are of particular interest for a number of reasons. Like Ryan, Jean for the moment appears barred from the human race—but only for the moment. Like Ryan, she assumes the posture of being victimized by the social dictates and policy measures associated with antiracism. For her, discrimination dissolves into discretion: if avoidance of certain groups is enacted through rational application of generalizations backed by statistical evidence about the dangers associated with those populations, then it can't be racist (Goldberg, 2002). Jean's fear at the very sight of two young black men proves justified. Before they do anything, they are guilty of blackness; they "are" a crime. Now she is angry at being thought of as a racist, and as such exhibits what Jean-Paul Sartre called "bad faith."[2] Jean throws herself into an emotional fit in order to take on an identity that enables her to evade herself. She presents herself manifestly as what she is (i.e., a racist) in order ironically to evade what she is.

We have spent some time unpacking these scenes because, we argue, they reveal in synecdochal form the broader politics of the film. In similar ways, *Crash* ironically evades the question of racism in all its sociohistorical force and political consequence while at the same time seeming to face it. It simultaneously insists that we take account of our own prejudices and their hateful, material consequences and unburdens us of the task at the same time by rendering such dispositions as timeless, universal attributes of a flawed humanity. Complicating further the gesture

[2]For an elaborated discussion of this concept as an analytic for understanding racism, see Lewis Gordon, *Bad Faith and Antiblack Racism* (Atlantic Highlands, NJ: Humanities Press, 1995).

toward responsibility is the relatively unthinking way in which characters engage in racist verbal assault or physical violence and the frequently unthinking way in which they act humanely toward others. By rendering such actions as reflexive, emotive, or generally prereflective, *Crash* calls on us to become consciously reflective and responsible for how we negotiate a post-9/11 urban context rife with fearsome strangers—a positioning it simultaneously undermines.

The apparent answer to the problem of white victimization is tolerance—and for their later expressions of tolerance, both Jean (by the simple gesture of affirming her Latina maid) and Officer Ryan are cinematically redeemed. The effort to humanize stark racists has earned Haggis critical accolades. Critics like David Denby (2005) were quick to seize on this alleged insight on the part of the film:

> *Crash* is the first movie I know of to acknowledge not only that the intolerant are also human but, further, that something like white fear of black street crime . . . *isn't always irrational.*[italics added] . . . In Haggis's Los Angeles, the tangle of mistrust, misunderstanding, and foul temper envelops everyone; no one is entirely innocent or entirely guilty. (p. 110)

Within this equal-opportunity view of racism, the primary "insight" of the film trades in the worst banalities: good and bad exists in everyone. In such a context, the question of responsibility for the violence that racism inevitably produces becomes free-floating. It is a logic sadly reminiscent of the "banality of evil" that Hannah Arendt (2003) discovered during the Adolf Eichmann trial in Jerusalem; everyone, she recalled, looked for "Eichmann the monster," only to find a man very much like themselves. That such attempts at "humanization" should now be equated with redemption is a tragic denial of that history. The racism both on display and normalized in *Crash* is banal in Arendt's terms precisely because it is so thoughtless, or as she later put it a "curious, quite authentic inability to think" (p. 159). A willful forgetting of such banality cannot elude the question of responsibility or forget that justice "'makes sense' only as a protest against injustice" (Bauman & Tester, 2001, p. 63). Any notion of humanization based on such forgetting is apparently the very precondition for the noncruel to do cruel things.[3]

Crash's delineation of racial conflict and its alteration of the concept of racism from a power-laden mode of exclusion into a clash of individual prejudices both privatizes and depoliticizes race, drowning out those discourses that reveal how it is mobilized "around material resources regarding education, employment conditions, and political power" (Goldberg, 1994, p. 13). When the conditions that produce racist exclusions are rendered invisible, as they are in *Crash,* politics and social responsibility dissolve into either privatized guilt (one feels bad and helpless) or disdain (victims become perpetrators responsible for their own plight). In a universe in which we are all racist pawns, it becomes difficult to talk about responsibility—let alone the conditions that actually produce enduring racist representations,

[3]This theme is taken up brilliantly in Zygmunt Bauman, *Life in Fragments* (Malden, MA: Blackwell, 1995).

injustices, and violence, the effects of which are experienced in vastly different and iniquitous ways by distinct groups. The universalizing gesture implicit in Haggis's theory of racism cannot address the dramatic impact of racism on individuals and families marginalized by class and color, particularly the incarceration of extraordinary numbers of young black and brown men and the growth of the prison-industrial complex; a spiraling health crisis that excludes large numbers of minorities from health insurance or adequate medical care; crumbling city infrastructures; segregated housing; soaring unemployment among youth of color; exorbitant school dropout rates among black and Latino youth, coupled with the realities of failing schools more generally; and deepening inequalities of incomes and wealth between blacks and whites.[4] Nor can it grasp that the enduring inequality that centuries of racist state policy has produced and is still, as Supreme Court Justice Ruth Bader Ginsburg observes, "evident in our workplaces, markets and neighborhoods"("Race on screen," 1997). It is also evident in child poverty levels, which statistics show are "24.7 percent among African Americans, 21.9 percent among Hispanics, and 8.6 percent among non-Hispanic whites" (Chelala, 2006).

David Shipler (1998) argues powerfully that race and class are the two most powerful determinants shaping an allegedly postracist, post–civil rights society. After interviewing hundreds of people over a 5-year period, Shipler wrote in *A Country of Strangers* that he bore witness to a racism that "is a bit subtler in expression, more cleverly coded in public, but essentially unchanged as one of the 'deep abiding currents' in everyday life, in both the simplest and the most complex interactions of whites and blacks" (p. 27). Positioned against civil rights reform and racial justice are reactionary and moderate positions ranging from the extremism of right-wing skinheads and Jesse Helms–like conservatism to the moderate "color-blind" positions of liberals such as Randall Kennedy, to tepid forms of multiculturalism that serve to vacuously celebrate diversity while undermining and containing any critical discourse of difference.[5] But beneath its changing veneers and expressions, racism is fundamentally about the relationship between politics and power—a historical past and a living present where racist exclusions appear "calculated, brutally rational, and profitable" (Goldberg, 1993, p. 105). It is precisely this analysis of politics, power, and history that *Crash* leaves largely unacknowledged and unexamined—with the single exception of a cheap dismissal by the Assistant District Attorney Flanagan (William Fichtner).

At the same time, we recognize that part of the popularity of *Crash* is due to its neorealistic efforts to make the new post-9/11 racial realities visible in American cities, especially in light of a pervasive ideology of race transcendence that refuses to acknowledge the profound influence that race continues to exert on how most people

---

[4]For a compilation of figures suggesting the ongoing presence of racism in American society, see Ronald Walters, "The Criticality of Racism," *Black Scholar, 26*(1) (Winter 1996), pp. 2–8; and Children's Defense Fund, *The State of Children in America's Union: A 2002 Action Guide to Leave No Child Behind* (Washington, DC: Children's Defense Fund Publication, 2002), p. xvii.

[5]For a devastating critique of Randall Kennedy's move to the right, see Derrick Bell, "The Strange Career of Randall Kennedy," *New Politics, 7*(1) (Summer 1998), pp. 55–69.

experience their everyday lives and their relationship with the rest of the world. *Crash* brings to the attention of the audience how racial identities are played out under the pressures of class, violence, and displacement. *Crash* also pluralizes the American racial landscape, making it clear that racism not only affects black populations. It is also refreshing to recognize that the Los Angeles that *Crash* portrays (in contrast to, say, the L.A. of *Short Cuts*) is not entirely white, and that the public sphere is a diverse one that reflects a cosmopolitan American audience. Mostly, *Crash* explodes the assumption that racism is a thing of the past in America, but then blunts the insight by denying both power and history. Instead, Haggis implies that racism may be a public toxin, but we are all touched by it because we are all flawed. But drawing attention to race is not enough, especially when racism is depicted in utterly depoliticized terms, sliding into an expression of individualized rage far removed from hierarchies and structures of power. *Crash* seduces its audiences with edgy emotive force, but renders the truth of racism comfortable because it is removed from the realm of responsibility or judgment. It enacts the evasion of collective responsibility in the face of a pervasive system of racism bounded by relations of power and structures of inequality that encourage such failures. This is not a film about how racism undermines the social fabric of democracy; it is a film about how racism gets expressed by a disparate group of often angry, alienated, and confused—but often decent—individuals.

Racism in America has an enduring, centuries-old history that has generated a set of economic conditions, structural problems, and exclusions that cannot be reduced to forms of individual prejudice prevalent across a racially and ethnically mixed polity in equal measure. While the expression of racism and its burdens cannot be reduced to specific groups, it is politically and ethically irresponsible to overlook how some groups bear the burden of racism much more than others. How does one theorize the concept of individual responsibility, character, or equal-opportunity intolerance within a social order in which the national jobless rate is about 6%, but unemployment rates for young men of color in places such as south central Los Angeles have topped 50%? How does one ignore the fact that while it is widely recognized that a high school diploma is essential to getting a job when more than "half of all black men still do not finish high school" (Eckholm, 2006, p. A1). Law professor David Cole (1999, p. 144) points out in *No Equal Justice* that while "76 percent of illicit drug users were white, 14 percent black, and 8 percent Hispanic—figures which roughly match each group's share of the general population," African Americans constitute "35 percent of all drug arrests, 55 percent of all drug convictions, and 74 percent of all sentences for drug offences." Within such a context, the possibilities for treating a generation of young people of color with respect, dignity, and support vanish, and with it the hope of overcoming a racial abyss that makes a mockery out of justice and a travesty of democracy.

In addition, it is crucial to point out that *Crash* apologizes for the growing violence and militarization of urban public space that are part of a "war on crime" largely waged against black and brown youth by what David Theo Goldberg (2002, p. 4) has called "the racial state." As the state is stripped of its welfare functions and it negates any commitment to the social contract, Goldberg argues that its priorities shift from social investment to racial containment, and its militarizing functions begin to function more visibly as a state apparatus through its control over the modes of rule and representation that it employs. Unfortunately, the film's commitment to privatized understandings of racism leaves intact the myth that collective problems can only be addressed as tales of individual plight that reduce structural inequality to individual pathologies—fear, alienation, selfishness, laziness, or violent predisposition. But the visibility of racism is not simply an outcome of people randomly crashing into each other. The harsh and relentless consequences of racism are not merely present when individuals collide. Racism structures everyday life and for most people is suffered often in silence, outside of the sparks of unintended crashes (Essed, 1991). Only white people have the privilege of becoming aware of racism as a result of serendipitous encounters with the Other. The fight against racism will not be successfully waged simply through the inane recognition that we are all racists.

The popularity of a film as deeply contradictory and often reactionary as *Crash* must be understood in the context of the growing backlash against people of color, immigration policy, the ongoing assault on the welfare state, the undermining of civil liberties, and the concerted attempts on the part of the U.S. government and others to undermine civil rights. *Crash* misrecognizes the politics of racism and refashions it as a new age bromide, a matter of inner angst and prejudices that simply need to be recognized and transcended, an outgrowth of rage waiting to be overcome through conquering our own anxieties and our fears of the other. We need to do much more to challenge racism in its newest incarnation. We should therefore be very attentive not to fight ancillary battles, viewing racism as merely an individual pathology with terrible consequences. *Crash* is not about viewing the crisis of racism as part of the crisis of democracy, but the crisis of the alienated and isolated self in a hostile (because increasingly diverse) urban environment. Consequently, it offers no solutions for addressing the most important challenge confronting an inclusive democracy—to critically engage and eliminate the conditions that not only produce the deep structures of racism but also destroy the possibility for a truly democratic politics. The struggle against racism is not a struggle to be waged through guilt or a retreat into racially homogenous enclaves; it is a struggle for the best that democracy can offer, which, as Bill Moyers (2007) points out, means putting into place the material and symbolic resources that constitute the "the means of dignifying people so they become fully free to claim their moral and political agency" (p. 1). It is a struggle that should be waged in the media as part of a politics of cultural representation; it is a struggle that needs to be waged against the neoliberal and racializing state and its failure both to equitably distribute power, resources, and social provisions and to create the basic conditions of engaged citizenship. It is also a struggle to be waged in neighborhoods, schools, and all of those places where people meet, talk, interact, sometimes colliding but mostly trying to build a new sense of political community where racist exclusion rather than difference is viewed as the enemy of democracy.

References

Allen, J. (2005, October 31). *The black and white of public opinion: Did the racial divide in attitudes about Katrina mislead us?* [Press release]. Washington, DC: The Pew Research Center. Retrieved April 23, 2009, from http://people-press.org/commentary/pdf/121.pdf

Arendt, H. (2003). *Responsibility and judgment.* (J. Kohn, Ed.). New York: Schocken Books.

Bauman, Z., & Tester, K. (2001). *Conversations with Zygmunt Bauman.* Cambridge, UK: Polity Press.

Brown, W. (2006). *Regulating aversion: Tolerance in the age of identity and empire.* Princeton, NJ: Princeton University Press.

Chelala, C. (2006). Rich man, poor man: Hungry children in America. *Seattle Times.* Retrieved January 2, 2006, from http://www.bread.org/press-room/news/page.jsp?item ID= 28514640

Cole, D. (1999). *No equal justice: Race and class in the American criminal justice system.* New York: New Press.

Denby, D. (2005, May 2). Angry people—Crash. *The New Yorker*, pp. 110–111.

Eckholm, E. (2006, March 20). Plight deepens for black men, studies warn. *New York Times*, p. A1.

Essed, P. (1991). *Understanding everyday racism.* Newbury Park, CA: Sage.

Feeney, F. X. (2005, May 5). Million dollar boomer. *LA Weekly.* Retrieved April 23, 2009, from http://www.laweekly.com/2005-05-05/film-tv/million-dollar-boomer/1

Goldberg, D. (1993). *Racist culture.* Cambridge, MA: Basil Blackwell.

Goldberg, D. (1994). *Multiculturalism: A critical reader.* Cambridge, MA: Blackwell.

Goldberg, D. (2002). *The racial state.* Oxford, UK: Blackwell.

Moyers, B. (2007). *A time for anger, a call to action.* Common Dreams. Retrieved May 1, 2009, from http://www.departments.oxy.edu/orsl/pdfs/MoyersSpeechOccidentalCollege.pdf

Race on screen and off. (1997, December 29). *The Nation*, p. 6.

Shipler, D. (1998). *A country of strangers: Blacks and whites in America.* New York: Vintage.

Reading 3.3

Latinos/as Through the Lens

Carleen R. Basler

The strange power of attraction possessed by motion pictures lies in the semblance of reality that the pictures convey. The impression of reality a film exerts over the minds of spectators is a visual suggestion. Through the medium of motion pictures, film is capable of creating an absolute realism with the potential for manipulating attitudes on the basis of visual images. Images of the social class, cultural, and political lives of Latinos/as are abundant in past and present film and filmmaking. But are they accurate, authentic, or merely stereotypes held by the "mainstream"? Does it influence what a viewer sees of Latinos/as on film if a Latino/a is behind the camera?

Using film as a medium for sociological analysis provides a way to connect film criticism with real life experience and empirical data. Sociological research is empirical; its conclusions are based on the experiences of individuals and groups. Film theory and criticism are not empirical in a scientific sense, thus failing to connect philosophical and critical speculation with the reality of lived experiences—especially for people considered to be outside of the supposed "American mainstream." Film stereotypes operate by gathering certain traits and assembling them into a particular image. A mediated stereotype that exists on the public screen runs the risk of making "real" the depiction of the "Other." Through the lens of sociology, film can represent a graphic manifestation of mainstream social attitudes about certain groups. Using film as a medium of study, one begins to understand that film representations of Latinos/as are part of a social conversation that reveals American mainstream attitudes about Others.

There are several key film stereotypes of Latinos/as as individuals and groups. Sometimes the stereotypes are combined, sometimes they are altered superficially, but their core characteristics have remained consistent for more than a century and are still evident today, even as American society becomes more racially and ethnically sensitive. Film depictions of Latinos/as have the power to "other" even when people censor themselves publicly. Negative stereotyping of Latinos/as has been challenged in many ways: studio-made films that attempt to accurately portray certain slices of Latino life, actors who manage to portray Latinos/as with integrity despite a filmmaking system that profits from stereotyping, and, more recently, a growing number of Latino/a filmmakers who reject the stereotyping paradigm of mainstream America. How Latinos/as absorb popular images of their ethnic group, and whether it influences their personal identity, is of great sociological concern considering the power of the film medium.

One of the most significant reasons to undertake a sociological study of Latinos/as through the medium of film is the U.S. film industry's general indifference toward the treatment of Latino themes and the presence of Latino/a participants in the film industry itself. The sometimes benign, but more often intentional neglect of Latino/a experiences in real life and the misrepresentation of Latino/a characters and culture in films has led to the perseverance of negative and racist stereotypes that, in turn, affect Latinos in their true everyday lives. Using film as a tool of analysis is powerful not only for the sociological interrogation of Latinos/as and film, but also in understanding the creation, maintenance, and motives behind general racial, ethnic, class, and gender stereotypes employed in films.

This reading begins with an overview of the history of the Latino/a image in the film industry, then analyzes how researchers have attempted to study and deconstruct many of the stereotypes the industry has helped solidify in the viewers' mind. Last, the reading addresses the various ways Latinos/as attempt to subvert and resist negative stereotypes, typecasting, and status quo filmmaking, thus challenging earlier film images and general attitudes regarding Latinos/as in the United States. Although the sociological examination of Latinos/as in film is relatively new, scholars already debate which film theories best explain the experience of Latinos/as in the film industry and the impact of projected Latino/a images through the film experience. A chronological examination of American cinema highlights how portrayals of Latinos/as have changed over time, often reflecting the larger sociopolitical processes of the day. A thorough content analysis prepares film scholars and researchers for a rigorous sociological analysis by focusing on both negative and positive stereotypes that have been created and reinforced about Latinos/as through film from the early days of filmmaking to the present day. When utilizing established film theory and bolstering the key concepts with sociological analyses, it is clear Latinos/as are often misrepresented in numerous ways.

Latinos/as and American Film

Latinos/as have a complex sociological relationship with American cinema that is complicated by several factors, most notably language, history, and stereotype. The word *Hispanic*, like the terms *Latino*, *Chicano*, *Mestizo*, *Mexican*, and *Spanish-American*, contains multiple meanings, including those connected to such identity markers as race, ethnicity, gender, sexuality, and politics (Noriega & Lopez, 1996).

The two most common terms used to describe people and cultures of Latin America in the United States are *Hispanic* and *Latino*. *Hispanic*, a label used by the U.S. government, infers some European (Spanish) heritage. *Latino* is a term that is more encompassing (combining those of Mexican, Cuban, Puerto Rican, and other Latin American descent under one umbrella term) and has been used with greater frequency since the mid 20th century. Both terms vastly condense and oversimplify the wide range of languages, histories, and cultures of diverse groups of people. Some Latinos trace their ethnic origins back to indigenous populations (such as the Incas, the Mayans, and the Aztecs) who occupied the lands before the arrival of the Europeans. For others, Latino ancestry and culture includes African customs and heritages, as Central and South America were involved in the African slave trade. Still others trace their lineage to European colonization. Concepts of *race* thus intersect with concepts of *nationality* and *ethnicity* in multifaceted ways. Racist assumptions about the superiority of whiteness (or European heritage) still linger even as the "hybridity" of Latino cultures would seemingly help to break it apart. Thus, the country of one's ancestry has tangible consequences in the linguistic tension between *Hispanic* and *Latino*, and is further complicated with the racialized term *Afro-Latino.*

From the very beginning of the American film industry, individuals and groups of Latin American descent, and more specifically, Mexicans and Mexican Americans, were stereotyped in ways that exposed the national consciousness in both foreign policy and the *sociological imagination* of what it meant "to be an American." Just as the culture in the United States defined its national identity by claiming to be "not Indian," the country's westward expansion also mandated another opposition: being "not Mexican." One of the methods used was usurping the term *American* to refer to U.S. citizens and to white citizens most specifically.

Marking U.S. racial identity and Mexican racial identity as distinctly separate was extremely important in the visual fulfillment of Manifest Destiny, the idea that Anglo-Americans were "destined" by God to own, inhabit, and control all the lands between the Atlantic and Pacific oceans in North America. America's aggressive westward expansion wrested control from Mexico the territories that became Texas, Arizona, Nevada, California, and New Mexico. Once those lands were part of the "new" American territory, many of the individuals who had lived in them for generations were directly treated as foreigners—as Others. In fact, Mexicans were made into an "inverted image of the Euro-Americans . . . in negative and barbaric terms, with the social problems of the times attributed to them" (McDonough & Korte, 2002, p. 252). Stereotypical images of how those Others looked, spoke, and acted became important tools for distinguishing them from the white Americans, hence, "real" Americans. Those stereotypes were generalized and used to describe almost anyone of Latin American ancestry.

Early silent movies such as *Martyrs of the Alamo* (1915) identified for (primarily white) American audiences the "good guys" (the white Texans) and the villains (the brown Mexicans) in the historical struggle over the Alamo. D. W. Griffith produced *Martyrs of the Alamo,* or *The Birth of Texas,* as it was also known, employing the same story line as *The Birth of a Nation* of "morally reprehensible" men of color (played by white men in blackface or brownface) posing a threat to white womanhood and, by extension, the white nation. The struggle over nation or state in this version of history was rooted in the efforts of white/Anglo men to protect white women from "disrespectful, uncivil, promiscuous, and sexually dangerous" (nonwhite) men (Flores, 2002, p. 103).

Absent from this film, and many other Alamo movies of the 20th century, was the political economy of land, white colonialism, and slavery, which the Mexican government had banned. By exploiting nativist and miscegenation themes, Griffith's work became central to the cinematic construction of the American racial order (Denzin, 2002). The message ultimately was that whites were superior and the United States should be reserved for white, native-born "Americans." The final scene of *Martyrs of the Alamo* showed a succession of flags: The Mexican Lone Star, the Confederate Flag, and finally the U.S. flag blowing in the breeze (Flores, 2002).

In early to mid 20th-century filmmaking, Latinos/as within the United States and persons of Latin American descent were racialized in ways that clearly reflected the mainstream cinema mind-set. Typically, pre–World War II Latinos were racially aligned with conceptions of goodness and badness based on stereotypical characters such as bandits, greasers, or Latin lovers.

During the governmental propaganda campaign in the early 1940s known as the Good Neighbor Policy, there was a much kinder and almost romanticized depiction of Latin Americans and those nations' ties with the United States. Family films made by Disney Studios such as *Saludos Amigos* (1943) and *The Three Caballeros* (1945) are examples of Hollywood's participation in the government's attempt to create films and images that celebrated the cultures of and a U.S. friendship with Latin America. During the same period there was a rise of Latin American themed musicals including *Down Argentine Way* (1940), *That Night in Rio* (1941), and *Week-End in Havana* (1941), all inspiring positive feelings about U.S. foreign policy and relations in Latin America.

While the Good Neighbor Policy films praised and idealized Latin Americans of other countries, Hollywood films rarely positively represented Latino people within the United States. Rather the early social problem films were produced to address much of the discrimination Latinos faced in America. For example, *A Medal for Benny* (1945) was a story about a Mexican American war hero who was awarded the Congressional Medal of Honor posthumously. The film explored the social injustices faced by many Mexican Americans, but it faltered by creating homogenized portraits of a variety of Latino people. Other social problem films included *The Lawless* (1950), *Right Cross* (1950), and *My Man and I* (1952), but this genre quickly dissipated as Hollywood became consumed by the Red Scare and many filmmakers were afraid of being blacklisted for producing any material that could be considered as critiquing the U.S. government.

Between the 1950s and 1970s, Hollywood's approach to Latino issues and Latino/a characters returned to patterns and practices employed in the decades before the Good Neighbor Policy. For example, the screen adaptation of the musical *West Side Story* (1961) racialized Latinos as non-American Others, even though the Latinos in the film are Puerto Ricans who have birthright American citizenship. Telling the story of a young Puerto Rican immigrant named Maria (portrayed by Anglo-American Natalie Wood) and her Polish American beau, the film ends on a tragic note, as the bigotry and prejudice of gang life results in the death of the white male lead. None of the main actors in the film were of Latin American descent. Employing Anglos or other non-Latinos to play Latino/a roles has been a continual criticism of the American film industry.

Throughout much of U.S. history, Latinos were often physically cordoned off into their own neighborhoods (barrios), discriminated against in the workplace, and verbally and physically harassed. Many Latino individuals and families who lived in the United States before and after annexation were actually deported as "aliens" during the Great Depression. Attempts to physically, linguistically, and metaphorically separate the United States from Latin America (particularly Mexico) continue to this day, as evidenced by the enormous amount of resources devoted to policing the border between Mexico and the United States, legislation that is distinguished by "English-only" rhetoric, and the tacit assumptions that only white individuals are unquestionably

"American" citizens. Over the past century, American films have often reinforced a sense of difference and distance between "them" (Latinos) and "us" (white Americans) illustrated in films such as *The Border* (1982), *Lone Star* (1996), and *Bordertown* (2007). Intriguingly though, some American movies have demonstrated a more complex cultural ambivalence toward Latinos, often based on perceived notions of class and race (for example, see *I Like It Like That,* 1994). While many Latino/a characters in American films have been treated as racialized stereotypes, others seem to be treated as members of an ethnic group assimilating into whiteness, such as *Real Women Have Curves* (2002). The dominant cultural conception of Latinos in film has been consistently negotiated along these ethnic and racial lines.

Ethnic stereotyping in movies certainly does occur, but much of the literature about how certain ethnic groups have been depicted actually creates and often reinforces its own distortions, creating "stereotypes of movie stereotypes" (Cortés, 1992, p. 75). Movies are part of the societal curriculum, the continual, informal curriculum of family, peer groups, neighborhoods, churches, organizations, institutions, and other societal influences that "educate" all of us throughout our lives.[1] Within the societal curriculum, the media serve as lifelong educators. Movies teach by disseminating information about a myriad of topics, including race, ethnicity, culture, and nationality. Intentional or not, movies contribute to intercultural, interracial, and interethnic understanding or misunderstanding.

The Complexity of Latino/a Stereotypes in American Film

All people stereotype. Stereotyping is a psychological mechanism that creates categories, allowing people to manage the massive amounts of data they encounter every day. Even though stereotyping may be a natural human process, it is how people use the process that can turn the "natural" into a manufactured reality, used for creating a superior social location of one group over another. It is important to account for the ways that dominant groups assign selective characteristics to people—social, cultural, political, sexual, racial, class, and ethnic—to create Others. Films have great potential to transmit both negative and positive stereotypes to mass audiences, even if the viewer has no real contact with any individual or groups that are being stereotyped in the film. Thus, a rigorous sociological examination is requisite of how movies do indeed stereotype Latinos/as.

Stereotypical depictions of various outcast races, ethnicities, and cultures, often excruciatingly derogatory by contemporary standards, were commonplace in American popular culture before the invention of motion pictures.[2] Early films'

---

[1]For an extended discussion of the concept "societal curriculum," see Carlos E. Cortés, "The Societal Curriculum: Implications for a Multiethnic Education," in *Education in the 80's: Multiethnic Education,* James A. Banks, Ed. (Washington, DC: National Education Association, 1981), pp. 24–32.

[2]See Cecil Robinson, *Mexico and the Hispanic Southwest in American Literature* (Tucson: University of Arizona Press, 1977), and Arthur G. Pettit, *Images of the Mexican American in Fiction and Film* (College Station: Texas A&M University Press, 1980).

racial and ethnic topics stemmed from other popular forms of entertainment that predated cinema, including comic strips, vaudeville, and novels. In turn, film escalated these depictions to new and often more extreme levels because of its capacity to heighten emotions as well as its ability to communicate to mass audiences. One side effect of the power of the American cinema was often crushingly brutal portrayals of other races and cultures, negative depictions that spread to larger audiences than ever before possible.

One sociological perspective is that stereotypes are preexisting categories in any society or culture. They are learned in the socialization process (Lippmann, 1941; Miller, 1982). There are two important features of stereotyping. First, when learned stereotypes are expressed, they are reinforced, and thus validated and perpetuated. Second, validation solidifies attitudes that suggest how certain individual and groups should be treated (Miller, 1982). Stereotypes persist not only because they are category labels, but because they are implied programs for action (Royce, 1982). Different stereotyping scenarios can be delineated depending on the power relationships between groups; they may be stratified, oppositional, or cooperative. If the groups are stratified, the dominant group creates subdominant stereotypes endowed with two sets of characteristics: *harmless* (with out-group members portrayed as childlike, irrational, and emotional) when they pose no threat, or *dangerous* (treacherous, deceitful, cunning) when they do. A systematic examination of film "curriculum" requires combining the rigor of content analysis, such as identifying, categorizing, and evaluating ethnic characters,[3] and relevant theories of social interactionism to help deconstruct stereotypes and properly ascertain the social intention of the film.

According to Berg (2002), there are few nonstereotypical portrayals of Latinos/as in Hollywood cinema. Before *Zoot Suit* (1981), *La Bamba* (1987), and *Stand and Deliver* (1988), most films with Latinos as key or even central characters were mostly one-dimensional, stereotypical projections of Latinos/as who often lacked self-determination or any real agency that was directly the consequence of or influenced by the Anglo characters in the film.[4] Berg (2002) delineates six classic Latino stereotypes based on American racial, ethnic, class, and linguistic stratification found in American films that persist today.

Bad Bad vs. Good Bad: El Bandito/Greaser vs. the Latin Lover

One of the earliest and most common stereotypes representing Latinos in U.S. cinema is the *greaser* or *bandit.* The Mexican male bandit stereotype has its roots in the silent "greaser" films and has continued to appear in many contemporary

---

[3]For more, see Carlos E. Cortés, "Who Is Maria? What Is Juan? Dilemmas of Analyzing the Chicano Image in U.S. Feature Films," in C. E. Rodríguez (Ed.), *Latin Looks: Images of Latinas and Latinos in the U.S. Media* (Boulder, CO: Westview, 1997).

[4]Anthony Quinn's dignified, defiant vaquero in *The Ox-Bow Incident* (1943), Ricardo Montalban's Mexican government agent in *Border Incident* (1949), and Katy Jurado's role as a resourceful businesswoman in *High Noon* (1952) are exceptions to the oversimplified Latino characters in U.S. films.

westerns. Typically, he is treacherous, shifty, and dishonest. His reactions are emotional, irrational, and unusually violent. His intelligence is limited, resulting in flawed strategies. He is dirty and unkempt—usually displaying an unshaven face, missing teeth, and disheveled oily hair. According to Berg (1997), a modern incarnation of the greaser type, the Latin American drug runner, shows superficial changes in the stereotype without altering the essence. "He has traded his black hat for a white suit and his tired horse for a glitzy Porsche, yet he is still driven to satisfy base cravings for money, power, and sexual pleasure—and routinely employs viscous and illegal means to obtain them" (p. 113). From the half-breed villain in *Broncho Billy and the Greaser* (1914) to a cocaine-addicted, bloodthirsty powermonger Tony Montana in *Scarface* (1983), the movie Latino bandit is a demented, despicable creature who must be punished for his brutal behavior and usually ends up dying in a hail of bullets.[5] One could suggest that the comedy team of Cheech Marin and Tommy Chong linked Chicanos to a counterculture that was more American mainstream (*Up In Smoke,* 1978). However, it is also possible to see Marin as an updated version of the greaser stereotype, lazy, not intelligent, and deviant (drug abusing).

The *Latin lover* is juxtaposed to the bandit/greaser stereotype. During the 1920s, a new kind of screen lover was introduced—the Other as attractive, seductive, romantic. According to this stereotype, men (and women) of Latin American descent are depicted as more sensual and sexual than their Anglo North American counterparts, furthering assumptions that Latinos are emotional and "hot-blooded."[6] The stereotype expresses the image of an alluring, darker-skinned sex object whose cultural difference hinted at exotic and erotic secrets, perhaps tinged with violence and sadomasochism. The Latin lover has continued to be a consistent screen figure, played by a number of Latino actors, including Cesar Romero, Ricardo Montalbán, Gilbert Roland, Fernando Lamas, Jimmy Smits, and Antonio Banderas. Ironically, the Latin lover stereotype can be solely attributed to the screen work of Italian immigrant Rudolph Valentino (for example, *Son of the Sheik,* 1926).

The Latin lover image has very different connotations than the greaser stereotype. Whereas the greaser is an overtly racialized Other, the Latin lover is more of an ethnic type, usually depicting a lighter-skinned individual that could potentially be assimilated into whiteness. Sexual relations between a white woman and a greaser character are considered unnatural or off limits. Latin lovers are romantic leading men who regularly succeed in winning the hearts of their white female leads. It is important to note that actors cast as Latin lovers have significantly fairer complexions than those of the actors hired to play greasers. These racialized codes of skin color were enforced and in some cases created by Hollywood makeup artists, who regularly darkened the complexions of greaser characters, or even slightly

---

[5]The 1932 version of *Scarface* featured an Italian immigrant character. The fact that the current version employs a Cuban immigrant (more specifically a *Marielito* second-wave Cuban immigrant) reflects the sociopolitical change of fear from early 20th-century Italian immigrants to contemporary anxieties surrounding Latin American immigration.

[6]Making such stereotypes apparent, California governor Arnold Schwarzenegger apologized in September 2006 for saying of Latina California congresswoman Bonnie Garcia that her "black blood" mixed with "Latino blood" equals "hot."

darkened white actors who were cast to play Latin lovers or "good" Latino characters, such as the "browning" of Charlton Heston in *A Touch of Evil* (1958) or Paul Muni in *Bordertown* (1935). The "not-quite-whiteness" of Latin lovers made them exotic, but acceptable and appealing to many moviegoers. Whereas the greaser/bandit image provided viewers with a villainous stereotype fueling Anglo audiences' fear of miscegenation and immigration.

Whore vs. Madonna: The Half-Breed Harlot vs. the Dark Lady

The corresponding Latina stereotype to the bandito or greaser is the *half-breed harlot.* She is a familiar stock figure in American cinema, particularly westerns. Like the bandit, she is a secondary character, and not always a half-breed. She is lusty and hot-tempered, and her main function is to provide as much sexual titillation within industry standards (Pettit, 1980). For example, Doc Holliday's woman Chihuahua (Linda Darnell) in *My Darling Clementine* (1946) is a classic example. The half-breed harlot is a slave to her passions; her character is based on the premise that she is sexually driven. Stereotypically, she is a prostitute because she enjoys the work and male attention, not because social or economic forces have shaped her life and opportunity choices.

Opposite the harlot is the *dark lady.* She is mysterious, virginal, inscrutable, and alluring because of these characteristics. Her cool distance is what makes her so fascinating to Anglo males. She is circumspect and aloof while her Anglo sister is direct and forthright. She is reserved and the Anglo is boisterous. She is as opaque as the Anglo woman is transparent. One classic example of the dark lady stereotype was played by actress Dolores Del Río—she portrayed the fascinating Latin woman who aroused American leading men's appetites the way no other Anglo woman could in the films *Flying Down to Rio* (1933) and *In Caliente* (1935).

By infusing the Latin lover and the dark lady with overtly sexual characteristics manifestly lacking in their Anglo counterparts, American cinema has stereotyped and marginalized Latinos/as through racial idealization and sexual fantasy.

Not So Funny: The Male Buffoon and Female Clown

The *male buffoon,* according to Berg (2002), is the second banana comic relief, such as the characters of Pancho in *The Cisco Kid,* Sgt. Garcia in Walt Disney's *Zorro,* and Ricky Ricardo in *I Love Lucy.* Many of the same stereotypical characteristics that are threatening in the bandit or greaser are targets of ridicule with the male buffoon. Comedy is used to deal with the Latino's accentuated differences, a way of taming his fearful qualities. Comedy was supposedly found in his simple-mindedness (the bumbling antics of Sgt. Garcia), his failure to master standard English ("Let's went, Cisco!"), and his childish regression into emotionality (Ricky's frustrated verbal explosions in Spanish). The reincarnation of this stereotype can be seen in the *Cisco Kid* films (1931–1945) and even the *Cisco Kid* television movie in 1994 starring actors Jimmy Smits and Cheech Marin.

The *female clown* neutralizes the overt sexual threat posed by the half-breed harlot. The strategy is to mitigate the Latina's eroticism by making her an object of comic derision. A key example is the film career of the beautiful Mexican actress Lupe Vélez, the comic star of the 1940s *Mexican Spitfire* movies. Though very physically attractive, the female clown is often portrayed as a dizzy yet alluring dingbat. The stereotype is often expressed by the actress portraying childlike emotionality, and irrationality, and as a comic foil. Other examples include Carmen Miranda as the "Lady in the Tutti-Frutti Hat" in *The Gang's All Here* (1943). An updated version of the female clown stereotype is the character Rosario on the popular television series *Will & Grace.*

Since the 1980s, there has been a gradual but notable change in the images of Latinos/as in U.S. films. The change may be attributed to certain Latino/a filmmakers whose concerted efforts have produced films that more accurately reflect the reality of Latino life in the United States. Some examples of the more successful directors are Luis Valdez (*Zoot Suit, La Bamba*), Ramón Menéndez (*Stand and Deliver*), Edward James Olmos (*American Me,* 1992), and Robert Rodríguez (*Spy Kids* trilogy). Yet, stereotypical images of Latinos/as abound across the large and small screen even in the 21st century. Images of Latino "gangbangers" are a mainstay on television crime shows and the narco-drug wars in Latin America are continually fodder for major action films.[7]

Identifying stereotypes of Latinos/as is not a sufficient examination of the larger social problem in the film industry. A comprehensive sociological analysis of image and content must also combine with an analysis of *control*—*why* films were made that utilize these stereotypes—and an analysis of *impact*—*how* the films actually influence viewers. A sociological understanding of film content, control, and impact allows the researcher to establish a direct causal link between the media and society. Identification of these links enables an examination of their utility in larger society. Such an interrogation begs the important questions: Why must Latinos/as play roles that overgeneralize minor or even fictionalized aspects of their cultural backgrounds? Why are they often confined to playing "ethnic" or racialized characters? How can films accurately represent Latino culture and retard the misrepresentations that are perpetuated by past negative stereotypes?

Conclusion

Film representation needs to be understood within a social and historical context. The images of Latinos/as in American film do not exist in a vacuum but as a part of the larger discourse on Otherness in the United States. Beyond their existence as

[7]One need look no further than the 2006 production *Bandidas* starring Selma Hayek and Penélope Cruz to see how Berg's six stereotypes have morphed for current racialized gender stereotypes about Latinas.

mental constructs or film images, stereotypes are part of the social conversation that reveals the mainstream's attitudes about Others. It is necessary to investigate how standardized cinematic techniques and the accepted norms of "good" filmmaking (including the star system, casting, screenwriting, camera angles, shot selection, direction, production design, editing, acting conventions, lighting, framing, makeup, costuming, and *mise-en-scène*) contribute to the totality of a stereotype image. A solid sociological analysis will investigate the narrative function that stereotypes play within classical American films and their purpose in various popular genres.

References

Berg, C. R. (1997). Stereotyping in films in general and of the Hispanic in particular. In C. E. Rodríguez (Ed.), *Latin looks: Images of Latinas and Latinos in the U.S. media.* Boulder, CO: Westview.

Berg, C. R. (2002). *Latino images in film: Stereotypes, subversion, resistance.* Austin: University of Texas Press.

Cortés, C. E. (1992). Who is Maria? What is Juan? Dilemmas of analyzing the Chicano image in U.S. feature films. In C. A. Noriega (Ed.), *Chicanos and film: Essays on Chicano representation and resistance* (pp. 74–93). New York: Garland.

Denzin, N. K. (2002). *Reading race: Hollywood and the cinema of racial violence.* Thousand Oaks, CA: Sage.

Flores, R. R. (2002). *Remembering the Alamo: Memory, modernity, and the master symbol.* Austin: University of Texas Press.

Lippmann, W. (1941). *Public opinion.* New York: Macmillan.

McDonough, K., & Korte, A. (2002). Hispanics and the social welfare system. In P. S. J. Cafferty & D. W. Engstrom (Eds.), *Hispanics in the United States: An agenda for the twenty-first century* (pp. 237–276). New Brunswick, NJ: Transaction.

Miller, A. G. (1982). Historical and contemporary perspectives on stereotyping. In A. G. Miller (Ed.), *In the eye of the beholder: Contemporary issues in stereotyping* (p. 27). New York: Praeger.

Noriega, C. A., & Lopez, A. M. (Eds.). (1996). *The ethnic eye: Latino media arts.* Minneapolis, MN: University of Minnesota Press.

Pettit, A. G. (1980). *Images of the Mexican American in fiction and film.* College Station: Texas A&M University Press.

Royce, A. P. (1982). *Ethnic identity: Strategies of diversity.* Bloomington: Indiana University Press.

CHAPTER 4

Gender and Sexuality

Ariel: But without my voice, how can I . . .

Ursula: You'll have your looks . . . your pretty face . . . and don't underestimate the importance of "bo-dy lan-guage." Ha! [singing]

Ursula: The men up there don't like a lot of blabber / They think a girl who gossips is a bore / Yes, on land it's much preferred / for ladies not to say a word / After all, dear, what is idle prattle for? / Come on, they're not all that impressed with conversation / True gentlemen avoid it when they can / But they dote and swoon and fawn / On a lady who's withdrawn / It's she who holds her tongue who gets her man.

The Little Mermaid (1989)

David: You're gay for saying that.

Cal: I'm gay for saying that?

David: You know how I know you're gay?

Cal: How? How do you know I'm gay?

David: Because you macramed yourself a pair of jean shorts.

Cal: You know how I know *you're* gay? You just told me you're not sleeping with women any more.

David: You know how I know that you're gay?

Cal: How? Cuz you're gay? And you can tell who other gay people are.

David: You know how I know you're gay?

Cal: How?

David: You like Coldplay.

40-Year-Old Virgin (2005)

We go to the movies and we see tales of men and women, sex and sexuality. In film after film we see men and women coming together against remarkable odds. She's hardheaded; he's a slob (*The Break-Up,* 2006). She's desperate; he's a cad (*Bridget Jones's Diary,* 2001). She's liberal, he's conservative (*He Said, She Said,* 1991). She's driven; he's demeaned (*Why Did I Get Married?,* 2007). Given these differences, it is a wonder men and women ever get together. After all, men are from Mars and women are from Venus. We are so profoundly different, we hail from opposing planets, right? Many Hollywood films suggest as much. Film representations of relationships often reinforce popular notions concerning gender; that is, men and women are more different than we are alike, and this fact of difference is a "natural" part of our genetic makeup. Of course, sociologists know that almost nothing is more *social* than gender, and they have clearly established that "differences" between men and women are both socially constructed and greatly exaggerated (Kimmel, 2007).

Gender, according to sociologists, refers to the meanings a society gives to masculinity and femininity. In daily interactions, we are called on to perform as either masculine or feminine. Gender then is not just an identity or a status; it is a continual process of negotiation. Gender is something we "do" (West & Zimmerman, 1987). Men are expected to maintain a masculinity that includes toughness, bravado, strength, assurance, and confidence, all with little display of emotion. Women are expected to perform a femininity that exudes beauty, caring, nurturance, neediness, and compassion, complete with every range of possible emotion.

If we play our roles wrongly, we often face social sanctioning. Of course, the price is much steeper for the man who drifts toward femininity, as we are a culture in which the traits of masculinity are more highly valued. We see these roles strictly enforced in children's films. In the scene from *The Little Mermaid,* Ariel bemoans her fate if she loses her voice, but recognizes that a woman who "holds her tongue" is preferred by men when it comes to romance. In Disney classics we find variations on this theme with repetitive images of the hapless princess, lacking any autonomy while in great need (of being rescued and of finding a husband/prince). The prince arrives, fully confident and capable of saving her from whatever forces bind her. They marry.

Research concerning these "differences" has shown that men and women are more alike than we tend to believe. For instance, many studies have found that *men* are the ones who hold strong ideological beliefs about romantic love (e.g., believe in love at first sight, fall in love more quickly) (see Kimmel, 2007). The problem is, when we assume men and women to be somehow *innately* different, we fall into denial of culpability. We don't really have to work at communication when we assume that men and women simply *can't* communicate. In this case, films presenting gendered relationships in this Mars/Venus manner both reflect the commonsense assumptions of gender and contribute to a culture of gendered understandings of interaction. As Lorber (1994) points out, gendered social arrangements are justified by social institutions (e.g., religion) and cultural productions (e.g., films).

The relationships featured in film are based on the characters doing gender appropriately—that is, in ways that are congruent with their assumed biological sex. Further, sexuality is linked to the sex/gender binary such that "the possession of erotic desire for the feminine object is constructed as masculine and being the object of masculine desire is feminine" (Schippers, 2007, p. 90). Accordingly, heterosexuality is based on the presumed differences between and the complementariness of masculine and feminine. The scene from *40-Year-Old Virgin* captures the banter of two straight guys playing a video game. We can see the length that men will go in order to *not* appear as anything other than masculine and, since sexuality is linked to gender, heterosexual. It is important to maintain the image of manhood—especially in the company of one's guy friends. As Kimmel (1994) notes, often the performance of masculinity (and heterosexuality) happens *among* men *for* men.

The readings in this chapter explore the social construction of gender and sexuality. In the first reading, Jean-Anne Sutherland examines how women with power are portrayed in film. Unpacking the concept of power, she identifies three forms of power and explores the meaning of these for women and the goals of feminism using examples from film. Women are using the power-over model when they enact masculine values and behaviors, as we see in *She Hate Me* (2004) when the women objectify and humiliate Jack on the basis of sex. This model is also found in the film *The Devil Wears Prada* (2006), with a female boss who is "chillingly mean, emotionless, and incapable of maintaining relationships." Another version of power, power-to, is more positive and life-affirming, and it involves an awakening, or raised consciousness, as seen in *The Color Purple* (1985), *Real Women Have Curves* (2002), and *Waitress* (2007). But the move to organized action to create social change is rooted in the power-with model, as seen in the historic discrimination case represented in the film *North Country* (2005).

Masculinity in film is the subject of the second reading, in which Michael Messner considers how cinematic representations of hegemonic masculinity have seeped into the "theatre" that is U.S. politics. Using the case of movie star turned California governor, Arnold Schwarzenegger, Messner explores the versions of masculinity Schwarzenegger's film roles captured, linking these to his "performances" as state governor. He focuses on the ways that Schwarzenegger, and the media, presented his political persona as the "Terminator" divorced from human compassion and emotions, and the "Kindergarten Commando" who is still a tough guy, but one who is vulnerable, compassionate, and caring (especially about children). Messner argues that film images take on lives of their own—that power is rooted in these symbolic images and often the symbolic becomes "real" as they are played out not by actors on the screen, but by politicians on the stage of government.

"Doing gender" is elaborated on in Besty Lucal and Andrea Miller's reading, in which they use a feminist constructionist analysis to explore the binary world of gender and sexuality. We assume that men and women are opposites (Mars and Venus), and we assume a gendered duality that forces people to be "one or the other"—you are man or woman, and you are heterosexual or homosexual. What is missing in this dualistic formula is the reality of many lives, that gender and sexuality are not fixed identities or experiences. However, with only two choices on both dimensions, the boundaries are made clear, at the same time that lived experiences are covered up or erased. Lucal and Miller explore the meaning of sexual desire and sexual identity, and the "either/or" constraints placed on changing and fluid human experiences in the film *Chasing Amy.* With *Transamerica* they draw our attention to the fixed categories of male/female and masculine/feminine, as the main character seeks to bring sex (through surgery) into alignment with (gender).

Exploring women and power, masculinity, and bisexual and transgender identities, the readings in this chapter invite you to consider the ways that gender and sexuality are socially constructed in film and in life. As sociologists, we begin with the understanding that gender is learned and then performed in interaction with others. Film is one of the mediums that provide us with the "meanings" that make up our gendered realities.

References

Kimmel, M. (1994). Masculinity as homophobia: Fear, shame and silence in the construction of gender identity. In H. Brod (Ed.), *Theorizing masculinity* (pp. 119–141). Thousand Oaks, CA: Sage.

Kimmel, M. (2007). *The gendered society* (3rd ed.). New York: Oxford University Press.

Lorber, J. (1994). *The paradoxes of gender.* New Haven, CT: Yale University Press.

Schippers, M. (2007). Recovering the feminine other: Masculinity, femininity, and gender hegemony. *Theory & Society, 36,* 85–102.

West, C., & Zimmerman, D. (1987). Doing gender. *Gender & Society, 1,* 125–151.

Reading 4.1

Constructing Empowered Women

Cinematic Images of Power and Powerful Women

Jean-Anne Sutherland

If you were asked to think of a powerful woman in film, who might come to mind? Perhaps Sigourney Weaver in *Alien* (1979)? Maybe *Thelma & Louise* (1991)? Or Jennifer Lopez after her karate classes in *Enough* (2002)? Perhaps Nurse Ratchet in *One Flew Over the Cuckoo's Nest* (1975) struck you as powerful, or *Norma Rae* (1979) when she stands on the table in the factory, holding the "UNION" sign as the loud machines slowly click off in support of her and the union. How many women would we come up with, and what would they look like?

As a starting point, it is important to observe the complexity of the presentation of powerful women in film. It is not straightforward. For instance, oftentimes when we see a powerful woman in a film, her power is the problem to be explained, overcome, or destroyed. Power is not the solution, as it is for men in film, but rather a flaw that must be rectified. Think of the two women in *Fatal Attraction* (1987). We are given Alex (Glenn Close), the confident, aggressive woman. She uses her sexual powers to lure Dan (Michael Douglas) into an affair. Her power is "bad." In fact, it is "mad," as in insane. Thus, the dutiful wife (Anne Archer) kills the seductress, quite legitimately (Alex killed their daughter's bunny, after all). Similarly, Sharon Stone in *Basic Instinct* (1992) may have indeed been powerful and clever, but her power, also highly sexualized, was to be overcome lest she continue to slaughter men. Annette Bening's character in *American Beauty* (1999) had many characteristics of a powerful male: she was passionately driven to succeed in her career, she wanted to maintain the lifestyle of the financially successful, she even engaged in an extramarital affair as many "successful" male characters have done in countless films. Yet, she is thought of as "the bitch," the relentless nag who "caused" poor Lester's (Kevin Spacey) breakdown.

If asked to think of powerful men in film, the question is less complicated. We might begin with Douglas Fairbanks, Edward G. Robinson, and Humphrey Bogart, covering several decades of early film. Then we could move to John Wayne, Clint Eastwood, Al Pacino, Sylvester Stallone, Michael Douglas, Arnold Schwarzenegger, Bruce Willis, Dwayne Johnson (aka "The Rock"), or Vin Diesel.[1] To be asked about "powerful men in film" might even strike us as a somewhat redundant question (that

---

[1]Noting the whiteness of this list of powerful men, we can consider the successful black actors that come to mind: James Earl Jones, Denzel Washington, Will Smith, Forest Whitaker, Don Cheadle, to name a few. Racial differences in terms of the representations of men and power in film are beyond the scope of this chapter, but necessary to note.

is, most men in film tend to be powerful—there's nothing remarkable about that). We've seen men exert power in film for so long that it seems odd to draw attention to it. The representation of men and power is rather straightforward—their "power" is more often than not the solution to whatever crisis the film revolves around.

Thus, we should pause and consider our ideas about power and women in film. Does Thelma's power (the kind that provoked her to run away from her constrained life) register as the same power Jennifer Lopez's character managed to muster (the kind that motivated her to kill her abusive husband)? Does either of these look like the kind of power Norma Rae had to gather to stand up for nonunionized factory workers? What does it mean to say that a woman is "powerful" in a film?

From 1929, when sound entered movies, until July 1, 1934, it was not at all unusual to see a film where the central character was a sexualized, self-sufficient woman. This period is what film critics have dubbed the "pre-Code" era of Hollywood. According to Mick LaSalle (2000), this era was dominated by films depicting multifaceted women who held jobs, took lovers, committed crimes, struggled with loss and the complexity of emotion, and claimed their sexuality. These films presented powerful women. They worked, they had sex, and they crossed boundaries. They acted a bit like men. These were not "women's films." Rather, these were the "movies the general public flocked to see." Alas, all of that came to an end—not gradually, but in one signing on the line. The so-called Legion of Decency, an organization of Catholic clergy, made their demands to the heads of studios: stop the indecency or we'll tell Catholics to stop seeing movies! An agreement was struck. The "codes" were enforced. Films no longer looked the same and neither did the women.

After the establishment of the Production Code, women returned home from work, denied their sexuality, sought marriage and motherhood above all else, and in general became more "ladylike." While we can find exceptions, as LaSalle points out, the change in women in film was startling and abrupt. But, we might ask if these pre-Code women were really powerful or whether they reinforced stereotypes of women using their sexuality as "vixens." We could then look at the films of the 1960s and 1970s that began to chisel away at the mostly dichotomous presentations of women in film: the virgin (Doris Day) and the whore (Sophia Loren). Was the critique of patriarchy in *The Stepford Wives* (1975) a step forward? Did Katherine Ross's character have power or did her eventual demise into robot perfection of femininity symbolize the futility of fighting against male expectations of women?

In this reading, I first discuss the concept of power. In order to decide who and what is powerful, it is necessary to first consider power and the multiple ways in which the term might be defined. I then discuss what those kinds of power tend to look like in film. Feminist scholarship acknowledges that some kind of acquisition of power is a necessary step in terms of women overcoming myriad layers of oppression. Thus, I also consider the extent to which these images of powerful women suggest social change for women—some crack in the traditions of sexual oppression and inequality. I will consider (1) what is power, (2) what do these varieties of power look like in film, and (3) do any of these representations actually offer a challenge to patriarchy?

What Is Power?

When we refer to power, it may well be that we take for granted what power is, without further consideration of the word. *Power*, after all, is a word we toss around in our lives almost daily. If we are asked to define it, perhaps we assume it looks only one way. Across disciplines, scholars have defined and conceptualized power in multiple ways. I have chosen three definitions most prominent in the literature, beginning with the most commonly understood notion of power.

Power-Over

The definition of power with perhaps the longest history in sociology, philosophy, and psychology is *power-over* (Nash, 2000; Yoder & Kahn, 1992). Power-over is the manner in which Weber defined power: the ability of one actor to carry out his (*sic*) will against another. Weber described it this way, "The chance of man or a number of men to realize their own will in a communal action even against the resistance of others who are participating in the action" (Nash, 2000). Chances are, if a student took an introduction to sociology course, this is the definition that emerged from his or her text. In a 1981 study, Paap found that nearly two-thirds of introduction to sociology texts defined power in this Weberian fashion. Not only is it a definition that runs through sociology and across disciplines, it also registers as the most "commonsense" definition. If a group has power, group members can achieve their goals. A person who has power can achieve his or her goals, and perhaps punch someone in the process.

Amy Allen (2000) argues that this conceptualization takes the form of a dyadic, master/subject relation; one is powered (master) and one is not (subject). In Marxist terms, the bourgeoisie have power (masters) and the proletariats have none (subjects). In that sense, the fate of the subjects is fully determined by the power wielded by the masters. This is a kind of masculinized power. When placed within the context of masculinity, power is seen in terms of force, inequality, and the ability to impose not only physical strength over others, but meanings and ideas as well (Connell, 1987). As Kimmel (1994) notes, manhood itself is equated with power. Hegemonic masculinity (the dominant culture's construction of manhood) is an image of men *in, with,* and *of* power.

According to Allen (1998), on the continuum of feminist analyses of power, the two ends are represented by those who see power as *domination* or *empowerment.* Domination theorists have focused their critiques on the ways in which men dominate or oppress women. Their analyses of domination, inequality, and the subjugation of women work within this power-over conceptualization. Prominent feminists such as Andrea Dworkin and Catharine MacKinnon have, according to Allen, grounded their analysis of inequality in the notion that masculinity means domination and femininity means subjection. The domination view holds that "what it means to be a woman is powerless, and what it means to be a man is to be powerful" (p. 23). These theorists do acknowledge that different men have different levels of power—white men more than black men, for example. But this perspective represents only a portion of a complete conceptualization of power. With their focus on men's domination of women, these theorists do not account for the moments in which women *do* assert their power over men and other women.

Important to note as we consider the feminist scholarship on power is that feminists are not working toward power-over for women. The goal of feminism is not to transfer domination and oppression (and masculinity) into the hands of women. Rather, feminists have exposed this type of power in order to dismantle it.

Power-To

If you were to search for the word *power* in the dictionary, whether the analog version (*Oxford English Dictionary*) or, more conveniently, online (dictionary.com), you would see that the first definition concerns the idea of personal control: "the ability to do or act." This "ability" is itself gendered in that research reveals that men tend to report higher levels of mastery than do women (Ross & Mirowsky, 2002).[2] If the ability to act requires a sense of control (or high levels of mastery), it is no wonder that research finds this more so in men than women. After all, at an early age we socialize boys toward notions of ability and strength and girls with notions of passivity and neediness (through childhood toys as well as film). Boys are taught to do and act while girls are encouraged to nurture others. Boys are taught to compete, and their levels of self-esteem generally rise throughout childhood. Girls learn the importance of physical attractiveness, and their levels of self-esteem tend to drop off in early adolescence.[3]

In social psychological terms, we associate *power-to* with what Bandura calls "personal control" or "self-efficacy" (see Yoder & Kahn, 1992). Bandura (1997) described self-efficacy as the extent to which we believe we have the ability to achieve results from our actions. This kind of personal empowerment is often conceptualized as "mastery," or the sense of having control over the circumstances of your life. In borrowing from these social psychological concepts, I mean to suggest a kind of power whereby a woman recognizes her own abilities and sense of agency. That is, do women have the power to act? Under what circumstances do women grasp their sense of agency and realize the control they have over the consequences of their lives?

"Empowerment theorists," as Allen (1998) dubs them, focus on "women's power to transform themselves, others, and the world" (p. 26). Carol Gilligan, Sara Ruddick, and Virginia Held have focused on the ways in which women's skills (e.g., nurturing, care giving, relationship building) have been devalued. These works build on the notion that as a woman performs care duties, particularly the role of mother, she is not powerless, but in fact empowered. Allen notes that to have power

---

[2]As Ross and Mirowsky point out, the gender gap in sense of control is affected by education, physical functioning, household income, and work history (working for pay increases personal control). They also find that the gender gap widens with age: as women age, their sense of control drops.

[3]This decline in adolescent girls' self-esteem is less true of African American girls than for other racial/ethnic groups. See Tamara R. Buckley and Robert T. Carter, "Black Adolescent Girls: Do Gender Role and Racial Identity Impact Their Self-Esteem?" *Sex Roles*, Vol. 53 (9/10, 2005), pp. 647–661.

in the mothering role, according to Ruddick (1989, p. 37), is "to have the individual strength or the collective resources to pursue one's pleasure and projects." Nancy Hartsock makes a similar argument. By taking the standpoint of women, it is possible to create notions of power distinct from male analysis of power-over (in Sprague, 2005).

The empowerment theorists are working with a picture of power as power-to, the ability of a woman to recognize the control she has over her life, see the results of her actions, and recognize the power of the self. Empowerment theorists, like the domination theorists, lack a complete analysis. In other words, while noting that women draw power from previously undervalued roles (such as mothering), they fail to note the ways in which these roles are themselves determined by a gendered society.

For our purposes, these two definitions of power are incomplete. The first, power-over, while social is also hierarchical. It embraces a masculinist typology that equates power with domination. Domination theorists who work within this framework critique the ways in which men dominate women in society, but fail to account for ways in which women dominate or assert power. The second definition is also incomplete as it offers a view of power as individual and experiential. Utilizing this conceptualization takes power out of the social context. Empowerment theorists acknowledge feminine practices, but do so without consideration of male dominance. Neither of these two types of power offers any real "threat" to the patriarchal structure. In the first, we have women adopting the very modes of masculinity that feminists oppose. In the second, we have a step forward from that—women individually recognizing the ways in which they have personal power, but without the goal of *social* change. There is a third definition we must consider.

Power-With

A third, more integrative approach to power is needed in our analysis of powerful women in film. Allen (1998) calls this third approach *power-with.* She argues, "We must be able to think about the kind of power that diverse women can exercise collectively when we work together to define, and to strive to achieve, feminist aims" (p. 32). This kind of analysis would allow us to account for masculinist domination, female empowerment, and also address the coalition building that is necessary to fully address oppression and structured inequality.

This approach depicts power as a kind of solidarity. Beyond the masculine variety of power-over, and the social psychological power-to, power-with is the kind of power whereby women come together as a group to challenge systems of oppression and bring about social change. The masculinized power-over is not the sort of power feminists have strived for. And, while personal empowerment can mean movement from oppression, it is still that—personal. However, social change is possible when solidarity is formed and women work as a collective. It is the kind of power that has produced significant change in the lives of both women and men.

What do these forms of power look like in film?

Power-Over: Masculinity

Recall power-over in the Weberian sense, which asserts that one actor (or group of actors) has the ability to control the actions of another. Thus, A can get B to do what A desires, even if said action is not in the best interest of B. We see men exerting this sort of power in films all the time. We always have. Westerns, war films, gangster films, hero-conquers-all films—all of these depict a man (or group of men) with the power to control the actions of others. They are powerful because they are masculine. In these films, we have grown accustomed to seeing women as the barmaids, the prostitutes, the tragic vixens, the adoring wives, or the hapless beauties in need of rescue.

But wait! There has been some change, hasn't there? Women are increasingly playing roles that demonstrate a kind of power-over. Yet we have to ask, what does it look like when women use power-over? I argue that three things happen when women have this sort of power in film: (1) women become "powerful" by their adoption of masculine characteristics; (2) as masculine women they often engage in the exploitation of others; and (3) when physically strong, they are often highly sexualized. We find evidence of this through an analysis of three films: *She Hate Me, The Devil Wears Prada,* and *Lara Croft: Tomb Raider.*

She Hate Me (Spike Lee, 2004)

Spike Lee's *She Hate Me* is the story of Jack (Anthony Mackie), who has fallen on hard times. A formerly successful biotech executive, he finds himself penniless when his attempts to expose corruption at his corporation cost him his job. All of his assets frozen, he finds himself financially ruined. Enter his ex-fiancé Fatima (Kerry Washington), a stunningly gorgeous woman, now in a relationship with another stunningly gorgeous woman, Simona (Monica Bellucci). Fatima and Simona approach him with a business venture. They want to get pregnant. They ask if he will impregnate them for a very handsome fee. He rants and protests, but eventually agrees. Soon, all of their wealthy lesbian friends find out and a real business venture is under way—he gets them pregnant, they pay him lots of money. (Interestingly, all of these lesbian women, with the exception of Simona, choose the "old-fashioned method," as subsequent scenes show each having varieties of fairly aggressive intercourse with Jack.)

They have the money; he has the sperm. They have the power, thus they can get Jack to do what they want Jack to do. At first glance we might cheer just a little when we see successful women with substantial incomes aggressively getting want they want, and with such confidence. We might be inclined to chuckle at the reversal of roles: a man "putting out" to a group of demanding women. But, looking closely, we might be a bit more critical of the power we see. In fact, we see *women becoming "powerful" by their adoption of masculine characteristics.*

In one scene, the first group of women comes to Jack's home as Fatima and Simona explain the "deal." The women begin to agree to the terms and costs. One woman stops the conversation, asking how she can be sure of what she is "getting."

She sits with her arm resting on the back of the couch (a rather masculine pose), and says to Jack, "Strip, Bitch." In the moments that follow, Jack slowly begins to remove his clothes, starting with his shirt, until he stands naked before them. He is quiet, his face reflecting humiliation and shame. One of the women comments, "Now you know how it feels."

Is this a moment of powerful women in film? Or a moment when women are using the same kind of masculine power-over that results in the oppression of others? Later in the film one of the women says to Jack, "We see you as nothing but dick, sperm, and balls." To objectify a man as women have been long objectified is a seriously impoverished view of power. The goal of feminism is not to place the power of humiliation and abuse into the hands of women. Rather, that is the kind of oppressive power that feminists have challenged and sought to change.

The Devil Wears Prada (David Frankel, 2006)

Meryl Streep's performance as Miranda Priestly, the cold and cynical fashion magazine editor, was highly praised for its dead-on ruthlessness. Miranda is not just any magazine editor; she is THE editor whose decisions determine the fashion world. Hers is a powerful position, and she dominates those around her. When Andy (Anne Hathaway) lands a job as her personal assistant, she is unaware of the inner workings of the industry and thus unprepared for the scorn and cruelty she encounters at her new job. Her days are spent running, anxiety-ridden, here and there, trying desperately to satisfy a boss who only places further demands, mixed with cold criticism. We laugh at Andy's foibles and Miranda's tight-lipped criticisms. And we perhaps enjoy a film that centers on a woman as head of a powerful industry (albeit the fashion industry).

The Devil Wears Prada offers audiences a commanding and powerful woman. Instead of Michael Douglas in charge (as in *Wall Street*, 1987), we have Meryl Streep. The most powerful executive in the room is a woman. But, again we must ask what this really looks like. Miranda is a boss not unlike other bosses we've seen before. Cold, chillingly mean, emotionless, and incapable of maintaining relationships, Miranda could be any male executive that has come before.

Again, we see a woman with power, but it looks like the masculinized version of power-over that we have grown accustomed to, perhaps the only kind of power we recognize as "power." We simply cheer when we see women given a chance to do it. Is it any sign of progress when women perform as the kind of oppressive executives that feminists have criticized? Miranda represents yet another version of masculinized power that offers no alternative or dialogue concerning workplace dynamics (in which women are far more often the power*less*). In *The Devil Wears Prada,* we are presented with a masculinized woman, at least in the norms of bureaucratic leadership. And, in film, *as masculine women they often engage in the exploitation of others.*

Of course, another key to this film is the contrasting types of power between the two women. Miranda is presented as a negative role model of power—it looks too much like male power and denies her emotional connections to family and friends. Andy is driven toward a career, but not if the cost is her boyfriend or her relationship with friends. There is a certain "feminist/antifeminist" feel to this contrast. If women are to attain the power that Miranda holds, it must look different. And, yes, Andy should have a career, but not if it means she loses her adoring boyfriend.

Lara Croft: Tomb Raider (Simon West, 2001)

In *Lara Croft: Tomb Raider,* Angelina Jolie plays Lara Croft, the product of a wealthy British family, who has been expertly trained in weapons and combat. She goes about her adventures, collecting artifacts from tombs, temples, and ruins around the world. She stares down danger; she flips, fights, handles big guns, wears dark shades, and delivers wry, cold lines that reinforce her coolness. And she does it all while looking extraordinarily sexy. While we might have found Harrison Ford ruggedly attractive as Indiana Jones, Jolie is the very image of sexy in these films. Ford wore worn, baggy pants and jacket. Jolie is seen in tight-fitting clothes that accentuate her long legs, her shapely hips and her seemingly perfect breasts. Lara Croft is powerful, perhaps. But, make no mistake, she is HOT.

As in the *Kill Bill* (2003, 2004) series, *Lara Croft* presents us with a woman who can "fight like a man." In *Kill Bill: Vol. I* and *Vol. II,* Quentin Tarantino gave us sexualized women who could not only fight (some in little school-girl outfits), but chop off limbs with utter disregard. Lara Croft fights with impressive strength, but she too is highly sexualized while doing it. In these films, we are given strong women, but *when physically strong, they are often highly sexualized.*

We are offered images of women who can fight, and we may be inclined to see that as progress—as women gaining power in the form of muscle and strength. Of course, this too is masculine power of the power-over variety. As with the previous films, this kind of power offers no real movement from the forms of power feminists have critiqued. We can go as far as saying that power in this form is no threat at all to patriarchy. It either *is* patriarchy, redubbed with women, or it is physical power so heavily sexualized as to border on the comical. We can cheer for Lara Croft. We can even nudge our daughters and say, "See how strong she is?" But we must also recognize the limitations of such representations of power. First, they are based solidly on the sexuality of the woman. Her power is unconditionally linked to her physical attractiveness. Thus, our daughters might be motivated to lift weights, but quite possibly equally motivated to have breast implants. And second, the kind of power demonstrated in all three of these films—the kind that mimics masculinity, the variety that continues to oppress, and the physically strong yet highly sexual—none of these offers serious threat to the traditions of patriarchy. In fact, they are drawn within the very limits of patriarchy, by patriarchy. There will never be enough Lara Crofts to really scare anyone.

Power-To: Agency and Autonomy

The second definition of power we considered was power-to, a personal sense of control. With this power, a woman can recognize the extent to which her actions determine her fate. She can come to know her own sense of agency and recognize her autonomy.

What does power-to tend to look like for women in film? We find this form of power begins to surface in films that portray women who, recognizing the restrictions of the norms dictating their lives, begin a search for "more." Lacking any real knowledge of personal empowerment (power-to) at the film's beginning, as the story unfolds, she comes to know something core about herself. These films have three prominent features. The women in these films tend to (1) separate from their lives, their cultures, and traditions as they begin to experience them as restrictive; (2) "wake up" from years of unconscious living and find agency where there was little; and (3) discover themselves through a discovery of their sexual selves or through the recognition that their lives are no longer dependent on having a man present. We'll take a look at these three features as they appear in *Real Women Have Curves, The Color Purple,* and *Waitress.*

Real Women Have Curves (Patricia Cardoso, 2002)

In this film, set in East Los Angeles, Ana (America Ferrera), a first generation Mexican American, has just graduated from high school. Her exceptional performance in school resulted in a full scholarship to Columbia University in New York City. Her family, a traditional Mexican family, struggles to survive financially on the earnings of Ana's sister, Estela's (Ingrid Oliu) sewing factory. Having given up her hopes for Columbia, Ana spends the summer working in Estela's factory with her sister, her mother, and several women from the neighborhood. Ana's mother is firmly rooted in the traditions of her culture and wants the same for Ana. At 18 years old, it is time for Ana to learn to sew, raise kids, and take care of a husband. She tells Ana, "A mother knows the right man for their daughter." Her advice to Ana: work hard, lose weight, find a man. As hard as Ana tries, she just can't find this advice acceptable.

Throughout the film, Ana *begins to separate from her life, her culture, and traditions, and begins to experience them as restrictive.* With a deep love for her family and a newfound respect for her sister's work, she still longs for the opportunities college offers. Not only has her mother's adherence to traditional ways proven restrictive of her future goals, her mother's sense of what it means to be a woman (get married, raise kids) has placed tremendous emphasis on physical beauty. Ana rebels against the notion that she is the "fatty" her mother calls her, instead finding a way to love her body, just as it is. Eventually, Ana finds the courage to tell her father that she must leave Los Angeles and take advantage of the opportunities available to her at Columbia. Her mother never comes out of the bedroom to "give her blessing," a heartbreaking image of one generation's fear of letting go and another's inability to *not* go. The last scene of the film shows Ana emerging from the subway in New York City, a confident and empowered smile on her face.

The Color Purple (Steven Spielberg, 1985)

The Color Purple tells the story of Celie (Whoopie Goldberg), a young black girl living in rural Georgia in the early to mid 20th century. Celie's father (Leonard Jackson) sexually abuses her, resulting in the birth of two children that her father "took away." Celie is eventually married to a local farmer called Mister (Danny Glover) who soon begins to abuse her. Celia lives with Mister, his son Harpo (Willard E. Pugh), and Harpo's defiant wife Sofia (Oprah Winfrey). Mister's lover, Shug (Margaret Avery), comes to live in the house as well, and at first she is as hateful to Celie as is Mister. Eventually Sofia leaves the abuse of Harpo, taking her children (however, her spirit is eventually broken when she is separated from her children, and forced to work for the mayor's wife). Celie and Shug grow closer, with Celie finding herself attracted to Shug's tenacity and spirit. Together the women uncover secrets from the past: Celie's father was not in fact her biological father; her sister whom she thought dead was in fact alive; and the children she birthed were alive, well, and in her sister's care. In a powerful scene, Celie, finally *having found some agency where once there was none,* confronts Mister for his years of abuse. Celie eventually moves to Tennessee and builds her own successful business. In the end, she returns to Georgia, this time an empowered woman.

In this course of this film, Celie emerges from the lowest possible social position. Not only is she a woman, but she is a poor, uneducated African American woman living in the segregated southern United States in the early 20th century. It was rare for a woman such as Celie to question the powerlessness of her life. When women, such as Sofia, did stand up for themselves, they were beaten, imprisoned, and forced into subservience. But even this experience did not break Sofia's spirit, and she stands with Celie as she is transformed from a woman lacking power (autonomy, agency, sense of control) to a self-supporting woman in control of her own fate.

Waitress (Adrienne Shelly, 2007)

In *Waitress,* we again encounter a woman living in an abusive marriage. Jenna (Keri Russell) is married to Earl (Jeremy Sisto), but she is miserably unhappy and hides the earnings from her job at the diner, in hopes of leaving him. Meanwhile she discovers she is pregnant. As the months go by, she becomes involved with her kind, dashing, and married obstetrician-gynecologist (Nathan Fillion), which further complicates her already complicated life. Not wanting to be pregnant, feeling miserable and stuck with Earl, Jenna feels powerless and without options. A talented baker, she pours her emotions into ideas for pies such as, "I Don't Want Earl's Baby Pie." Other than her talent for baking, she has little else. While she feels loved and respected by Dr. Pomatter, his marriage prevents them from being together. She has very little money of her own. She lacks the physical strength to stand up to Earl. She pins her hopes on the chances of winning $25,000 in a pie-baking contest. With that she could leave Earl.

What makes this film remarkable sociologically is its ending. If given only the paragraph above, and knowing something about the genre of light-hearted American films, we might assume the ending. She has the baby and manages to leave abusive Earl. The gorgeous doctor whom she has grown to love, finds his way to her. But, in this film, that is not quite the case.

This is a film depicting a seemingly powerless woman, who comes to power through her questioning and search for empowerment. This movie ends with Jenna indeed having her baby, a daughter. Before leaving the hospital she tells Earl she has no intention of going home with him—their marriage is over. Gone is the abusive husband. Then the twist. She tells the gorgeous doctor, who appears shaken and a bit devastated, that they too are finished. Jenna inherits some money from a customer (Andy Griffith) who recognized her potential. The film ends with Jenna, her daughter by her side, in her own pie-baking business on the site where once the diner stood.

The kind of empowerment that Jenna found involved the slow awareness that she had agency. In the course of the film, Jenna found that she could control the outcome of her life and, unlike so many films with happy endings for women, *she found this in the recognition that her life no longer depended upon men.*[4] Empowerment theorists might argue that her power stemmed from the birth of her daughter—from the sense of strength and power that accompanies the mothering role. Whether due to motherhood or her months of reflection and struggle (the process of "waking up"), or some combination of both, Jenna emerges in the end a woman empowered.

All three of these films offer hope. All suggest a very important kind of power, the kind that moves women to question their taken-for-granted circumstances. When we see this form of power in film, we are moved by women who find a place of core strength and a sense of agency and autonomy. More so than with films projecting a masculinized sense of power, these films offer images of power that actually change lives and lead women to step out of the expectations of femininity and womanhood and choose their own paths. Certainly as we hope for films that inspire or reflect (or both) social change, this is movement.

But how much does it actually challenge the structure of patriarchy? I believe that it suggests a crack in the surface, that there is at least a consciousness of women and personal empowerment. That is an important image to portray in film as it offers women a picture of their gendered trappings (as *Kate Chopin: The Re-awakening,* 1998). But do these films offer critiques of the social context? Is the change suggested in these films social, or does it suggest a kind of experiential, personal change? While an important and valuable "change" occurs, it has yet to resemble *social* change. I believe that we see evidence of true social change in films that take power a step further.

Power-With: Collective Power of Women

Power-with involves women working together to define and achieve feminist goals. Consider some of the significant changes that have impacted women's lives over the past 50 years: women's suffrage, women's access to contraceptives, the right of women to initiate a divorce, protections against rape, laws against sexual harassment, equal pay laws, and women entering previously male-dominated jobs such as the police force, law, and medicine. While developing personal consciousness of these issues played a vital role in the lives of the women involved, it was the collective actions of women (and the men involved in the women's movement) that

[4]Other films, such as *How Stella Got Her Groove Back* (1998), depict the power that comes when a woman discovers her forgotten sexual self.

forced this consciousness into public social institutions, such as law, medicine, education and the media.

Films depicting women working together for change are far more common in the documentary genre (see the Film Index) than in Hollywood feature films When we see power-with demonstrated in Hollywood films, they are the stories of women (1) struggling within the constraints of an oppressive system, (2) coming to realize the extent of their oppression, and then (3) working together to confront the system that oppresses them. In these films, as in the film discussed below, an individual woman begins to awaken to the oppressiveness of the system in which women operate. Hard as she might try, her individual attempts at change fail until the community of women comes together as a group.

North Country (Niki Caro, 2005)

Based on the book by Clara Bingham and Laura Leedy Gansler, *Class Action: The Story of Lois Jensen and the Landmark Case That Changed Sexual Harassment Law, North Country* tells the story of women working in the male-dominated world of mining and their fight against the abuse and harassment they receive from their male coworkers and bosses. Josey (Charlize Theron), a single mother of two, has recently left her abusive husband. Seeking employment that will allow her to care for her children, she is thrilled to land a secure job that offers a steady income. In the days before sexual harassment laws, the women in the mines suffered myriad forms of abuse from male coworkers who resented their very presence in their masculinized world. One supervisor tells the women, "there are all sorts of things a woman shouldn't do," and working at the mine was certainly one of them. On Josey's first day, she tells this same supervisor that her paperwork is in order, including the results of a physical. The man sneers, "Doc says you look real good under those clothes. . . ." When the women are silent he retorts, "Sense of humor ladies! Rule numero uno."

The workplace harassment and abuse is extensive. It begins with the verbal, "Which one of these ladies will be my bitch?" "Cunts" is written on the door to their locker room. It is psychological warfare. One woman finds that a man has ejaculated in her locker, on her clothing. Another is taunted as she tries to use the outdoor port-o-pot. When the men eventually turn the toilet on its side, she climbs out, covered in urine and feces, clearly traumatized by the event. The abuse turns physical when Josey is attacked and nearly choked and raped as her attacker tells her, "you're gonna learn the goddamn rules if I have to beat them into you myself." Clearly these women are *struggling within the restraints of an oppressive system.* They have no recourse—they can "tough it out" or they can quit their jobs. Of course, living in an economically depressed region, the women often make the case that they "need this job."

The abuse these women sustain builds camaraderie. They *come to realize the extent of their oppression* as the abuse they endure increases in frequency and intensity. However, they also realize their powerlessness. When Josey takes her complaints to management she is told sternly, "I suggest you spend less time stirring up your female coworkers, less time in the beds of your married coworkers," and more

Photo 4.1 Collective power fuels social change in *North Country*. The case against sexual harassment in the workplace is successful once women *and* men take a stand in support of Josey.

time at her actual job. Their reaction to her attempt at presenting a legitimate set of complaints stuns Josey in its hostility, not to mention the blatant disregard for truth (she is not sleeping with the married coworker, the lone male who sympathizes with the women's experiences), and the denial of their reality (they can't possibly focus on their jobs when each workday is spent in constant self-defense). When Josey tries to induce change, the women at first turn against her. They fear the loss of their jobs and make it clear they will not support her in the legal actions for the rights of the women to work in harassment-free environments. She asks one woman, "What about what happened to *you*?" Her coworker replies, "It's my business." Josey tells her and the others, "Actually it's all of our business."

In the course of her court proceedings, it appears as though Josey's case is hopeless. She is attacked for her sexual history and the paternity of her children. She only needs three plaintiffs to come forward with similar grievances in order to make her individual charges into a class-action case, but the others remain fearful and sit quietly in court. In a pivotal scene, her lawyer (Woody Harrelson) demands that those present in the courtroom "stand up and tell the truth." Slowly, the women rise in support of Josey. First her coworkers, then her mother, then her father, and slowly other men in the room until most of the room stands in support. Her case is won and laws against sexual harassment and discrimination in the workplace slowly begin to change the lives of women.

Initially working alone, Josey had no chance of winning. But by *working together to confront the system that oppressed them,* social change became possible. It was not until the women stood together that life for women in the workplace would begin the slow process of change. The courtroom scene begins with the women rising to their feet and ends with first a few brave men, then slowly more and more men rising in support. This scene captures the kind of solidarity in the women's movement that includes men. Feminists don't want to turn the tables and oppress men, rather they count on men to rise to their feet, along with women, and support them in their call for social change. Real change is possible when the collective power of women (power-with) takes a stand.

Conclusion

We have considered cinematic images of power and powerful women. In the process we have come up with multiple definitions of power and multiple images of power across films. This discovery has various implications for women, conceptually and politically. Beginning with the first kind of power, power-over, we could come up with numerous films that display powerful women in this manner. Oftentimes, when women are seen as possessing power-over, it is seen as a problem, not a solution. She is too powerful, thus she is evil, a temptress, a black widow, or a bitch. If she mimics masculinized power, she is likely to be "the bitch," or we find that she is humbled (or killed!) in the end. Those physically strong women are so highly sexualized as to offer no real threat. In fact, that kind of individual power may actually reinforce patriarchy as a system. It reminds women that patriarchy itself is not so oppressive. It can't be if we keep seeing images of these powerful women. Thus, any lack of power one might "feel" is wrong—she only has herself to blame.

We could also come up with a fairly nice list of films that present power-to. When we contrast this image to power-over, we certainly find it more pleasing. We even consider it feminist. Women in these films are bristling under the weight of gendered expectations. They leave their abusive husbands or have adventures. They often reach out to another woman or group of women in the process. And this is good. Personal empowerment is to be cheered. Conceptually it is much preferred to the former. But what does it say about the politics of gender? Does the content or message of these films offer a threat to patriarchy? If we compared it to the films in the power-over category, we would say yes, somewhat. They present us with women who develop the power to seek and find change. However, this is not the kind of power necessary for social change.

Significantly fewer films portray the power-with model. Both conceptually and politically, these films offer an actual challenge to the definition of power and to its social foundations. We move beyond a power-over depiction of women that, in reality, is merely repeating the masculinist patterns of patriarchy. And we step beyond power-to, which depicts individual women coming to terms with their lives. Power-with films offer us images of women, coming together in solidarity, in numbers to do what one woman cannot do alone. As we move beyond masculine and individualized definitions of power, let's hope that more directors are brave enough to offer us cinematic images of the collective power of women.

References

Allen, A. (1998). Rethinking power. *Hypatia, 13*, 21–40.

Allen, A. (2000). *The power of feminist theory: Domination, resistance, solidarity.* Boulder, CO: Westview.

Bandura, A. (1997). *Self-efficacy: The exercise of control.* New York: W. H. Freeman.

Connell, R. W. (1987). *Gender and power: Society, the person and sexual politics.* Stanford, CA: Stanford University Press.

Kimmel, M. (1994). Masculinity as homophobia: Fear, shame and silence in the construction of gender identity. In H. Brod (Ed.), *Theorizing masculinity* (pp. 119–141). Thousand Oaks, CA: Sage.

LaSalle, M. (2000). *Complicated women: Sex and power in pre-code Hollywood.* New York: Thomas Dunn Books, St. Martin's Press.

Nash, K. (2000). *Contemporary political sociology: Globalization, politics, and power.* Malden, MA: Blackwell.

Paap, W. R. (1981). The concept of power: Treatment in fifty introductory sociology textbooks. *Teaching Sociology, 9*(1), 57–68.

Ross, K. E., & Mirowsky, J. (2002). Age and the gender gap in the personal sense of control. *Social Psychological Quarterly, 65*, 125–145.

Ruddick, S. (1989). *Maternal thinking: Towards a politics of peace.* New York: Ballantine.

Sprague, J. (2005). *Feminist methodologies for critical researchers.* Lanham, MD: AltaMira Press.

Yoder, J. D., & Kahn, A. S. (1992). Toward a feminist understanding of women and power. *Psychology of Women Quarterly, 16*(4), 381–388.

Reading 4.2

The Masculinity of the Governator[1]

Muscle and Compassion in American Politics[2]

Michael A. Messner

The big news story on November 7, 2006, was that voters had returned control of the U.S. Congress to the Democrats. This represented a dramatic turning of the electoral tide against the policies of Republican President George W. Bush—especially against his stubborn mantra to "stay the course" in the war on Iraq. But apparently swimming against this tide was another story. On that same day, Republican Arnold Schwarzenegger was reelected as California governor by a landslide, winning 56% of the vote over Democratic challenger Phil Angelides's mere 39%. During a year of resurgent Democratic strength nationally, in a solidly Democratic state, and only a year after his popularity had plummeted with voters, how do we explain Schwarzenegger's resounding victory? In this article, I will explore this question by examining Schwarzenegger's public masculine image.

A key aspect of Schwarzenegger's public image, of course, is his celebrity status, grounded first in his career as a world champion bodybuilder, and even more so in his fame as one of the most successful action film stars of his generation (Boyle, 2006). My aim here is not to analyze Arnold Schwarzenegger's biography. Nor do I intend to offer a critical analysis of his films—I confess, I have watched some of them and not others (and I enjoy the ones that I have seen). Instead, my aim is both practical and theoretical: I will outline the beginnings of a cultural analysis of how and why Schwarzenegger rose to political power, what his appeal was and is, and how some current debates in gender theory might be useful in informing these questions. I will consider what Schwarzenegger's deployment of a shifting configuration of masculine imagery tells us about the limits and possibilities in current U.S. electoral politics. And I will deploy the concept of "hegemonic masculinity" to suggest how Schwarzenegger's case illustrates connections between the cultural politics of gender with those of race, class, and nation. In particular I hope to show how, when symbolically deployed by an exemplar like Arnold Schwarzenegger, hegemonic masculinity is never an entirely stable, secure, finished product; rather, it is always shifting with changes in the social context. Hegemonic masculinity is "hegemonic" to the extent that it succeeds, at least temporarily, in serving as a

[1]The term *Governator* became a widely used way to refer to Arnold Schwarzenegger in the popular media and among Californians in the aftermath of his election as California governor. The term symbolically links his job as governor with his best-known film role as the Terminator, and speaks to his successful construction of a hybrid celebrity personality that I discuss in this paper.

[2]This article first appeared in *Gender & Society*, 21 (2007), pp. 461–480. Reprinted with the author's permission.

symbolic nexus around which a significant level of public consent coalesces. But as with all moments of hegemony, this consent is situational, always potentially unstable, existing in a dynamic tension with opposition.

Masculinities and Politics

Since the late 1980s, sociologists have tended to agree that we need to think of masculinity not as a singular "male sex role," but as multiple, contextual, and historically shifting configurations. At any given moment, a dominant—or hegemonic—form of masculinity exists in relation to other subordinated or marginalized forms of masculinity, and in relation to various forms of femininity (Connell, 1987). Very few men fully conform to hegemonic masculinity. In fact, it is nearly impossible for an individual man consistently to achieve and display the dominant conception of masculinity, and this is an important part of the psychological instability at the center of individual men's sense of their own masculinity. Instead, a few men (real or imagined) are positioned as symbolic exemplars for a hegemonic masculinity that legitimizes the global subordination of women and ensures men's access to privilege. What makes this masculinity hegemonic is not simply powerful men's displays of power, but also, crucially, less powerful men's (and many women's) consent and complicity with the institutions, social practices, and symbols that ensure some men's privileges (Messner, 2004). To adapt a term that is now popular in market-driven bureaucracies, hegemonic masculinity requires a *buy-in* by subordinated and marginalized men, and by many women, if it is to succeed as a strategy of domination.

Thus, the concept of hegemonic masculinity is most usefully deployed when we think of it not as something that an individual "has"—like big muscles, a large bank account, or an expensive car. But then, what is it? Where does it reside? Can we define it or is it something about which we simply say, "I know it when I see it"? To ask these kinds of questions, we need to develop ways of thinking about gender that are *global*, both in the geographic and in the conceptual sense of the word. Here, I want to explore the ways that we can think about hegemonic masculinity as a symbolically displayed "exemplar" of manhood around which power coalesces—and importantly, *not* just men's power over women, but also power in terms of race, class, and nation (Connell & Messerschmidt, 2005). I will suggest that it is in the symbolic realm where an apparently coherent, seemingly stable hegemonic masculinity can be forged (Gómez-Barris & Gray, 2006). We can track this symbolic masculinity as it reverberates into institutions—in the case of Arnold Schwarzenegger, into the realm of electoral politics—and we can see how hegemonic masculinity works in relation to what Collins (1990) calls a "matrix of domination," structured by race, class, gender, and sexuality.

Terminating the Feminized American Man

Arnold Schwarzenegger began his public career as a world champion bodybuilder. Many people mark his starring role in the award-winning 1977 documentary on bodybuilders, *Pumping Iron,* as his film debut. However, Schwarzenegger actually appeared in a few other television and B film roles before that, including a typecast role in the 1970 film *Hercules in New York.* Schwarzenegger's celebrity star rose rapidly in the early 1980s, with a series of films that featured his muscular body as the ultimate fighting machine: *Conan the Barbarian* (1982), *Conan the Destroyer* (1984), *Commando* (1985), *Red Sonja* (1985), and *Predator* (1987). Among these popular 1980s films, it is *The Terminator* (1984) that most firmly established Schwarzenegger as a major film star and as king of a particular genre.

Conan, John Matrix, and the Terminator appear in the 1980s at the same time that Rambo and other hard, men-as-weapon, men-as-machine images filled the nation's screens. Susan Jeffords (1989) calls this cultural moment a "remasculinization of America," when the idea of real men as decisive, strong, and courageous arose from the confusion and humiliation of the U.S. loss in the Vietnam War and against the challenges of feminism and gay liberation. Jeffords's analyses of popular Vietnam films are especially insightful. The major common theme in these films is the Vietnam veteran as victimized by his own government, the war, the Vietnamese, American protestors, and the women's movement—all of which are portrayed as feminizing forces that have shamed and humiliated American men.

Two factors were central to the symbolic remasculinization that followed: First, these film heroes of the 1980s were rugged individuals, who stoically and rigidly stood up against bureaucrats who were undermining American power and pride with their indecisiveness and softness. Second, the muscular male body, often with massive weapons added as appendages, was the major symbolic expression of remasculinization. These men wasted very few words; instead, they spoke through explosive and decisively violent bodily actions. Jeffords argues that the male body-as-weapon serves as the ultimate spectacle and locus of masculine regeneration in post–Vietnam era films of the 1980s. There is a common moment in many of these films: the male hero is seemingly destroyed in an explosion of flames, and as his enemies laugh, he miraculously rises (in slow motion) from under water, firing his weapon and destroying the enemy. Drawing from Klaus Theweleit's (1987) analysis of the "soldier-men" of Nazi Germany, Jeffords argues that this moment symbolizes a "purification through fire and rebirth through immersion in water" (Jeffords, 1989, p. 130).

During this historical moment of cultural remasculinization, Schwarzenegger was the right body at the right time. Muscular Arnold, as image, reaffirmed the idea of categorical sex difference, in an era where such difference had been challenged on multiple levels. In this historical moment, the Terminator's most famous sentence, "I'll be back," may have invoked an image of a remasculinized American man, "back" from the cultural feminization of the 1960s and 1970s, as well as a resurgence of American power in the world.

It's possible to look at this remasculinized male subject in 1980s films as a symbolic configuration of hegemonic masculinity that restabilizes the centrality of men's bodies, and thus men's (at least white U.S. men's) power and privilege. Indeed, Messerschmidt's (2000) statement that "Hegemonic masculinity . . . emphasizes practices toward authority, control, independence, competitive individualism, aggressiveness, and the capacity for violence" (p. 10) seems to describe precisely the masculinity displayed by Schwarzenegger and Stallone in these 1980s films. But we need to be cautious about coming up with such a fixed definition of hegemonic masculinity. Though it clearly provided symbolic support for the resurgent conservatism of the Reagan era, this simplistic reversion to an atavistic symbology of violent, stoic, and muscular masculinity probably fueled tensions in gender relations as much as it stabilized them. As Connell and Messerschmidt (2005) note:

> Gender relations are always arenas of tension. A given pattern of hegemonic masculinity is hegemonic to the extent that it provides a solution to these tensions, tending to stabilize patriarchal power or reconstitute it in new conditions. A pattern of practice (i.e., a version of masculinity) that provided such a solution in past conditions but not in new conditions is open to challenge—in fact is certain to be challenged. (p. 853)

There's plenty of evidence that by the end of the 1980s, the remasculinized muscular hero who wreaks havoc with his guns (biceps and bazookas), while keeping verbal expression down to a few grunts or occasional three-word sentences delivered in monotone, was not playing well in Peoria. The 1988 installment of the Rambo series (*Rambo III*) was listed by the 1990 *Guinness Book of World Records* as the most violent movie, with 221 acts of violence and over 108 deaths. Despite (or because of?) this carnage, the film did not do well. Its gross in the United States was $10 million lower than the film's overall budget, and Stallone's tired one-liners (Zaysen: "Who are you?" Rambo: "Your worst nightmare") reportedly left audiences laughing derisively. Other icons of heroic masculine invulnerability tumbled from their pedestals: one of the actors who played the Marlboro man in cigarette ads died from lung cancer. By the late 1980s and early 1990s, the idea of men as invulnerable, nonemotional, working and fighting machines was frequently caricatured in popular culture and made fun of in everyday life. Health advocates grabbed on to this caricature with "culture jamming" counteradvertisements aimed at improving men's health. For instance a counter ad distributed by the California Department of Health Services referenced years of "Marlboro Country" ads by depicting two rugged cowboys riding side-by-side on their horses with a caption that read, "Bob, I've got emphysema." These sorts of ads invert the intended meanings of the Marlboro Man, illustrating how narrow cultural conceptions of masculinity are unhealthy—even deadly—for the men who try to live up to them. This new cultural sensibility is direct a legacy of the feminist critique of hypermasculinity. By the 1990s, these kinds of counter ads could rely on readers to make the ironic connections, drawing on their own familiarity with the "straight" tobacco ads that were referenced, in addition to their familiarity with the increasingly prevalent cultural caricatures of hypermasculinity as dangerous, self-destructive, and (often) ridiculously laughable.

The Birth of the Kindergarten Commando

Many professional-class white men in the 1980s and 1990s began to symbolically distance themselves from this discredited view of "traditional masculinity," and forged new, more "sensitive" forms of masculinity. But this is not to say that successful and powerful men have fully swung toward an embrace of femininity and vulnerability. Some men's brief flirtations with "soft" "new man" styles in the 1970s—the actor Alan Alda comes to mind—were thoroughly discredited and marginalized. Instead, we have seen the emergence of a symbology of masculinity that is hybrid: toughness, decisiveness, and hardness is still central to hegemonic masculinity, but it is now normally linked with situationally appropriate moments of compassion and, sometimes, vulnerability. The 1980s and 1990s saw the increasingly common image of powerful men crying—not sobbing, but shedding a tear or two—in public: President Ronald Reagan in speaking of soldiers' sacrifices after the 1983 U.S. invasion of Grenada; General Norman Schwartzkopf at a press conference noting U.S. troops killed during the 1990–91 Gulf War; basketball player Michael Jordan in the immediate aftermath of winning his first NBA championship in 1991. These emotional displays may have been fully genuine, but I emphasize that they were not delivered in the aftermath of a loss, in a moment of vulnerability, failure, or humiliation. Try to imagine, for a moment, superstar NBA player Dirk Nowitzki after the Dallas Mavericks' 2006 loss in the NBA Finals, dropping to his knees at center court, overcome with grief, weeping openly with his face buried in his hands. That's not likely to happen. Tears are appropriate as public masculinity displays in the immediate aftermath of *winning* an NBA championship, or of having just successfully overrun a small third-world country with virtually no military. Powerful men have found it most safe to display public grief or compassion *not* in relation to their own failures, or to the pain of other men—this might be perceived as weakness—and *not* in terms of women's struggles for respect and equality—this might be perceived as being "pussywhipped" (a recently revived epithet in pop culture). Rather, the public compassion of this emergent masculinity is most often displayed as protective care—often for children—which brings us back to the Governator.

Schwarzenegger's original Terminator character was an unambiguously violent male-body-as-weapon, severed from any capacity for human compassion. But in the late 1980s, this image began to be rounded out by—not *replaced* with—a more compassionate persona. We can actually watch this transformation occur in *Terminator 2: Judgment Day* (1991). In this film, Schwarzenegger, though still a killing machine, becomes the good guy, even showing occasional glimpses of human compassion. And significantly, it is his connection with a young boy that begins to humanize him. Taken together, Schwarzenegger's films of the 1990s display a masculinity that oscillates between his more recognizable hard-guy image and an image of self-mocking vulnerability, compassion, and care, especially care for kids (e.g., *Kindergarten Cop,* 1990; *Jingle All the Way,* 1996). I call this emergent hybrid masculinity "The Kindergarten Commando." Indeed, in Schwarzenegger's first major foray into California politics in 2002, he plugged his ballot initiative for after-school activities for kids by saying that he had been "an action hero for kids in the movies; now I want to be an action hero for kids in real life." In the 1994 comedy *Junior,* Schwarzenegger appropriates an ultimate bodily sign of femaleness: pregnancy and childbirth. But Schwarzenegger's gender hybridity could never be mistaken as an embrace of a 1970s styled "androgyny." Instead, in the Kindergarten Commando masculinity of Arnold Schwarzenegger, we see the appropriation and situational display of particular aspects of femininity, strategically relocated within a powerfully masculine male body.

In his initial 2003 run for California governor, Schwarzenegger positioned himself as a centrist unifier, and his film-based masculine imagery supported the forging of this political image. Hardness and violence, plus compassion and care, is a potent equation for hegemonic masculinity in public symbology today. And what tethers these two seemingly opposed principles is *protection*—protection of children and women from bad guys, from evil robots from the future, or from faceless, violently irrational terrorists from outside our borders.

The post-9/11 world has provided an increasingly fertile ground for the ascent of the Kindergarten Commando as compassionate masculine protector. Iris Marion Young (2003) has argued that the emergent U.S. security state is founded on a renewed "logic of masculine protection." And as Stephan Ducat (2004) has argued in his book *The Wimp Factor,* right-wing movements have seized this moment to activate a fear among men of "the mommy state"—a bureaucratic state that embodies weakness, softness, and feminist values. The desire for a revived "daddy state" is activated through a culture of fear: only the man who really cares about us and is also tough enough to stand up to evil can be fully trusted to lead us in these dangerous times. The ascendance of this form of hegemonic masculinity is both a response to feminist and other critiques of the limits of a 1950s, John Wayne–style masculinity, *and* it thrives symbiotically with pervasive fears of threats by outsiders.

This is not a symmetrical symbiosis, though. In a male political leader today, compassion and care seem always to be subordinated to toughness, strength, and a single-minded resolve that is too often called "decisiveness," but that might otherwise accurately be characterized as stubborn narrow-mindedness. This asymmetry is reflected in Schwarzenegger's recent films, in which the violent, tough-guy hero has never been eclipsed by the vulnerable kid-loving guy: in *End of Days* (1999), Schwarzenegger saves the world from no less a force of evil than Satan himself. The emergent Kindergarten Commando masculinity is forged within a post-9/11 context in Schwarzenegger's 2002 film *Collateral Damage.* Here, firefighter Gordon Brewer is plunged into the complex and dangerous world of international terrorism after he loses his wife and child in a terrorist bombing. Frustrated with the official investigation and haunted by the thought that the man responsible for murdering his family might never be brought to justice, Brewer takes matters into his own hands and tracks his quarry ultimately to Colombia. By the time *Terminator 3: Rise of the Machines* hit the theaters in 2003, Schwarzenegger was governor of California. In *T3,* it was clear that the hero, though still an admirably efficient killing machine, had mobilized his human compassion to fight for humanity against the evil machines.

Hegemonic Masculinity in a Matrix of Domination

My argument thus far is that the currently ascendant hegemonic masculinity constructed through a combination of the film images of Arnold Schwarzenegger is neither the stoic masculine postwar hero image of John Wayne, nor is it the 1980s remasculinized man-as-machine image of Rambo or the Terminator. These one-dimensional masculine images, by the 1990s, were laughable. Instead, the ascendant hegemonic masculinity combines the kick-ass muscular heroic male body with situationally expressive moments of empathy, grounded in care for kids and a capacity to make us all feel safe. Feminism, antiwar movements, health advocates, and even modern business human relations management have delegitimized pure hypermasculinity. But many people still view effeminacy as illegitimate in men, especially those who are leaders. So, neither hard nor soft are fully legitimate, unless they are mixed, albeit with a much larger dose of the former than of the latter. And commercial interests have fruitfully taken up this hybrid masculine image: Heterosexual men, as we saw in the TV show "Queer Eye for the Straight Guy," are seen as more attractive to women when softened—provided they still have power, muscles, and the money to purchase the correct draperies, fine cuisine, clothing, cosmetics, and other body-management products. Arnold, of course, has all of this. His masculinity displays were effective in securing power. But toward what ends? What do we see in the play of "hard" and "soft," in "strength" and "compassion" in terms of what he *does* as governor? Three events are very revealing, and I will discuss them very briefly. First, the governor's playfully aggressive use of references to his film and bodybuilder careers in his ongoing budget battles with the Democrats. Second, his class politics—particularly in his dealings with business and labor interests in California. And third, the "woman problem" that emerged during his first election.

Girlie Men and Political Intertextuality

During his earliest days in office, Schwarzenegger famously mobilized the "girlie men" epithet and turned it on his Democratic opponents in the California legislature. In doing so, he deployed references to his own *Terminator* films (urging voters to "terminate" his Democratic opponents), and the "girlie men" comments referenced the "Saturday Night Live" skit that had originally spoofed him. This illustrates the often-noted fact that cultural symbols don't float free: they emerge from, and in turn enter into, social relations. Schwarzenegger strategically deployed the imagery of the Kindergarten Commando to get himself elected. But in the real life of governing, when push came to shove, he fell back on the Terminator, not the lovable Kindergarten Cop "Protector of Children," as a strategy for deploying power.

Schwarzenegger's "girlie man" taunt is not the first time that a politician has drawn from popular commercial culture to invoke an image aimed to undermine his opponents. Recall, for instance, Vice President Walter Mondale's 1984 attempt to attack his Republican presidential race opponent's lack of substantive ideas by humorously deploying a "Where's the beef?" chant that referenced the then-popular Wendy's hamburger commercial. However, by comparison, Schwarzenegger's "girlie men" insult is a rather unprecedented multilevel image: it is a veritable Möbius strip of meanings, with life imitating ironic schlock, imitating life, imitating more schlock.

Audiences get a sense of pleasure and power—a sense of authorship from being insiders as they participate in decoding familiar intertextual messages like "Where's the beef?" "girlie man," or "I'll be back!" And if we think of the electorate as an audience (and certainly political parties use all the advertising expertise that they can muster), an election can be seen as a sort of poll of the audience's preferences. The electorate is buying a particular candidate who has been sold to them. Schwarzenegger's reference to girlie men and to *The Terminator* appealed to his supporters, but it also set off a firestorm of criticism from feminists, gay/lesbian organizations, and Democratic legislators. The governor's plea to voters to "terminate" his Democratic foes in November not only disrupted his thus far carefully crafted image of the bipartisan get-it-done compromiser, it also indicated the reemergence of a gloves-off muscular masculinity behind which the kind and compassionate Kindergarten Cop receded into the background. And this spelled some trouble for the Governator.

The symbolic symmetry of the new man—the Kindergarten Commando—was broken, leaving Schwarzenegger once again vulnerable both to sarcastic media caricature and to open questioning about the misogyny and homophobia that might lie behind the warm smile. However, the fact that Schwarzenegger's power was anchored so much in the symbolic realm facilitated his ability to deploy his power in a form that allowed for humorous, ironic interpretation; the implied self-mocking in his girlie man comments gets him off the hook, perhaps, from otherwise coming across as a bully: Democrats who decry the sexism or homophobia embedded in the girlie man comment appear perhaps to have no sense of humor. In the vernacular of the shock radio so popular with many young white males today, people who object to Schwarzenegger's comments as sexist or homophobic are "feminazis"; they just don't get the joke (Benwell, 2004; Messner & Montez de Oca, 2005).

Hegemonic Masculinity and Class Politics

Meanwhile, though, it's clear that the joke was on some of the most vulnerable people of California. My second example concerns Schwarzenegger's class politics. In August 2004, he vetoed a minimum wage increase. Simultaneously, he supported Walmart's economic colonization of the California retail industry. Walmart's importation of notoriously low-waged jobs has been resisted by organized retail unions and by several California cities and towns, yet it was clear which side Schwarzenegger took in this struggle. This illustrates how hegemonic masculinity enters into class relations. If there was compassion here, it was compassion for big business; if there was muscle to be deployed, it was against the collective interests of working people, defined by the governor's business logic as "special interests."

Governor Schwarzenegger also attacked public employees unions in his effort to control state spending. A National Public Radio story on March 15, 2005, noted that members of the California Nurses Association were showing up and protesting in every public venue in which he appeared. As he was giving a speech to supporters, one could hear the voices of the nurses chanting something in the background. Schwarzenegger commented to a cheering crowd, "Pay no attention to them. They are the *Special Interests.* They don't like me in Sacramento, because I am always kicking their butts!" Indeed, earlier that year Schwarzenegger had vetoed the rule that mandated a lower nurse/doctor ratio, and also took $350,000 in campaign contributions from pharmaceutical companies while opposing a prescription drugs law that would have helped consumers.

The ideological basis of these class politics and their links to the politics of race and immigration were further demonstrated in Schwarzenegger's speech at the 2004 Republican National Convention. He told his own rugged individualist immigrant story, with clear pull-yourself-up-by-your-bootstraps Horatio Alger themes. While positioning himself on the surface as someone who cares about and understands immigrants, his story reiterated conservative themes that are grounded in the experience of white ethnics, rather than in that of the vast majority of California's current immigrant population who deal daily with poverty, institutionalized racism, and escalating xenophobia. Schwarzenegger's narrative thus helps to reconstruct a white male subject and demonstrates how hegemonic masculinity is never just about gender: it is also about race and nation (Montez de Oca, 2005).

Hegemonic Masculinity as Heterosexy

My third example concerns Schwarzenegger's "woman problem." During the final weeks of the 2004 election, the *Los Angeles Times* broke a series of stories indicating that several women had complained of Schwarzenegger's having sexually harassed and humiliated them in various ways over the years. The women's claims were quickly trivialized by being more benignly defined as "unwanted groping." Ironically, these accusations probably enhanced Schwarzenegger's status with many men, and may have helped to secure the complicity of many women, as evidenced by the "Arnold, Grope Me!" signs seen at some of his rallies in the final days of his run for governor. Here, we can see that the hegemonic masculinity created by Schwarzenegger's symbolic fusing of opposites also involves the construction of a particular form of masculine *heterosexuality.* We should not underestimate the extent to which the imagery of hegemonic masculinity is electrified with an erotic charge—a charge that serves as a powerful linking process in constructing dominant forms of femininity, and through that, the consent of many women. In fact, it is likely that while the "groping" charges solidified an already-existing opposition to Schwarzenegger, it also pulled some voters more solidly into his camp.

A comparison with former President Bill Clinton is useful in this regard. Stephen Ducat (2004) discusses how during his first term as president, Clinton was vulnerable to questions about his masculinity due to his lack of military service, his support for women's and gay issues, and especially to the perception that his wife Hillary "wore the pants" in his family. Clinton, according to Ducat, lacked symbolic ownership of the phallus, necessary for a man with power. Attempts by Clinton's handlers to symbolically masculinize Bill, and to feminize Hillary, did little to help either of their images.

However, in the aftermath of the scandals surrounding Bill Clinton's sexual relations with White House intern Monica Lewinsky, Clinton's poll numbers skyrocketed. A 1998 Gallup poll conducted after the scandal broke found that Americans saw him as the most admired man in the world. He had morphed, in Ducat's words, from "emasculated househusband to stud muffin," from "pussy" to "walking erection" (p. 150). Hillary Clinton's "stand by your man" posture apparently enhanced her popularity, too.

Hegemonic Masculinity and Women in Politics

To summarize, Arnold Schwarzenegger's sexy, hybrid mix of hardness and compassion is currently a configuration of symbols that forge a masculinity that is useful for securing power among men who already have it. But for a woman striving for power—at least in the context of the United States' current gender order—these opposites don't mesh as easily. Strength and compassion, when embodied in a woman leader still appear to clash in ways that set her up for public crucifixion. U.S. congresswoman Pat Schroeder's brief flirtation with a presidential run in 1988—derailed by a public tear—comes to mind. Though Schroeder's many successful years in Congress, and especially her position as head of the Congressional Armed Services Committee, might have made her seem a serious candidate for the presidency, one public tear made her seem perhaps too feminine to become president. By contrast, during her years as first lady, Hillary Clinton was pilloried for her supposed "ballbusting" of her husband, for having her own ambitions to gain political power—in short, for being too much like a man. Former British prime minister Margaret Thatcher is no exception; she proves the rule. Thatcher was notoriously conservative with respect to slashing the British welfare state, and she complemented U.S. President Ronald Reagan with her militaristic saber rattling. An individual woman *can* occasionally out-masculine the men and be a strong leader. But, as with Thatcher, she'd better leave compassion and caring for the poor, for the sick, and for the aged literally at home.

The Dangers of a Compassionate Masculinity

Arnold Schwarzenegger is not the first male politician to attempt to craft a post-feminist hybrid symbology of hegemonic masculinity. George Bush, Sr., battled his own reputation as a bureaucratic wimp with a masculinizing project of waging war against Saddam Hussein. He signaled his compassionate side with speeches encouraging others (instead of the government) to be "a thousand points of light" to help the poor and homeless. And the 2004 presidential election between Kerry and George W. Bush seemed to devolve into another old "Saturday Night Live" satire: *quien es mas macho?* This reveals something important about the ascendant hegemonic masculinity. It did not seem to matter to many voters that Bush had been a lousy student who partied his youth away, and only escaped the shame of a drunk driving conviction, Vietnam War service evasion, and possible desertion from his National Guard service through his born-with-a-silver-spoon family connections. Nor did it seem to matter to many voters that John Kerry had served willingly in Vietnam and had been honored for bravery and war wounds. Kerry was still—with enough success to neutralize this apparent war hero advantage—stained by his association with "elite liberalism."

This is nothing new. Adlai Stevenson's unsuccessful runs for the presidency in 1952 and 1956 offer a good case in point. The conservative attack on Stevenson can be seen as part of the postwar hysteria about "reds" and "homosexuals." Kimmel (1996, p. 237) notes not only was Stevenson labeled "soft" on communism, but he was the classic "egghead." The candidate whom the *New York Daily News* called

"Adelaide" used "tea cup words," which he "trilled" with his "fruity" voice, and was supported by "Harvard lace cuff liberals" and "lace panty diplomats."

A month before the 1952 Democratic Convention, FBI head, J. Edgar Hoover, ordered a "blind memorandum" be prepared on Stevenson. The "investigation" concluded that Stevenson was "one of the best known homosexuals" in Illinois (Theoharis, 2002, p. 180). Though the FBI memorandum was not made public at the time, attacks on Stevenson's masculinity (linked with his liberalism and intellectualism) formed a core of the contrast that Republicans successfully drew between Stevenson and war hero General Dwight D. Eisenhower, who handily won the election—and reelection four years later. For the past half-century, conservatives have used a version of this same gender strategy to wage a successful symbolic campaign that links liberalism with softness: book-learning and intellectual curiosity are viewed as a lack of inner strength and determination. Seeing the complications and gray areas in any public debate are viewed as signs of waffling and a lack of an inner values-based compass. And compassion for the pain of others is seen as weakness.

To be sure, as President George W. Bush's slogan about "compassionate conservatism" showed, conservatives have incorporated the language of care into their project. As I have argued, a leadership masculinity without compassion is now symbolically untenable. But the new hybrid hegemonic masculinity always leads with the muscle. Muscle must first and foremost be evident; compassion is displayed at appropriate symbolic moments, suggesting a human side to the man. Liberalism suffers from the fact that it seems too often to lead with compassion, not with muscle. So when liberals try to look muscular, they are much more easily subjected to ridicule, like they are in some sort of gender drag, as evidenced by the infamous moment when 1988 democratic presidential candidate Michael Dukakis tried to dress up like a commander-in-chief but ended up looking more like a schoolboy taking a joyride in a tank while wearing a too-large military costume.

George W. Bush's love of military dress-up did not draw the same kind of ridicule. Or, at least it can be said that this kind of ridicule does not seem to stick in the way it does when aimed at a supposed liberal. And so, very sadly I think, in recent presidential campaigns we saw both candidates trying their hardest to appear tough, strong, decisive, athletic, and militaristic, while suggesting—parenthetically, almost as an afterthought—that they care about all of us, that seniors should get prescription drugs, that no child should be left behind. The asymmetries in the ways that these two candidates were able to deploy this hybrid masculinity were apparent. When Kerry said that smart leadership would lead to a "more sensitive" waging of the war in Iraq, it was the only opening that any manly hunter would need; Vice President Dick Cheney needed no shotgun to jump right on this opportunity to blast the war hero with the feminized symbolism of weakness and liberalism.

Gender, Politics, and Justice

The accomplishment of a stable hegemonic masculinity by an individual man in daily interactions is nearly impossible. But what helps to anchor an otherwise unstable hegemonic masculinity is the play of masculine imagery in the symbolic realm. Today, for the moment, the gender imagery seen in the combined films of Arnold Schwarzenegger creates a hybrid masculinity I am calling the Kindergarten Commando. This image, when deployed in the realm of electoral politics, secures power and privilege in a moment of destabilized gender and race relations, economic insecurity, and concerns about immigration, all permeated with a culture of fear.

The widespread consent that accumulates around this form of masculinity is, I suggest, an example of hegemony at work. And—importantly—the power and privilege that this hegemonic masculinity secures is not necessarily or simply "men's power over women." The erotically charged masculinity of the famously cigar-loving Governator was effective in securing power in terms of race and nation, and in class relations in California. What I am suggesting here is that the public symbolism of hegemonic masculinity is a means of consolidating power in a matrix class, race, and international politics. For California in 2004, it was Arnold Schwarzenegger's combination of muscle, heterosexuality, and whiteness—particularly the way his story reiterated the white European melting pot story of individualism and upward mobility in a meritocratic America—that formed a successful symbolic package that enough voters liked. As governor, Schwarzenegger mobilized this package first and foremost to wage class war on California's public workers and poor. But as Connell and Messerschmidt (2005) point out, hegemonic masculinity is always contingent and contextual. As contexts change, challenges are possible, perhaps inevitable. And California over the past few years has certainly been a site of rapid shifts and conflicts.

Schwarzenegger's attack on the underprivileged left him open to criticism of his own privilege and the possible use of his office to further his own interests. Opposition to Schwarzenegger mounted in 2005, as organized California teachers, nurses, firefighters, and other public employees waged massive protests and media campaigns against the governor. His having flexed his Terminator muscles left him open to questioning about whether he really cared about the elderly. Health activists and advocates for the elderly railed at the large donations he had accepted from the insurance and pharmaceutical industries, and his decisions that reflected those links. Perhaps, at least in California, a less conservative state than most, many of the governor's constituents wanted to see care and compassion reflected in *actual policies* rather than simply in some of his movies.

And so, after the ballot initiatives that he had sponsored were soundly defeated in the 2005 special election that he had called for, Schwarzenegger immediately shifted his strategy, leaving his combative Terminator persona behind, and returning to the Kindergarten Commando. He began to promote some liberal issues, including signing landmark legislation to control global warming. He finally agreed to sign legislation to authorize a modest raise in California's minimum wage. But he also advocated cutting thousands of poor off of the welfare rolls and continued his ties with corporate elites in the pharmaceutical industry. Perhaps the new model for Schwarzenegger might be closer to the masculinity of Bill Clinton—combining a moderate liberalism on social issues like women's and gay rights with a fiscal conservatism that continues to enlarge the gap between rich and poor. The success of this new man leadership style is at once a visible sign of the ways that liberal feminist critiques of "traditional masculinity" have been incorporated and embodied into many professional-class men's interactional styles and displays. What results is a rounding of the hard edges off of hypermasculinity and a visible softening of powerful men's public styles and displays. But this should not be seen necessarily as a major victory for feminism. Rather, if I am correct that this more "sensitive" new man style tends to facilitate and legitimize privileged men's wielding of power over others, this is probably better seen as an example of feminism being co-opted into new forms of domination—in this case, class and race domination.

Schwarzenegger's return to Kindergarten Commando masculinity appears to have worked. His shift to more centrist stands—undoubtedly influenced by his more liberal wife, Maria Schriver—has calmed the anti-Arnold storms of 2005. And clearly, the muscle still matters: one of the largest advantages he had over his Democratic opponent in 2006 was that voters saw Schwarzenegger as a much stronger leader. A preelection poll conducted by the *Los Angeles Times* found that 60% of likely voters saw Schwarzenegger as a "strong leader," while only 20% viewed Phil Angelides as strong.

In short, I speculate that Governor Schwarzenegger's 2004–2005 rejection of the hybrid "Kindergarten Commando" masculine imagery that had gotten him elected contributed to a dramatic decline in his popularity, and to his thrashing in the special election of 2005. When he dropped the oppositional tough-guy approach and redeployed the Kindergarten Commando, his popularity again soared in 2006, contributing to his landslide reelection.

Conclusion

If we are to work toward economic justice for working people, immigrants, and the aged, toward equality for women and racial and sexual minorities, and if we are to work toward the creation of a more just and peaceful world, we need to tackle head-on the ways that dominant forms of masculinity—while always contested and shifting—continue to serve as a nexus of power that secures the privileges of the few, at the expense of the many. Governor Arnold Schwarzenegger's strategic shifting of his public persona from Kindergarten Commando to Terminator and then back to Kindergarten Commando illustrates how, in the realm of electoral politics, hegemonic masculinity is a malleable symbolic strategy for wielding power.

In electoral politics, men's militaristic muscular posturing in seeking office limits women's abilities to seek high office in much the same ways that narrow masculine displays of dress, demeanor, voice, and style narrow women's chances in corporate or professional occupations. Women's activism in public life challenges these limitations, but if meaningful change is to occur, male leaders must also stop conforming to a singular masculine style of dress, demeanor, or leadership style.

It is unlikely that new, expansive, and progressive imagery will emanate from top male politicians. It's more likely that some men who seek high office but have progressive (even feminist) values may try to be "stealth feminists"—while posturing in military garb (like Dukakis in 1988), downplaying a deep commitment to reversing the human destruction of the environment (Gore in 2000), or overemphasizing long past military accomplishments (Kerry in 2004) instead of focusing on issues and values that they cherish. This doesn't work, partly because it is bad theatre. Even if this strategy succeeds in getting someone elected, it's unlikely that stealth feminism will work; men who get to the top using masculine muscle will rightly assume that, once in office, their constituents expect them to flex those muscles (e.g., President Jimmy Carter's ill-fated use of military power in 1980 in his attempt to end the Iran hostage crisis).

What we need is a renewed movement of ordinary women and men working side-by-side to push assertively for an ideal of the public that is founded first and foremost on compassion and caring. The seeds of such a movement currently exist—in feminist organizations, the peace movement, religious-based immigrant rights organizations, and union-based organizing for the rights of workers. A coalition of these progressive organizations can succeed in infusing local and national politics with the values of public compassion. This won't happen easily, or without opposition. We need to expect that such a movement will have to be tough, will have to fight—against entrenched privilege and against the politics of fear—in order to place compassion and care at the top of the public agenda. Out of such a movement, we can generate and support women and men who will *lead* with love and compassion and *follow* with the muscle.

References

Benwell, B. (2004). Ironic discourse: Evasive masculinity in men's lifestyle magazines. *Men and Masculinities, 7,* 3–21.

Boyle, E. (2006, November). *Memorializing muscle in the auto/biography(ies) of Arnold Schwarzenegger.* Paper delivered at the annual meetings of the North American Society for the Sociology of Sport, Vancouver, BC.

Collins, P. H. (1990). *Black feminist thought: Knowledge, consciousness, and the politics of empowerment.* Boston: Unwin Hyman.

Connell, R. W. (1987). *Gender & power.* Stanford, CA: Stanford University Press.

Connell, R. W., & Messerschmidt, J. W. (2005). Hegemonic masculinity: Rethinking the concept. *Gender & Society, 19,* 829–859.

Ducat, S. J. (2004). *The wimp factor: Gender gaps, holy wars, and the politics of anxious masculinity.* Boston: Beacon Press.

Gómez-Barris, M., & Gray, H. (2006). Michael Jackson, television and post-op disasters. *Television and New Media, 7,* 40–51.

Jeffords, S. (1989). *The remasculinization of America: Gender and the Vietnam war.* Bloomington: Indiana University Press.

Kimmel, M. (1996). *Manhood in America: A cultural history.* New York: Free Press.

Messerschmidt, J. W. (2000). *Nine lives: Adolescent masculinities, the body, and violence.* Boulder, CO: Westview.

Messner, M. A. (2004). On patriarchs and losers: Rethinking men's interests. *Berkeley Journal of Sociology, 48,* 76–88.

Messner, M. A., & Montez de Oca, J. (2005). The male consumer as loser: Beer and liquor ads in mega sports media events. *Signs: Journal of Women in Culture and Society, 30,* 1879–1909.

Montez de Oca, J. (2005). As our muscles get softer, our missile race becomes harder: Cultural citizenship and the muscle gap. *Journal of Historical Sociology, 18,* 145–171.

Theoharis, A. (2002). *Chasing spies.* Chicago: Ivan R. Dee.

Theweleit, K. (1987). *Male fantasies, volume 1: Women, floods, bodies, history.* Minneapolis: University of Minnesota Press.

Young, I. M. (2003). The logic of masculinist protection: Reflections on the current security state. *Signs: Journal of Women in Culture and Society, 29,* 1–25.

Reading 4.3

Working the Boundaries

Bisexuality and Transgender on Film

Betsy Lucal and Andrea Miller

Stop for a moment and think about your most recent encounter with a stranger. How did you know if that person was a man or a woman? While you may not have been aware of it at the time, you most likely quickly made an assessment to allow you to proceed with the encounter under the assumption that you knew the individual's sex or gender. It may have been something about that person's appearance: the clothes he or she wore, the cut and style of hair. It may have been something he or she did: the timbre of voice that greeted you or the way that person walked.

Here's another question: Did your assessment include an attempt to determine whether that person was heterosexual or homosexual? Maybe not. Much of the time, we take for granted that the people we meet are heterosexual. In other words, our presumption of heterosexuality is so strong that we assume that others are heterosexual until something leads us to question that assumption. And, interestingly enough, the things that lead us to question such an assumption also relate to appearance and behavior. Perhaps it was a man who moved his hips a little too much when he walked or a woman who dressed in "masculine" clothing and had very short hair.

In our society, we make such assessments and behave based on those assessments all of the time. We also make decisions about our own appearance and behavior based on the assumption that others are making similar assessments of us.

Key to the sociological perspective is the understanding of gender and sexuality as social constructions, as social products. Gender and sexuality are not biologically determined; they are human creations, developed over a lifetime of interaction in the context of one's culture and society. We make a variety of assumptions about gender and sexuality that are a product of the social context in which we live.

First, there is our assumption about the relationship between sex and gender. Males are boys who grow up to be men; they look and act in "masculine" ways. Females are girls who grow up to be women; they look and act in "feminine" ways. Even though none of us looks or acts in exclusively masculine or feminine ways, we still tend to believe that men and women are "opposites." Similarly, masculine and feminine traits are conceptualized as opposites: boys are aggressive and girls are passive; men are prone to violence while women are prone to caretaking.

We grow up in a society that works very hard to ensure the persistence of these differences, yet, at the same time, we insist that the differences are "natural." Through the wonders of ultrasound technology, we need not wait until a baby is born to ask the most fundamental of all questions: "Is it a boy or a girl?" As soon as that question is answered, a pervasive process of gendering begins. We give different names, distinctive clothing, and different toys based on the answer. Throughout our lives, we expect men and women to see the world differently, to pursue different careers, and to have different priorities.

We train men and women to see each other as opposites and then expect them to fall in love, get married, and live happily ever after! For a society that provides so much support to heterosexual couples (e.g., all of the legal benefits of marriage), we also set them up for failure. How will they understand each other? Surely they won't have similar interests or like the same things. And, yet, people who couple with their opposites are the ones we see as normal—because, after all, *opposites attract!*

In the context of these assumptions, there is little room for the existence of complex identities, such as bisexuality and transgender. Dualistic gender and sexual identities are firmly rooted in our social landscape. However, this landscape is fraught with tensions and contradictions as it both *forces* and *relies on* social actors to imagine and experience their sexual and gendered identities as dichotomous (i.e., as being either one or the other). Even though lived experience shows us that sexual and gendered lives are rarely consistent, but usually dynamic (changing across contexts and over the course of a lifetime), these irregularities and shifts are ignored. Instead, they are subsumed, even hidden, under binary, dichotomous, "either/or" categories.

These dualistic categorizations are not mere conceptualizations existing only in the theoretical realm; they have real-world consequences for individuals who personally and socially identify as bisexual and/or transgender. For them, the dualistic conceptualizations of heterosexual/homosexual and man/woman are not enough because such conceptualizations do not reflect the reality of their experiences or identities. Identities are indeed complex; yet this complexity is wholly absent from the sexual duality of hetero/homo and the two-and-only-two gender categories that have come to serve as cornerstones of our social world.

Using a feminist social constructionist framework, in this chapter we examine how bisexuality and transgender can challenge either/or categorizations of sexual and/or gender identity, specifically as seen in the films *Chasing Amy* (1997) and *Transamerica* (2005). Using these films, we consider how sexual and gender categories have clear boundaries at the same time that people's experiences overflow and escape these strict categories.

Bisexuals and transgender individuals are oftentimes denied access to power and privileges otherwise granted to those who maintain "heterogender" (Ingraham, 1994) or "heterosexualized" patterns of behavior and interaction. This reading considers the depiction of bisexual and transgendered lives on film and reexamines *both* sexuality and gender identification categories through our analysis of *Chasing Amy* and *Transamerica.*

We begin by exploring the (in)visibility of bisexual and transgender identities. In *Transmen and FTMs,* Jason Cromwell (1999) argues that transgendered persons become "erased" when they "blend in and become unnoticeable and unremarkable as either a man or a woman" (p. 39). Likewise, when bisexuals are identified by social others as lesbian, gay, or straight, their identity as bisexual is erased. Thus, bisexuals, too, can appear "unremarkable." These identities become co-opted under the rubric of "heterogender" and homosexuality because they deviate from normative gender and sexual expectations. Thus, "doing bisexuality" (Miller, 2006) and "doing transgender" requires others to call their heteronormative assumptions into question—to rethink their own gender and sexuality assumptions. Using these two films we show how "changing gender [and/or sexuality] may be less empowering than changing how others *perceive* [italics added] our gender [and/or sexuality]" (Cromwell, 1999, p. 5).

Doing Gender/Doing Sexuality

With the groundbreaking work of West and Zimmerman (1987) in their article "Doing Gender," gender was repositioned as something that one accomplishes through his or her daily interactions with other people. The assumption is that individuals are constantly doing gender, whether they intend to or not. The social constructionist approach to gender reveals the potential individuals possess both to maintain and to deviate from what Lorber (1994) has referred to as "gendered processes," or those learned patterns that are enacted in order to conform, deviate from, or test gender-appropriate behavior.

What is perhaps most important with regard to this conceptual framework is the understanding of gender as a product of social interaction. One does not have to consciously perform his or her gender, for others will "do gender" for him or her regardless of what gender the social actor intended to project (Lucal, 1999). For example, those who participate in androgynous presentations of self (Goffman, 1959, 1976) to break down strict gender polarizations tend to be placed by others into one of the two available gender categories rather than into the gender-neutral category the androgynous actor may have intended (Frye, 1983; Lorber, 1994). Lucal (1999) has written about her experiences of being mistaken for a man, showing how important appearance cues are to people's assessment of gender. While Lucal is not trying to self-present as "masculine," her size, very short hair, nonfeminine clothing, and other cues lead people to read her as a "man" even though she is female.

When people are doing gender, they are also doing sexuality. A female who is feminine in her appearance and behavior is assumed to be "doing heterosexuality." On the other hand, a male whose appearance and behavior are deemed too feminine is assumed to be "doing homosexuality." We are surprised, for example, to find that a feminine female is a lesbian or that a masculine male is gay.

It is essential also to recognize that sexuality is not accomplished in isolation from one's gender (or race or social class) (Miller, 2006). By using the "doing gender" and "doing sexuality" approach, we emphasize that neither gender nor sexuality is muted. Instead, individual actors "do sexuality" similar to the ways in which they do gender. Recall here Garfinkel's (1967) assertion that, at the end of the day, we see persons as either male or female, masculine or feminine, and, consequently, as heterosexual or homosexual.

In the context of this schema, there is no room, it turns out, for an individual to "do bisexuality" (Miller, 2006). If conforming to gender norms means doing heterosexuality and deviating from gender norms is assumed to mean that one is gay or lesbian, then there is no way for individuals to signal their bisexuality, nor is there a way for others to attribute bisexuality to them. Next, we analyze the film *Chasing Amy* as an example of doing sexuality in the context of doing gender and of the (in)visibility associated with bisexuality.

Chasing Amy (Kevin Smith, 1997)

Bisexuality calls into question the either/or dualism of sex, gender, and sexuality, especially for those self-identified bisexuals in primary other-gender relationships, such as the character Alyssa Jones (Joey Lauren Adams) in the film *Chasing Amy.* In an attempt to deconstruct the rubric that supports either/or categories, we examine them in the context of biological essentialism and social constructionism. Using the film *Chasing Amy* provides us with a lens for considering how these competing theories impact "sexual truths," allowing us to see the processes involved with the erasure and delegitimization of bisexual identity in U.S. society.

Essentialist theories regarding sexuality are informed by a rubric that emphasizes the "natural" and biological aspects of sexuality. According to this model, dichotomous and mutually exclusive sex/gender categories (male/masculine and female/feminine) are associated with dichotomous, mutually exclusive sexuality categories (heterosexual/homosexual). In this context, bisexual identities are erased because their sexual identity does not exist within the heterosexual/homosexual binary—the only two possibilities recognized by this model.

Such assumptions are woven throughout the opening scene of the movie when Holden (Ben Affleck) and Alyssa, both comic book producers, meet at a comic book convention. She is sitting at the "minority voices" table—as the only woman on a panel, the audience is left to guess whether her minority status is as a "woman" or as a "lesbian." After a casual introduction, Holden immediately takes a liking to Alyssa and heterosexual assumptions prevail as Holden agrees to meet Alyssa at a club called Meow Mix in New York City. Upon arriving, Holden's business partner and best friend, Banky (Jason Lee), points out that there are a lot of women in the club. Holden notices Alyssa dancing and starts to approach her when his friend, Hooper (who is a black gay male played by Dwight Ewell), says, "There is something you ought to know." Holden responds: "What? Does she have a boyfriend?" Hooper replies, "No." Holden smugly says, "Then what's to know, my friend? What's to know?" The assumption is that Alyssa is straight and, as long as she doesn't have a boyfriend, and is therefore available, there is nothing more to know.

As Holden and Alyssa reminisce about their common acquaintances (they grew up in the same town, but went to different high schools), Alyssa is invited up on stage to sing. Alyssa agrees and begins to sing a throaty love song. Holden believes she is singing this throaty love song for him and never notices the woman who has made her way through the crowd to get closer to the stage. When Alyssa finishes singing, she points to this woman and beckons her closer. Holden, of course, believes this summons is for him. Only when Alyssa walks off stage and dodges past him to kiss the woman does he finally understand what his friend wanted him to know. As the women continue to kiss, Holden stares in astonishment as his heteronormative assumptions begin to crumble.

Alyssa realizes that Holden is "weirded out," so she begins to "prove" her gay identity. She shares "war stories" with Banky, who seems more troubled with the techniques she uses to "fuck girls" than with her "gay" identity. For example, Banky and Alyssa talk about the permanent injuries they have received from performing oral sex on women. Here Alyssa continues to legitimate her identity as "gay" by talking about her encounters with other women, to the dismay of Holden, who quietly sits and listens.

Turning our attention to Alyssa and Holden's developing relationship, we see that their experiences challenge the "two-and-only-two" gender and sexuality paradigm and call into question the assumption that sex, gender, and sexuality are not stable and static categories, but instead fluid and ever-changing.

This contradiction of essentialism is evident when Alyssa, who refers to herself as "gay," contemplates her attraction to Holden, a self-declared heterosexual.[1] In the following scene, we see Alyssa grapple with her "true" sexual identity once Holden reveals his love for her while driving home from what their friends have come to call their "pseudo-dates."

Holden continues to declare his love for Alyssa and asks her if she has any hesitation about loving him. Alyssa sits quietly through Holden's soliloquy, but when he stops the car, she jumps out into the pouring rain. Holden follows after her and says, "Aren't you at least going to comment?" Alyssa responds, "Here's my comment—Fuck you!" Angrily, Alyssa verbally assaults Holden, asking him, "Do you remember for one second who I am?" Holden replies, "So? People change." Alyssa counters, "Oh, so it's that simple? You fall in love with me and want a romantic relationship? Nothing changes for you. But what about me, Holden?" Alyssa is astonished by this trivialization of her gay identity and she shouts: "There is no period of adjustment, Holden, I am fucking gay!" Alyssa and Holden walk away from each other, but once Holden reaches his car, Alyssa jumps on him and they desperately kiss one another.

Despite what she says, Holden seems to assume that, based on her behavior, Alyssa has decided that she can in fact merely "change," with little consequence to her identity. Because Holden is a self-declared heterosexual, he does not have to contemplate a change in his sexual identity; his love for Alyssa only reconfirms the reality of his heterosexuality. We never really see Holden question his heterosexual identity because his partner, Alyssa, meets the standard relationship form. For Holden, his relationship with Alyssa supports the assumption that one's sex, gender, and sexuality remain both congruent and unchanged throughout one's life (Lorber, 1994).

Holden entirely dismisses Alyssa's past relationships with women and expects her attraction to him ultimately to override her other experiences. It seems that, in Holden's eyes, Alyssa has simply failed to find her "true" sexual identity until now. Alyssa herself can manage little more than an angry "fuck you" because she, too, feels that her attraction to Holden is an example of how she has failed at "gayness."

[1]It is interesting to note that throughout the movie, while other characters refer to Alyssa as a "dyke" or "lesbian," Alyssa only refers to herself as "gay."

We see Alyssa confront essentialist notions about her "gay" identity in a scene with her all-woman friendship group. When Alyssa and her (apparently) lesbian friends get her comic book ready for mailing, the obfuscation begins. As Alyssa elusively attempts to describe her new lover, one of her friends accuses her of "playing the pronoun game." Alyssa denies this accusation; but, when her friends continue to confront her, she says, "Holden. His name is Holden." Finally, one woman lifts her wine glass as if to toast and says, "Here's to the both of you. Another one [lesbian] bites the dust."

The assumption here is that Alyssa must "pick a side." She is either gay (read lesbian) or straight. We see here that biphobia does not just occur in heterosexual society; in happens with great regularity in the so-called safe confines of lesbian and gay space (Miller, 2006). As a result, bisexuals participating in similar-gender or other-gender relationships are often mistaken for or misidentified as gay or straight. For example, recall the scene at Meow Mix when Holden assumes Alyssa is straight. Because Alyssa projects appropriate femininity, she is able to conform to heterogendered attributes (Ingraham, 1994). Hence, Holden's obvious bewilderment when he realizes Alyssa is gay. The feminine gender cues Holden read were not just that "Alyssa is a feminine woman" but also that "Alyssa is a feminine and heterosexual woman."

Social constructionist theory argues that social actors possess the agency to construct and participate in their own meaning-making systems. This approach calls into question the essentialist thesis that absolute sexual truths are necessary, as illustrated in a lovemaking scene. Holden, lying in bed with Alyssa asks, "Why me? Why now?" She responds:

> I've given that a lot of thought, you know. . . . I'm not with you because of what family, society, life tried to instill in me from day one. The way the world is; how seldom it is to meet that one person who just gets you. It's so rare. . . . And to cut oneself off from finding that person, to immediately halve your options by eliminating the possibility of finding that one person within your own gender, that just seems stupid to me. So I didn't. But then you came along—you—the one least likely. I mean, you were a guy. And while I was falling for you, I put a ceiling on that because you were a guy, until I remembered why I opened the door to women in the first place—to not limit the likelihood of finding that one person that would complement me so completely. So, here we are. I was thorough when I looked for you and I feel justified lying in your arms, 'cause I got here on my own terms and I have no question there wasn't someplace I didn't look. And for me that makes all of the difference.

Here Alyssa is not necessarily searching for the essentialist "truth" about her sexual identity, as she allows herself to experience her sexual activity and identity as being in flux. She is ready to confront the assumed impermeability of sexual identity categories, and challenges Holden to move beyond the categories of "hetero" and "homo," to question the idea that sexuality is "fixed" or "exclusive."

For both Alyssa and Holden, traditional sexual identity categories cannot explain their *lived* experience. Alyssa, the "gay" woman, is in love with Holden, the "straight" man. Is she still gay? Is she bisexual? What is her sexual truth? What about Holden? Who is he if he sleeps with a gay woman? What becomes of his identity status as a straight man? Is he still straight when he has sex with someone who openly identifies as gay? Essentialist thinking does not allow for such lived experience, which challenges binary sexual categories. Such categories are inadequate, and people whose experiences confound them must try to "fit" their identities and sexual activities within the confines of categories that cannot contain their lives.

While Alyssa allows room for a fluid sexual identity, we never see this epiphany with the character development of Holden. In fact, instead of contemplating how his sexual identity may be destabilized by his relationship with Alyssa, he trivializes Alyssa's own identity negotiations by asking, "But can I at least tell people that all you needed was some serious deep dicking?" The chance for social-sexual reflection is abruptly interrupted and Holden never has to problematize his own identity. His homophobic response that all Alyssa needed was heterosexual intercourse reifies Holden's heteronormativity.

Indeed, Holden's relationship with Alyssa has no consequences for his identity, as he maintains his heterosexual privilege by using essentialist assumptions to his advantage. In other words, because others know Holden to be masculine and straight, his sexuality does not come under scrutiny when he develops an intimate relationship with Alyssa. In fact, this relationship only works to reaffirm sexual binary categories, since Alyssa's involvement with Holden proves that she must *really* be straight.

In keeping with a definition of sexuality based on what one does (and with whom), Holden's friend, Banky, cannot fathom why Holden would continue this relationship with Alyssa, since the fact that she is a lesbian means that Holden has no chance of fucking her. Banky pointedly questions Alyssa about the number of times that she has been in their apartment, asking: "Isn't that grounds enough for the pink mafia [i.e., lesbians] to throw you out of their club?" The attempts made by both Holden and Alyssa to date and eventually become intimate may be real on the surface but, as Banky predicts, their relationship is bound to end badly because Alyssa's previous relationships with women are the ultimate indicator of her true (i.e., essential) sexual identity.

To disregard the notion of any single sexual identity seems impractical to Banky. Arguing that sexualities are fluid and dynamic may make sense theoretically, but provides little justification for Alyssa's sexual identity. In other words, her varying experiences and relationships with similar and same-gendered persons do not necessarily place her sexual identity in flux. The viewer is still led to decide only whether Alyssa is straight or gay. For example, while Alyssa never invokes the term *bisexual* (or, for that matter, *lesbian*) to describe herself, the viewer can see Alyssa's identity-questioning as an opportunity to examine the usefulness of stable categories like "hetero" and "homo."

We are not suggesting that the viewer should conclude that Alyssa is really bisexual—another absolute truth. But we do think her questioning helps to make the point that many people's sexual identities are in flux at some point in time, whether that flexibility is about attractions, preferences, fantasies, or actual relationships. Thus, the point is not to claim Alyssa as a true bisexual. Yet her character emphasizes that gender and sexuality are never determined once and for all and never function as mutually exclusive. Thus the question for the audience can become, "Why doesn't Alyssa use the term *bisexual* to describe her sexual identity?"

We can see Alyssa's character as an exemplar for rethinking the assumption that sexuality is some static personal attribute, property, or essence. Instead, Alyssa helps the viewer imagine bisexuality as one of the several available cultural categories from which she might choose. Through Alyssa's character development, we see that conventional categories that attempt to describe an individual's sexuality are usually misleading, for one category (e.g., homosexual, heterosexual, bisexual, asexual) does not adequately describe the variability of one's sexuality (Stein & Plummer, 1996).

The way in which categories are misleading is at the crux of *Chasing Amy.* For example, in a pivotal scene, the audience sees Banky throwing Alyssa's high school yearbook to Holden and pointing out Alyssa's senior picture with the words *finger cuffs* printed below it. Banky goes on to recount several rumors about Alyssa that he has heard from her sexual exploits with high-school boyfriends. Holden dismisses these rumors and tells Banky that, "She's never even been with a guy."

Instead of looking at her past and concluding that it is evidence of her bisexuality, Holden cannot look past masculinist and heterosexist assumptions that suggest Alyssa's relationships with women were an acceptable part of her past, something that can easily be forgotten and forgiven once she cements her relationship with Holden. However, her sexual activities with men cannot simply be explained away as meaningless youthful indiscretions or some sort of "bi curiosity." Here, bisexuality is not only dismissed, but is actually erased because it is not positioned as a legitimate sexual identification category. Again, Alyssa never names herself as "bisexual." The assumption is that her sexual identity remains intelligible within the binary—she is lesbian/gay when she is partnered with a woman and heterosexual when she is with Holden. By excluding *bisexual* from her language, Alyssa provides a type of binary normalcy—a guise under which heteronormativity is perpetuated. In other words, *normal* is defined as two-and-only-two gender categories, justifying a binary logic that includes us all (and always excludes bisexuals).

Even though Alyssa does not claim a bisexual identity, her lived experiences at least allow viewers to witness sexual identity categories in flux, as well as to arrive at their own conclusions with regards to why Alyssa erases herself as bisexual. For example, even though a bisexual identity remains unspoken, unclaimed, and unremarked upon in *Chasing Amy,* its presence can be felt throughout the film. In Alyssa Jones, we have a character who is doing bisexuality without anyone—including herself—naming it as such. While she is unapologetic about any of her relationships, it is the assumptions of heteronormativity to which she ultimately refuses to conform. Alyssa and Holden do not live happily ever after.

The final scene of the movie occurs "one year later." Alyssa and Holden have not seen each other in a while when they meet at a comic book convention. Holden approaches Alyssa, who is sitting with another woman, promoting her comic. It is unclear whether this woman is a friend or a lover. The woman excuses herself from the table as Holden begins to make small talk.

When Alyssa's woman friend/lover returns to her side and asks, "Who was that?" Alyssa responds, "Oh, just some guy I knew." Here, Alyssa's sexual identity seems to be confirmed as really lesbian, since she does not say that she and Holden had a sexual relationship. Ultimately, any attempt to pin down a bisexual reading of Alyssa's character is thwarted.

In *Transamerica,* on the other hand, the visibility of transgender and the instability of gender categories permeate the film from the opening scene. In spite of the story of gender transition that is at the center of the movie, by the end of the film, the partitioning of gender categories remains in flux.

Transamerica (Duncan Tucker, 2005)

While all of us deviate from strict gender categorizations in some way, consider the lives of people who experience their sex and gender as incongruent, as not matching up in the expected ways. This describes the experiences of many *transgender* individuals. But this umbrella term can be used to refer to a whole spectrum of people who are more different from each other than we might expect, given their placement in a single category. Some identify as *genderqueers,* for example, and do not wish to change their bodies, just to live (full or part time) in a gender somehow different from their assigned sex. Others, for various reasons, seek some medical and/or surgical intervention; most commonly, they take hormones to change their bodies. Some others desire a full regimen of sex reassignment surgery (SRS).

Like bisexual identities, transgender subjectivities call into question the "natural" sex/gender/sexuality binaries. From an essentialist perspective, there are males and females, whose masculine and feminine genders, respectively, follow naturally from their sex. Transgender identities, however, raise the possibility that the relationship between sex and gender is, instead, a social construction and need not take the form of a mutually exclusive dichotomy.[2]

The opening scene of *Transamerica* leaves the viewer with no doubt about its subject matter. The film begins with a clip from *Finding Your Female Voice,* with Bree (Felicity Huffman) practicing a female voice along with the coach by repeating the mantra, "This is the voice I want to use." As Bree prepares for the day, we watch her dress, putting on heavy, opaque panty hose; shaping garments (a padded girdle and fake breasts), and makeup. Bree leaves the house decked out in pink; there is no doubt that she is very good at doing femininity.

As Bree walks to the bus stop, we hear a voiceover of her meeting with a doctor who must sign the consent form that will allow her to have SRS. The doctor asks which procedures she's already had, immediately signaling to the viewer that SRS involves much more than a single surgical procedure. Bree recounts three years of hormone therapy, electrolysis, facial feminization surgery, a tracheal shave (to make her Adam's apple less prominent), jaw recontouring, and a brow lift. The doctor tells her that she looks "very authentic."

---

[2]It should be noted that some people consider transgender to be a biological phenomenon.

Photo 4.2 Challenging categorizations of sexual and gender identities in *Transamerica.* Bree (Felicity Huffman) tries to look "authentic" at the bus stop by crouching down so as to appear the appropriate height of a "real" woman.

Though she may look authentic, as Bree waits at the bus stop, she crouches down when she notices that she is taller than the men waiting for the bus. It seems unusual that a transwoman just a week away from her final surgery would still be so self-conscious. But perhaps the director is simply reinforcing the point that Bree is concerned about fitting in and being normal. As she tells the doctor, "I believe the term is 'being stealth.'" (Like the B-1 Stealth Bomber, a trans person who is stealth cannot be detected [or "read" or "clocked"].)

When the doctor reminds Bree that "gender dysphoria" is classified as a "very serious mental disorder," Bree quips, "Don't you find it odd that plastic surgery can cure a mental disorder?" This plastic surgery will rid Bree of the penis she finds so disgusting and will, in her eyes—and the eyes of society—complete her transition to full womanhood and allow her to fit easily into the binary. The result of this surgery, as she tells the doctor, will be female genitals that appear normal even to a gynecologist. Before she goes to bed that night, Bree tucks her penis between her legs, allowing her to envision herself as completely female (and as completely feminine in her gauzy night gown/robe). Bree's disgust for her penis reinforces the belief that, even for trans people, sex and gender are either/or. In a society that insists individuals must be either men or women—never both, let alone neither—it is, of course, not surprising that some trans people internalize these same beliefs.

Later in the film, we see another example of Bree's desire to live life as a "normal" woman. As she travels cross-country with Toby (Kevin Zegers), the son she fathered as a result of a "tragically lesbian" affair in college, Margaret (Elizabeth Peña), her therapist, has arranged for her to spend the night at the home of another transwoman, Mary Ellen (Bianca Leigh). Toby does not yet know that Bree is trans and she is horrified to find that the woman and her transgender friends, who are in varying stages of transition, are in the midst of a party. It is not surprising that she panics and tells her hostess that she and Toby must leave. Toby, however, wants to join the party, later reporting to Bree that the people there were "nice."

Among the guests at the party are well-known transgender people like Calpernia Addams and David Harrison (the only transman in the film). David tells Toby that trans people are "gender-gifted" rather than "gender-challenged." When Toby says to him, "Dude, I thought you were a real guy," David's response is "We walk among you." In a memorable scene, Bree tells her hostess that one of the women there could not pass "on a dark night at two hundred yards"; but it turns out that this Mary Kay cosmetics representative is a "G.G." or "genuine girl" (or "genetic girl," as this abbreviation is often defined). Mary Ellen says that Bree needs to "check her T-dar." (In other words, Bree is not as good at spotting other transgendered people as she might think.) Here, we get the sense that gender is more complicated than the sex equals gender, either/or dichotomy might suggest.

Bree's concerns about what Toby thinks of her as a result of her association with the other trans people are soon overshadowed by him finding out that she is not a "real" woman. Toby is driving when they stop along the roadside late at night so Bree can urinate. When he looks in the rearview mirror, he sees Bree's penis. When she returns to the car, he says nothing but speeds off. Soon, they have a confrontation at Sammy's Wigwam, a roadside souvenir stand. Having mistaken Bree for Toby's mother (something that happens several times in the film), Toby tells Sammy (Forrie Smith) that "she's not even a real woman; she's got a dick" and goes on to call her a "fucking lying freak."

Here we see another example of the assumption that, to be a "real" woman, Bree must get rid of her penis (which we know she wants to do, given the disgust she expressed about it to the doctor at the beginning of the film). Despite the fact that Bree has lived as a woman throughout the film to this point, her womanhood is immediately called into question upon Toby's sighting of her penis. In his eyes—*and* in hers—only by "cutting it" off can she become a complete woman.

This belief is reinforced when Bree enters her parents' home later in the film. Bree's mother's first question is whether Bree is still a "boy." To find out, she grabs Bree's crotch and announces that, indeed, she is still male. But Bree grabs her mother's hand and forces her to touch Bree's breast, showing her that she is female, too. While this scene can be taken to counter the taken-for-granted assumption that male and female, masculine and feminine, cannot exist in the same body, the viewer already knows that this situation is temporary. Bree hates her penis and wants to have it turned into female genitals. There is no acknowledgment of the reality that some transwomen (by choice or because of limited resources) keep their penises (and even continue to use them as sexual organs).

Bree's mother's refusal to see her as a woman continues. When Bree comes into the living room wearing a flowing, pink chiffon dress, her mother tells her that she looks "perfectly ridiculous." When the family goes out to dinner, Bree's mother forces her to act like a man and pull her chair out for her so she can sit down. This action takes on a whole different meaning, however, when Toby pulls Bree's chair out for her, too. Without saying anything, Toby shows that he has a renewed belief in Bree's womanhood.

Bree finally returns to Los Angeles and has her surgery. When we see her in the bathtub touching her new genitals, obviously pleased with them, it is clear that her transformation into a "real" woman is complete. Her penis is gone, fashioned into female genitals, and she can now begin her new life. But the final scenes, when Toby reappears in her life, both reinforce and call into question the sex-gender dichotomy of either/or and the naturalness of these distinctions.

Bree rises from the sofa and starts walking out of the room to get Toby the beer he has requested. As she walks into the kitchen to get it, she stops, turns around and tells Toby that he will not be allowed to put his feet up on her brand new coffee table. The fact that she has seen what he is doing despite not watching him can be viewed as an example of her motherly (feminine) attributes. We usually think of mothers, not fathers, as having "eyes in the back of their heads," seeing everything a child does. It is evident that Bree's role in Toby's life will be a maternal and feminine one.

But the final scene (viewed from outside the apartment, through appropriately feminine gauzy curtains) is this: when Bree hands Toby a beer, he cannot open the bottle, so he hands it to her. She exchanges it for the one she has already cracked open and proceeds to open the other bottle as well. As viewers, then, we are left with a reminder of Bree's former life as Stanley; men, not women, are usually the ones with the strength to open a stubborn lid.

However, the impression here is not that Bree is still really Stanley. Instead, it appears that, because her transition is now complete, Bree can be comfortable mixing genders, something she consciously avoided before her final surgery. Indeed, in this scene, she is still wearing pink—but now her pink sleeveless blouse is paired with loose-fitting, flowered pants. The new Bree can show more skin (with a sleeveless shirt) and can wear pants (for the first time in the film). Freed from her concerns about being read, Bree can now do both femininity and masculinity. If Toby needs help with a "masculine" task, she can now provide it. The old Bree would have asked Toby to open *her* bottle.

On one hand, finalizing her transition with surgery makes Bree invisible as a trans person—much like Alyssa erases herself as bisexual in *Chasing Amy.* On the other hand, however, the ending of the film makes her visible for the first time as someone "gender-gifted," as transman David Harrison characterized transgender people earlier in the film. She can do both femininity and masculinity, thus questioning the natural, mutually exclusive status of these categories.

Conclusion

Many people experience their sex, gender, and sexuality as existing in the realm of the taken for granted. They learn the gendered appearance and behavior expectations associated with their sex category and do gender in ways that mostly conform to those expectations. Their common, everyday deviations from such norms and expectations are considered to be temporary and mundane, and do not call their sex, gender, or sexuality into question (Garfinkel, 1967; Kessler & McKenna, 1978). Heterosexual men can wear pink shirts and take care of babies, while heterosexual women can enjoy watching football (but probably not playing it) and wear T-shirts and jeans without leading to any questioning of their identities.

Other people, however, cannot take for granted such fundamental components of their identities. Feminine lesbians risk getting misread as heterosexual and being invisible to other lesbians (Maltz, 2002), while transwomen may face the prejudice and discrimination that they experience because our society is patriarchal and denigrates femininity (Serano, 2007). Like openly gay men, they risk violent reactions to their deviations from gender and sexual norms. Some heterosexual women's sexuality is called into question by those people who believe they "look like lesbians," while the "Effeminate Heterosexual Man" was the title of a "Saturday Night Live" skit in the 1990s.

And, as the films analyzed here show, there are the issues and dilemmas of a transgendered and/or bisexual existence. As Cromwell (1999, p. 122) notes, "Transgender people and people with nonheterosexual identities queer the Western binaries of body-equals-sex-equals-gender-equals-identity as well as the binary of heterosexual and homosexual." Along with such queering, however, often comes invisibility.

In *Chasing Amy,* Alyssa's experiences can easily be read as evidence that she is bisexual and that she is doing bisexuality; we see her kissing a woman (and know that she is known as someone who has sexual/romantic relationships with women) *and* we see her sexual/romantic relationship with a man. But Alyssa identifies as gay, does not continue her relationship with Holden, and, in the end, chooses not to go public with the fact that she had a relationship with him. The fact that Alyssa's prior sexual relationships with men, rather than those with women (the ones that were clearly most important to her identity to that point), are most troubling to Holden further queers her identity.

In *Transamerica,* on the other hand, the viewer is left with little doubt about Bree's transgender status. But it is also made clear that her transgender existence is temporary and will be resolved by the end of the film. Before her final surgery, Bree is depicted several times as a woman (given her feminine appearance—she always wears pastel colors—and behavior) with a penis. It is also clear that she does not want to continue to have a penis, supporting the notion that, ultimately, male and female/woman and man cannot exist simultaneously. Her reluctance to be in the company of other trans people (seen at the party in Mary Ellen's house) also supports the idea that her transgender status is temporary and uncomfortable for her. Unlike these transgender people, Bree's preference is, as she told the doctor at the beginning of the film, to "try to fit in" and to be "stealth."

Ultimately, however, even Bree queers the gender binary. While the viewer is left with no doubt that Bree is happier after her final surgery, it is also made clear that she is also more comfortable mixing masculinity and femininity. Presurgery, Bree always wore skirts and was always self-conscious about her appearance. Postsurgery, however, she wears pants for the first time in the film and opens Toby's beer bottle, something presurgery Bree never would have done.

In the final analysis of *Chasing Amy* and *Transamerica,* what remains problematic is not necessarily that the characters do not name themselves as "bisexual" or "trans," but the ease with which Alyssa and, to a lesser extent, Bree erase or at least make more difficult the possibility of such naming. Alyssa never legitimizes a bisexual identity; her validation comes from retaining her gay identity, such as when we see her dismissing Holden as simply "some guy she once knew." Bree does not claim a trans identity; she now considers herself authentically female and a woman. This erasure of identity is blinding—for both characters, previous lives, lovers, practices, and interactions are dismissed and instead replaced and validated by the static, binary categories of homosexual and woman.

In this respect, *Chasing Amy* is ultimately more conservative than *Transamerica.* The viewer is left with validation of the belief that static, dualistic categories are the best for everyone involved. The final images of *Transamerica,* on the other hand, leave the viewer more optimistic about the future prospects of gender fluidity and flux. Ironically, as a result of her final SRS, Bree appears to have been freed, at least a bit, from her rigidly gendered self, leaving the viewer to ponder the possibilities of "working the boundaries" between categories.

References

Cromwell, J. (1999). *Transmen and FTMs: Identities, bodies, genders, and sexualities.* Urbana: University of Illinois Press.

Frye, M. (1983). *The politics of reality: Essays in feminist theory.* New York: Crossing Press.

Garfinkel, H. (1967). *Studies in ethnomethodology.* Englewood Cliffs, NJ: Prentice Hall.

Goffman, E. (1959). *The presentation of self in everyday life.* Garden City, NY: Doubleday.

Goffman, E. (1976). Gender display. *Studies in the Anthropology of Visual Communication, 3,* 69–77.

Ingraham, C. (1994). The heterosexual imaginary: Feminist sociology and theories of gender. *Sociological Theory, 12*(2), 203–219.

Kessler, S. J., & McKenna, W. (1978). *Gender: An ethnomethodological approach.* Chicago: University of Chicago Press.

Lorber, J. (1994). *Paradoxes of gender.* New Haven, CT: Yale University Press.

Lucal, B. (1999). What it means to be gendered me: Life on the boundaries of a dichotomous gender system. *Gender & Society, 13,* 781–797.

Maltz, R. (2002). Fading to pink. In J. Nestle, C. Howell, & R. Wilchins (Eds.), *Genderqueer: Voices from beyond the sexual binary* (pp. 161–165). Los Angeles: Alyson Books.

Miller, A. (2006). *Voices on binary objections: Binary identity misappropriation and bisexual resistance.* Unpublished dissertation, American University, Washington, DC.

Serano, J. (2007). *Whipping girl: A transsexual woman on sexism and the scapegoating of femininity.* Emeryville, CA: Seal Press.

Stein, A., & Plummer, K. (1996). I can't even think straight: Queer theory and the missing sexual revolution in sociology. In S. Seidman (Ed.), *Queer theory/sociology* (pp. 129–144). Cambridge, MA: Blackwell.

West, C., & Zimmerman, D. (1987). Doing gender. *Gender & Society, 1,* 125–151.

CHAPTER 5

Work and Family

Buddy: What I am concerned with is detail. I asked you go get me a packet of Sweet'N Low. You bring me back Equal. That isn't what I asked for. That isn't what I wanted. That isn't what I needed and that shit isn't going to work around here.

Guy: I, I just thought . . .

Buddy: You thought. Do me a fucking favor. Shut up, listen, and learn. Look, I know that this is your first day and you don't really know how things work around here, so I will tell you. You have no brain. No judgment calls are necessary. What you think means nothing. What you feel means nothing. You are here for me. You are here to protect my interests and to serve my needs. So, while it may look like a little thing to you, when I ask for a packet of Sweet'N Low, that's what I want. And it's your responsibility to see that I get what I want.

Swimming With Sharks (1994)

Francesca Johnson: When a woman makes a choice to marry, to have children, in one way her life begins but in another way it stops. You build a life of details. You become a mother, a wife, and you stop and stay steady so that your children can move. And when they leave, they take your life of details with them. And then you're expected to move again only you don't remember what moves you because no one has asked you in so long. Not even yourself.

The Bridges of Madison County (1995)

Sociology as a discipline developed in the 19th century, a time of rapid social, political, and economic transformation. The processes of industrialization and urbanization were reflected in the growth of capitalism and the factory system, and the separation of home from work, production from consumption and reproduction. The separation of home and family into the private sphere and the economic and political into the public sphere became a dominant model of social life based on gender. Women's participation in the domestic realm became life-defining, as did men's participation in the political economy. Regardless of women's waged productive labor, women's primary role became that of wife and mother, while, in Marxist terms, men's position was defined by their relationship to the means of production (e.g., as workers in a waged economy).

Historically, sociology treated the private realm of home and family as separate and apart from the public realms of work, politics, and the economy. Further, the spaces we inhabit as social actors were conceptualized as not just physically separate, but as also involving separate functions and processes in our lives. Nowhere does this become more obvious than in the very narrow construction of normative family life in the mid-20th-century United States. "Bedroom communities" were built in urban areas, and workers commuted from the privatized life of home and family to the public domain of work. Home was a place where our bodies were nourished and cared for, our emotional well-being nurtured, and we found meaning in what sociologists call primary group relationships. Leaving home, we entered the public domain of secondary group relationships, engaging in contract-based exchange relationships and the life of the larger community.

Movies reflect this division as well, and we often follow characters from one realm into the other, understanding that the shift from public into private provides a frame for interpreting the actions of people engaged in social relationships and situations. In this chapter, the readings examine the private/public domains of social life through a sociological lens. In the first reading, Karla Erickson turns to the public sphere of social life, specifically the world of work, and applies sociological questions usually reserved for research on family life to the workplace. Using classical sociology, Erickson frames the transformation of work in a service economy with the concepts of community (*gemeinshaft*) and society (*gesellshaft*). She points out that more of our human interactions are taking place within the marketplace, at the same time that more of our physical and emotional needs are being met by services for pay.

Using the film *As Good As It Gets* (1997), Erickson explores connection, intimacy, and identity in the public space of an urban restaurant as a site of dining and social relationships. The second film, *Office Space* (1999), provides a view of a suburban restaurant as one setting where workers are pitted against not just management, but the dehumanizing effects of corporate control of the self. As Buddy, who represents the corporate position, so harshly explains to Guy in *Swimming With Sharks* (1994), "What you think means nothing. What you feel means nothing. You are here for me. You are here to protect my interests and to serve my needs." Yet people do more than serve the corporation in the context of work; they develop relationships and construct communities.

In the second reading, we move into the private realm of family, as Janet Cosbey explores different dimensions of family life and experience, including gender, race/ethnicity, social class, and family violence. She begins with two films to explore the gendered model of family life as separate from the world of economy and work. Using *Mona Lisa Smile* (2003), Cosbey examines the construct of the "ideal family," consisting of a homemaker wife, breadwinner husband, and their children, and the reproduction of this ideal in the historical anomaly of the 1950s era. Despite the fact that this ideal never represented the way we were (Coontz, 1992), it influenced the development of public and economic policies, as well as individual choices and experiences into the present day. Cosbey uses a gendered twist on the ideal family with the film *Mr. Mom* (1983), demonstrating the gendered division of labor between home and work, as well as inside the home, and the inequalities embedded in this social arrangement.

Race, ethnicity, and social class are key sociological dimensions of inquiry, and Cosbey turns to the intersection of these in family life with two films, *What's Cooking?* (2000) and *The Pursuit of Happyness* (2006). Last, she considers what sociologists have called the "dark side of families" (Finkelhor, Gelles, Hotaling, & Straus, 1983), family violence that often is hidden by the privacy of the family, with two films, *The Great Santini* (1979) and *Sleeping With the Enemy* (1991).

Exploring the private and public domains, the readings in this chapter invite you to consider the ways that we live in the *gemeinschaft* of our homes and families, as well as survive in the *gesellschaft* of work and commerce. As sociologists, we are interested in the ways our identities and experiences are shaped by gendered arrangements intersecting with race, ethnicity, and social class at home and at work. Going to the movies is one way to learn about social life in private and public, as well as the connections in between.

References

Coontz, S. (1992). *The way we never were: American families and the nostalgia trap.* New York: Basic Books.

Finkelhor, D., Gelles, R., Hotaling, G., & Straus, M. (1983). *The dark side of families: Current family violence research.* Thousand Oaks, CA: Sage.

Reading 5.1

Service, Smiles, and Selves

Film Representations of Labor and the Sociology of Work

Karla A. Erickson

Workplaces are sites in our society where people are rejected, promoted, discriminated against, sexually harassed, politically awakened, and intellectually challenged. What humans can do, they do at work: fall in love, cheat, laugh, cry, scream, bleed, build, plan, present, think, and talk. Despite the richness of activities that take place within work, workplaces have historically been studied through a set of questions that situate work as distinct from private life. When studying private life, scholars frequently pursue questions about emotion, connection, intimacy, and conflict. Yet, when scholars turn their curiosity toward work, the tendency has often been to ask a distinct set of questions about control, constraint, resistance, organization, power, and efficiency. Like many contemporary sociologists of work, I am compelled to apply those questions previously reserved for private life to public life. How does work shape our sense of self? How does our work and the places where we labor influence our ideas about emotion, friendship, and even intimacy? What do we give to our work emotionally, intellectually, socially, and spiritually, and what do we receive in return?

As scholars of labor, when we apply questions about trust, pleasure, intimacy, and care to work processes and workplaces we begin to treat work as a location and an experience that can be transformative, for better and for worse. Work is a powerful site for sociological inquiry because work can lead individuals to feel cheapened, work can act as a socializing agent, encouraging individuals to treat others with trust or, alternatively, with suspicion, and finally, work can profoundly influence individual ideas about other people's capacities, intentions, wants, needs, and skills. These questions, then, are important for studying contemporary social life in general, and are particularly important for scholars who seek to study service work. In this reading, I focus on scenes from two popular films that depict a series of social interactions in restaurants, to introduce some of the concerns and theories that are central to the study of work, occupations and organizations. A close analysis of scenes from *As Good As It Gets* (1997), directed by James L. Brooks, and *Office Space* (1999), directed by Mike Judge, provides a starting point from which to consider a series of questions about workers and consumers in the service sector: What kinds of connections are available to workers and consumers in the global economy? How do workers negotiate the terms of their labor? What do customers want and what do they get emotionally and socially through for-pay interactions with service workers? In what follows, I focus on scenes in *As Good As It Gets* and *Office Space* that use restaurant interactions as commentary on social life, paying particular attention to the social meanings that emerge around work and identity.

Restaurants as Scenes of Connection

In movies, the social interactions highlighted in a scene can make use of the partial privacy of a particular table nestled in the hustle and activity of the marketplace. The sound of people at other tables talking, the clink and crash of work in the kitchen, and the buzz and aura produced by many bodies, smells, and sounds in one place can provide a rich backdrop for plot development. In *As Good As It Gets,* the restaurant provides much more than mere backdrop; the restaurant becomes a site where workers and consumers navigate the boundaries between individuals working and living in a service society.

In *As Good As It Gets,* viewers get to know two of the three main characters through their points of contact within a restaurant in which one dines and the other works. Melvin Udall (Jack Nicholson) is an obsessive-compulsive man who spends most of his time self-sequestered in his well-appointed apartment writing romance novels. Melvin is a bigot who is riddled with compulsions from handwashing to checking the locks, and his inability to step on cracks in the sidewalk or eat with silverware washed by someone else. Despite his rather noteworthy limitations as a social actor, Melvin Udall somehow manages to write romantic fiction that is widely read. His abrasiveness toward others is routine. For example, when his editor's assistant, who has appreciatively read all of his many books, decides one day to eagerly inquire, "Mr. Udall, I just *have* to ask, how do you write women so well?" Udall responds without skipping a beat, "I think of a man and I take away reason and accountability." Like this exchange, most of Melvin Udall's points of contact with other people are characterized by fear, hurled insults, and abrasive attempts at achieving distance from others. Melvin spends his days trying to get other people to leave him alone, and he is largely successful through a combination of ornery behavior, obsessive-compulsive habits, and forceful insults that combine the worst of misogyny, racism, homophobia, and anti-Semitism.

When we first meet the character of Melvin Udall, there is reason to believe that the film will be taken up with his battle against others in a world, specifically New York City in the 1990s, that requires contact with other people, both wanted and unwanted. Yet as the film progresses, Udall's battle to remain alone is replaced by a growing need to be connected to others, first Carol Connelly (Helen Hunt), his waitress at the restaurant he frequents; next a dog, Verdell, that he takes care of while his neighbor recovers from a violent attack; and finally the neighbor himself, Simon Bishop (Greg Kinnear), who is trying to regain his foothold in life after his body has been brutalized and his spirits destroyed as a result of the attack against him. The movie is categorized as a romantic comedy, but it has none of the lightness often associated with that genre. It *is* funny, but often at the expense of the pain and absurdity of daily life including children who are very sick but cannot receive consistent medical care, and the diffuse costs of homophobia within families, communities, and between neighbors. For example, the title of the movie, *As Good As It Gets,* comes from a scene in which Melvin has "courageously" burst into his therapist's office (his self-proclaimed courage comes from the fact that he had to step on lines in the floor to get to the therapist, an act he is strongly compelled to avoid) only to be thrown out because he does not have an appointment. In one of the great comic lines of the movie, Melvin exasperatedly shouts, "How can you diagnose me with obsessive compulsion and then act like I had some kind of choice in the matter?" After his therapist kicks him out, an irritated and disappointed Melvin surveys the room of people also waiting for his therapist and asks in a tone that betrays both complacency and horror, "What if this is as good as it gets?" Like him, his audience of fellow patients also looks horrified, but too beaten down to respond at all. It is against this backdrop of a contemporary urban environment that is taxing at best, and frequently demoralizing, that Melvin, beleaguered and exhausted, heads to the same restaurant each day to eat his meal with disposable silverware that he brings with him in a plastic sandwich bag.

Melvin's routine patronage at this particular restaurant is a striking contrast with the rest of his routines that seek to isolate him and protect him from contact with others. As he enters the restaurant for the first time, nothing is particularly noteworthy, charming, or extraordinary about the restaurant itself that might inspire such loyalty. Like many regular customers, Melvin has a favorite table, and he is outraged to discover other people sitting in "his" booth. His behavior reveals his willingness to verbally accost strangers in an attempt to get what he wants, and introduces viewers to "his" waitress, Carol Connelly.

Carol is portrayed as the last bastion of goodness in a world gone wrong. As viewers, we get to know Carol through Melvin's contact with her at the restaurant. From watching her move quickly and effectively between tables, chatting with customers and coworkers, we learn that Carol is the mother of a chronically ill son. She lives with her mother and focuses all her energy and attention on her son, forming her own sort of obsession. Carol knows the names of her customers, seems respected by her coworkers, and has considerable influence with her manager, which she uses in the first scene to save Melvin from getting kicked out of the restaurant.

Scene 1: Melvin's Table

During the first scene staged at the restaurant, Carol tries to calm Melvin down when he finds that "his" table is occupied.

Melvin: (following Carol into the service area) I'm starving.

Carol: Go on, sit down. You know better than to be back here.

Melvin: There are Jews at my table.

Carol: It's not your table. It's the place's table. Wait your turn. Behave. Maybe you can sit in someone else's section. (all the other waitresses gasp)

Melvin: (to the customers at "his" table) How much more do you have to eat, your appetites aren't as big as your noses, huh?

Manager: (preparing to move toward Melvin) That is it! Forget it! I don't care how many people buy his books . . .

Carol: (holding out her hand to stop her manager) I know, I know. Just give me one more chance, I'll talk to him. (walks over to talk to Melvin)

Melvin: (shrugs, self-satisfied) They left.

Carol: Yeah, what do you know? Brian says he doesn't care how long you've been coming, you ever act like this again, you're barred for life. I'm going to miss the excitement, but I'll handle it.

Melvin: (nods assent) Three eggs over easy, two sausage, six strips of bacon with fries . . .

Carol: Fries today!

Melvin: . . . a short stack, coffee with cream and sweetener.

Carol: You're going to die soon with that diet, you know that.

Melvin: We're all going to die someday. I will, you will, and it sure sounds like your son will.

Carol: (face drops, becomes silent; after a pause . . .) If you ever mention my son again, you will never be able to eat here again, do you understand? Do you understand? Give me some sign you understand or leave now. Do you understand me, you crazy fuck? Do you?

Melvin: (nervously moving around his plastic silverware) Yes. (quietly) Yes.

Photo 5.1 The restaurant as the site of emotional labor and social connection in *As Good As It Gets.* Carol (Helen Hunt) and Melvin (Jack Nicholson) must negotiate boundaries in the unequal structure of a service exchange.

This first interaction between Carol and Melvin is richly staged emotionally, from the irritation of the manager and offense of the other customers, to Carol's tightly contained rage when Melvin insinuates that her son is dying. This scene quickly establishes Carol as a deft manager of her feelings and those of other people. For her part, we see Carol as a skilled worker who is valued enough by her boss to be able to step in and protect a regular customer against her manager's desire to throw him out.

This first exchange between Carol and Melvin also displays both the potential and danger of service exchanges as social interactions. Carol is portrayed as being in control of her work environment. She chooses to share aspects of her personal life with her coworkers and customers. Yet despite her ability to control some aspects of her work, Melvin still can use the knowledge he has overheard about her personal life to hurt her. Melvin is able to receive recognition and tolerance of his atypical, politically incorrect, and offensive behavior by dining out. By contrast, Carol is vulnerable to the risks of opening herself up to the emotional aspects of her job.

Like most service workers, Carol must learn how to put on multiple emotional performances simultaneously. As servers' bodies move between locations and slip between tables, they must also adjust their faces and affect to react to a range of customers. Sociologist Arlie Russell Hochschild (1983) defines these adjustments of feeling as "emotional labor." Hochschild uses the term *emotional labor* to describe work that "requires one to induce or suppress feeling in order to sustain the outward countenance that produces the proper state of mind in others" (p. 7). Hochschild and subsequent researchers who have followed in her steps to develop a sociology of emotion have directed attention to how service work makes instrumental use of workers' emotions on the job, the costs to workers of altering their emotional production for a wage, and the broader implications of moving through a service economy in which consumers know that feeling is being strategically altered and managed. Just as physical labor can tire and wear out the body, emotional labor can estrange workers from the part of their selves they use at work—their emotions (Hochschild, 1983, p. 7).

In *As Good As It Gets,* Carol and Melvin's interactions provide a context for examining the emotional stakes in service exchanges. Despite Carol's far superior emotional skill and appeal, Carol remains structurally vulnerable to Melvin's poor social skills. The power dynamics inherent in their commercial relationship set up unequal rights to feeling. As Hochschild (1983) explains, servers are situated to absorb their customers' emotional displays, both appropriate and inappropriate, while customers are not obliged to return the favor. In the emotional subtext of the film, Carol is a form of medicine for Melvin, inspiring him and soothing him, but as the one who is paid to serve rather than be served, Carol's emotional returns are constrained by the circumstances of her employment. By virtue of her position as service provider, Carol assumes a weaker right to feeling than does Melvin.

While previous studies of emotional labor have focused on worker strategies for protecting themselves from emotional demands (Paules, 1991; Pierce, 1995), I argue that part of what lubricates the "crucial steadying effect" of emotional labor in our culture is that the workers who provide a warm meal, a haircut, a smile, and a token of recognition in the marketplace often come to enjoy the emotional demands of their work. In my study of a restaurant called the Hungry Cowboy, both customers and coworkers describe the restaurant as a scene of familiarity and connection (Erickson, 2004a; Erickson, 2004b; Erickson & Pierce, 2005). Customers say the restaurant provides them with recognition. "Everyone knows me here," says one regular customer. A waitress named Alex goes further, comparing Friday to a reunion:

> I look forward to working on Friday night, I like to come in on Friday nights because it's like family reunion time. All of these customers that I haven't seen for awhile, I can give them a hug, I can say it's good to see you, oh my god, you're getting so grown up. I have always looked forward to Friday nights. A good night is lot of familiar faces, not a lot of money. Just to have fun and enjoy the people I'm waiting on. (Erickson, 2004b, p. 562)

Similarly, in *As Good As It Gets,* as the film develops viewers come to realize that as the single mom of a sick child who lives with her mother, Carol gets the majority of her adult interaction at work. We see Carol telling sweet stories about her son to her coworkers and customers, exchanging resources with her coworkers, and being recognized for the quality of her service by her customers and boss. Carol is routinely connected to others, and Carol relies on the restaurant as a significant site of connection as well. Carol is intent, even insistent, on producing meaningful interactions through her labor, so much so that she does not even give up on Melvin. The allegorical aspects of Melvin and Carol's relationship as a venue for commentary on contemporary life expand midway through the film when their relationship moves outside the restaurant on a day when Carol cannot come to work because her son is ill.

Scene 2: Carol Misses Work

As Americans consume more services and more Americans work in the service sector, many more of our points of contact with other people take place within the context of a service exchange. In *As Good As It Gets,* it is this same point of connection between customer and server that proves transformative for both Carol and Melvin. Unlike most consumer/provider relationships, Carol and Melvin's relationship is extended beyond the restaurant stage when Carol's son, Spencer, is ill and she misses work to care for him.

Melvin's day is ruined when she is not there to wait on him. Melvin pays the busboy for Carol's last name, finds her address, and goes to her house. She opens the door.

Melvin: I'm hungry. You have ruined my whole day. I haven't eaten.

Carol: (shocked) What are you doing here?

Melvin: This is not a sexist thing. If you were a waiter, I'd be doing the same—

Carol: (interrupting, exasperated) What are you doing? Are you totally gone? This is my private home!

Melvin: I'm trying to keep emotion out of this, even though it's an important issue to me and I have strong feelings on the subject.

Carol: On what "subject"? That I wasn't there to take crap from you and bring you some eggs? Do you have any control over how creepy you allow yourself to get?

Melvin: (hurt) Yes, I do, as a matter of fact, and to prove it I have not gotten personal and you have. Why aren't you at work? Are you sick, you don't look sick, just tired and bitter.

Carol: (noticeably relaxing) My son is sick, OK?

Melvin: What about your mother?

Carol: How do you know about my mother?

Melvin: I hear you talking while I wait.

This scene is important to the development of both characters precisely because they have moved from the ordinary stage for their contact—tableside—to Carol's doorway. In Carol's understandable surprise at finding Melvin at her door, she reveals a secondary surprise at discovering that Melvin has in fact been listening to her while she talks to other people at the restaurant. Having put a moratorium on references to her son, she seems to assume that Melvin does not know much (or care much) about her except for her utility to him, specifically, her usefulness as the one who knows how to serve him his "one warm meal" in a way he finds soothing. It is Melvin, then, who is reaching beyond the parameters of the service exchange in two ways: first, by showing up on her home territory rather than the middle ground of the restaurant where they ordinarily meet, and second, by revealing an interest in her life that extends beyond the predictability of the service she provides. This scene in the doorway of Carol's apartment also becomes the catalyst for the next stage of their connection. Because Spencer's health prevents Carol from getting to work, Melvin uses his wealth to ensure that she can be there when he needs her. He pays for Spencer's medical care, and in doing so, pushes far beyond the preliminary point of contact at the table in the restaurant, catapulting their relationship into uncharted waters.

Service Interactions and Social Connections

In *As Good As It Gets,* contemporary social life is portrayed as lonesome, dangerous, and potentially crippling. Carol is at once in need of healing herself while also the source of that healing for others. Carol, like many waitresses who have turned up on the silver screen before her, is also a master of emotional management. Carol's tenacity, her emotional resilience in regards to Melvin's frequent verbal assaults on her character and her importance, inspires him, in his words, "to want to be a better man." Her emotional resiliency operates as an antidote to the emotional defeat in which Melvin and his neighbor, Simon Bishop, are tempted to indulge.

Melvin, the unredeemable, is redeemed by a waitress. Not a high-paid therapist, a doctor, or a friend, but his favorite waitress. This outcome is not very surprising if we are at all familiar with the romantic comedy genre, in which unlikely lovers find each other in unusual locations. However, what is quite unusual is the ways in which Carol's work persona remains central to Melvin's understanding of her, even as his affection for and attraction to her deepen. For example, Melvin tells Carol how he marvels that all her customers cannot see what he sees:

> I might be the only person on the face of the earth that knows you're the greatest woman on earth. I might be the only one who appreciates how amazing you are in every single thing that you do, and how you are with Spencer—"Spence," and in every single thought that you have, and how you say what you mean, and how you almost always mean something that's all about being straight and good. I think most people miss that about you, and I watch them, wondering how they can watch you bring their food, and clear their tables and never get that they just met the greatest woman alive. And the fact that I get it makes me feel good, about me.

The manner in which Carol does her work is central to what Melvin believes he knows about her. This attentiveness to what her work communicates and reveals is in tension with a more generalized attitude that waiting tables is not an important means of making a living. In fact, the view that waiting tables is a superfluous or inconsequential form of labor is the necessary context for a funny scene later in the film. When Melvin arranges for a top physician, Dr. Martin Bettes (Harold Ramis), to care for Carol's son at Melvin's expense, the doctor is surprised to discover what Carol's "indispensable" job entails.

Doctor Bettes: My wife is Melvin Udall's publisher. She said that I was to take excellent care of this little guy because you are urgently needed back at work. What kind of work do you do?

Carol: I'm a waitress.

What makes this exchange funny is that waiting tables is viewed as unessential and unimportant labor. Being a doctor is important; being a waitress is not. And yet, within the context of the film, Carol's work is what brings Melvin into contact with her.

Thinking beyond the film, in the move from an economy dominated by industrial production to an economy dominated by services, our service interactions with others have become much more central to daily life. The sheer frequency of our market-based interactions with others inspires sociologists of labor to return to long-running questions about the nature of connections between individuals. Similarly, Carol and Melvin's relationship raises questions about the nature of market-initiated relations: Can relationships that begin in market exchanges transform into relationships characterized by more familial or intimate ties? What does it require to move a relationship that originates in the unequal structure of a service exchange that is characterized by emotional scripts and restraint on intimacy into a relationship that is expansive and personal? Thinking through these and similar questions encourages a return to the distinctions made by German theorist Ferdinand Toennies (1957/1912), who differentiated between two distinct types of relationships between people: *gemeinshaft*, which roughly translated means community, and *gesellshaft*, which means society. *Gemeinshaft* is characterized by close, intimate kin or kinlike relations, while *gesellshaft* is characterized by impersonal and sometimes instrumental relations between people.

Toennies's ideas continue to be relevant today when we, as participants in social life, and as scholars, attempt to differentiate between the mood, tone, or substance of interactions between two or more people. The concepts *gemeinshaft* and *gesellshaft*—or community and society—provide us with a vocabulary to distinguish between voluntary connections between similar individuals who support one another and means-to-an-end connections between heterogeneous social actors who have limited connections with or responsibilities for one another. During his lifetime (1855–1936), Toennies was attempting to make sense of the rapid cultural changes wrought by industrialization. Today, the rapid expansion of the service economy means that more and more of our human interactions take place within the marketplace, while more and more of our needs are fulfilled by services for pay.

As Good As It Gets is a movie about connections and emotions that uses neighbors in an upscale apartment complex and customers and food servers in a neighborhood restaurant as two aspects of social life that thrust us toward each other, for better and for worse. Ultimately, the movie puts Melvin, Carol, and Simon—three characters connected through unlikely means—on a road trip together to Baltimore. The message of that road trip is that they are all (and perhaps we, the viewers, are, too) hurt emotionally by the cumulative damage of their lives. Their emotional needs, which arise from different events and ailments, are ultimately treated as *shared* needs. The connections between their needs are made explicit in a conversation that takes place between Carol and Simon, that Melvin interrupts. Simon is explaining to Carol why he has not been in touch with his parents for many years. Carol replies, "OK, we all have these terrible stories to get over, and you—" she is interrupted by Melvin who spouts off from the back seat:

> It's not true. Some have great stories, pretty stories that take place at lakes with boats and friends, and noodle salad. Just no one in this car. But, a lot of people, that's their story. Good times, noodle salad. What makes it so hard is not that you had it bad, but that you're pissed that so many others had it good.

This outburst from Melvin is significant because it connects to the title of the film and the abiding fear Melvin revealed earlier, "What if this is as good as it gets?" Melvin's will never be a life of noodle salad and boat ride memories, and yet he delivers this speech from the backseat of a convertible driven by a woman he is attracted to and whom he respects, and shared by his neighbor, toward whom he has, despite his homophobic tendencies, grown increasingly fond. If as viewers we believe that connections with other people are generally good, then for Melvin, for Carol, and arguably even for Simon, their former lives were not as good as it gets.

In this film, the neighborhood and the restaurant give rise to new connections between people. And the connections between the three people in the car are situated as more real than the often accidental and unwanted ways in which we touch each other, talk to one another, and are forced to interact with people as we make our way through late capitalism. In fact, the relations between the three people in the convertible are situated as curative and healing, in contrast with many other social relations that are largely disinterested and disheartening. When Simon's "party friends" and even his family abandon him, when Carol cannot find men to date who can handle the complexity of her life, and when Melvin cannot even get the attention of the therapist he pays to treat his illness, these three become unexpectedly connected. Melvin's reference to "noodle salad" paints a picture of lives of ease, lives that do not require careful navigation to find one another and to meaningfully connect with each other. In contrast, the connections between the three main characters in this film are anything but easy; they are forged through trial and error, but emerge quite strong. As viewers, it is clear that Melvin's speech about noodle salad is itself mythological and nostalgic; he refers to a world as pretend as the romantic fiction he writes. Melvin longingly imagines a life that no one in the car, but also no one watching the film either, ever *really* experiences. In contrast to a fantasy about noodle salads and boat rides, the connections between Melvin, Carol, and Simon are situated as ultimately superior to connections to some family and some friends, many of whom cannot be counted on to stand by when disaster or illness strikes. In ending on this note, the film offers one vantage point from which to think about how to apply Toennies's terms to contemporary social life. These three people forge *gemeinshaft*-like relations out of *gesellshaft* points of connection. They build community amidst the rubble of kinship.

I turn next to *Office Space,* a movie that situates the restaurant as one of many sites in which workers must struggle against the dehumanizing effects of corporate control of the self at work.

The Ironies of Cubicle Culture

Office Space, as the title suggests, is a movie about work. The primary plotline involves Peter (Ron Livingston), a burnt-out office worker who has reached the limits of his tolerance for his work at a technology corporation called Initech. The perilous and mind-numbing conditions of work at Initech, where workers are faced daily with the threat of downsizing, are juxtaposed with the work conditions of Peter's love interest, Joanna (Jennifer Aniston), who works at a TGIFriday's-look-alike restaurant called Chotchkie's. Chotchkie's is located between Chili's and Flinger's, in a strip mall situated in the concrete landscape of the suburb where Initech is located. The viewer first encounters Chotchkie's when the bored-to-death cubicle workers, Peter and his friends, Samir (Ajay Naidu), and Michael (David Herman), go for a coffee break after just 30 minutes of work on a particularly irritating Monday morning.

Scene 1: Brian, the Superserver

Unfortunately for Peter, Chotchkie's does not provide the refuge from corporate culture he had hoped. At the restaurant, Peter, Samir, and Michael are accosted by a male waiter named Brian who epitomizes the ideal server Chotchkie's seeks to mold. Brian is aggressive, trying to sell dessert to the three men at nine o'clock in the morning, and portrays an enthusiasm that is both sickeningly sweet and utterly unrealistic given the confines of his role and the time of day.

Brian: So can I get you gentlemen something more to drink? Or maybe something to nibble on? Some Pizza Shooters, Shrimp Poppers, or Extreme Fajitas?

Peter: Just coffee.

Brian: Okay. Sounds like a case of the Mondays.

Writer and director Mike Judge, who also created the animated series *Beavis and Butthead* and *King of the Hill,* has captured with painful accuracy the irrationalities of consumer experiences in chain restaurants. In Brian's performance, the corporate script trumps the context. If shrimp poppers are what he's been instructed to sell, he'll suggest them to customers, even at nine o'clock in the morning. Like Judge's previous work, *Office Space* is cynically focused on the minutiae of suburban life, and here, in his first full-length motion picture, he has mapped the continuities between cubicle culture and corporate chain restaurants.

Office Space is a modern day rant about the alienating effects of work, about the unacceptable costs to the self extracted by jobs that belittle us, leaving us dreaming of a different life, and cubicles that are too small to contain us. For example, in one of the culminating, most famous scenes of the film, Peter steals the fax machine that Samir and Michael have struggled with for all the time they have worked at Initech. The three men take the machine out into a vacant field and proceed to brutalize the machine with baseball bats and fists like a group of gangsters beating up an enemy. The scene is set to the tune of "Still" by the Geto Boys, so as they punch, kick, and mutilate the machine, the Geto Boys chant, "Cause it's die muthafuckas, die mutherfuckas" (Geto Boys, 1996). The critique that *Office Space* offers of the unacceptable costs and dehumanizing effects of work is not subtle.

Unlike films that tangentially comment on the effects of business practices in late capitalism, here the criticism is centered in the dark and biting humor that emerges out of circumstances that have become all too common in the new economy. Throughout the film, Peter's struggle to escape the constraints of his work amidst the continuous threat of downsizing is compared with Joanna's struggle to retain her dignity in her job at Chotchkie's. Like Peter, Joanna is portrayed as resisting the character-shaping demands of her work. While Joanna's character and struggle is substantially underplayed compared to the screen time devoted to Peter's struggle, the parallel structure of the two narratives is noteworthy if only for the way it treats waiting tables as work that is similar to other forms of work, notably white collar work. By contrast, back in 1974, in *Alice Doesn't Live Here Anymore,*

Alice's job as a waitress is portrayed as the ultimate failure and insult to herself. In that film, waiting tables is presented as overtly shameful work that is the domain of lower-class people who are destined to toil, or of "fallen" middle-class women. Some 30 years later, in *Office Space* the central waitress figure is depicted as a smart working-class heroine whose compassion and charm make her a sought-after love interest in part because she possesses enough street smarts to assert her independence and to resist degradation in her job. Much like the character of Carol in *As Good As It Gets,* in *Office Space,* Joanna can be trusted as a solid, honest, truth-telling woman who provides a healthy alternative to a world riddled by people who have been beaten down and compromised by their work. In fact, unlike the vast majority of people at Initech who have lost all trace of resistance, and certainly unlike the superserver, Brian, who has given himself entirely to a corporately sponsored presentation of self, Joanna still has substantial fight left in her.

Joanna's specific struggle is with her officious manager who seems eternally disappointed by her reluctance to enthusiastically display what he calls "flair." At Chotchkie's, "flair" refers to the corporate-required buttons with funny, ironic statements on them that are provided by the company to "individualize the workers." During Joanna and Peter's first lunch date, Peter accidentally stumbles on the delicate matter of the flair.

Peter: We're not in Kansas anymore.

Joanna: (laughs) I know, really.

Peter: (indicating the buttons on her chest) It's on your . . .

Joanna: (looking down at the buttons she is wearing on her suspenders and uniform shirt) Oh right. That's one of my pieces of flair.

Peter: What's a piece of flair?

Joanna: Oh it's. . . . We're required to. . . . You know, the suspenders and these buttons and stuff, we're actually required to wear 15 pieces of flair. It's really stupid actually.

Peter: Do you get to pick them out yourself?

Joanna: Yes, we are, although I didn't actually choose these, I just sort of grabbed 15 buttons. I don't even know what they say. (uncomfortable) I don't really like talking about my flair. OK?

If we think of what occurs initially between Melvin and Carol as a form of commercialized intimacy, then what Joanna is struggling with is the corporate management of the self. In service work, the very character and emotions of service workers become the terrain of struggle in manager and worker debates.

Joanna's arguments with her boss and her refusal to conduct herself like Brian the superserver reveal ambiguities and tensions central to training service workers. Brian the superserver and Stan the manager both seem to believe in servers' ability to perform a script that does not sound rehearsed, to reproduce a smile, again and again without ever losing the sparkle. Within this framework, getting to know customers is a purely instrumental activity in that a connection with customers might provide servers with insights into how to get more money out of customers. According to the corporate philosophy then, more flair equals more fun for customers, which equals more money for the company. Joanna pushes back against this outlook, arguing against more flair. From her vantage point, being forced to go over the required minimum of flair would falsely convey support for and enthusiasm about corporate mandates that she thinks are "stupid."

In short, what Joanna's manager is attempting to convince her to do is "speed-up" her performance of self within the job. Sociologist Robin Leidner (1993) describes some of the tensions that arise when companies attempt to "speed up" human interactions:

> Organizations that routinize service interactions are acting on contradictory impulses. They want to treat customers as interchangeable units, but they also want to make the customers feel that they are receiving personal service. The tension inherent in this project was apparent when I asked one of the trainers at Hamburger University about McDonald's goals for customer service. He told me quite sincerely, "We want to treat each customer as an individual, in sixty seconds or less." (p. 179)

Joanna's manager wants her to act as if she wants to go above and beyond the minimum. He wants her to want what the company wants, and masks this behind a request for her to "express herself" by giving in and displaying more than the minimum 15 pieces of flair on her body. While Joanna's character is really tangential to the Herculean struggle that Peter and his friends at Initech undertake, she, too, struggles to retain an inviolable sense of self at work.

Scene 2: Scripted Selves

To understand the significance of the "flair wars" that Joanna and her manager play out, it is important to note that Chotchkie's is likely a play on words to refer to the Yiddish word *tchotchke,* which means "an inexpensive trinket" *(Oxford English Dictionary,* n.d.). Writer and director Mike Judge cleverly names the restaurant a word that refers to something worth collecting, but not antique or particularly valuable. Combined with the fact that Judge plays the unflinchingly corporate manager of Chotchkie's who relentlessly attempts to get Joanna to "express herself" with flair, the significance of the restaurant scenes to the larger meaning of the film becomes clear. Later in the film, in a conversation with her manager, Joanna debates the necessity of "speeding up" the personality demands of her job.

Stan (Chotchkie's manager):	We need to talk about your flair.
Joanna:	Really? I have 15 pieces on.
Stan:	Well, 15 is the minimum, okay?

Joanna: Oh, okay.

Stan: Now, you know, it's up to you whether or not you wanna just do the bare minimum or, uh— Well, like Brian, for example, has 37 pieces of flair on today. And a terrific smile. People can get a cheeseburger anywhere. They come to Chotchkie's for the attitude and the atmosphere, and the flair is part of it.

Joanna: So you want me to wear more?

Manager: (rolls his eyes) Look, we want you to express yourself.

Her struggles demonstrate the central tension between individualizing and management control. The manager's task is to encourage individuals to be who the company needs them to be under the guise of "expressing themselves," but Joanna refuses to play the game.

Later in the film, Joanna triumphs over her sniveling boss, even though to do so requires quitting her job.

Joanna: You know what, Stan, if you want me to wear 37 pieces of flair, like your pretty boy over there, Brian, why don't you just make the minimum 37 pieces of flair?

Stan: Well, I thought I remembered you saying that you wanted to express yourself.

Joanna: Yeah. You know what, yeah, I do. I do want to express myself, okay. And I don't need 37 pieces of flair to do it. (holds up both middle fingers and walks out)

Joanna's victory is satisfying, if fleeting. By the end of the film, she is working for another corporate restaurant in the same strip mall.

Popular Culture and the Sociology of Work

What can the movies teach us about work in the new economy? Films can be used as a common text through which to discuss and debate changes in the conditions of work and the implications of work on how we construct selves and relate to each other under conditions of late capitalism. Familiarity with sociological theories and research on the conditions of work leads to making the connections between what we get paid to do and who we think we are, in our own lives and in the larger society.

Occasionally movies and other media circulate back and get incorporated somehow into work itself. Witness how many workers have Dilbert cartoons up in their cubicles, for example. When I worked as a waitress, I was subject daily to the power of pop culture to inform our vantage point on work, because since my name is Karla, customers constantly reminded me of the character of Carla Tortelli from the incredibly popular 1980s television sitcom *Cheers*. *Cheers* depicted a Boston bar in which the same customers interacted with the same workers every night, forming a tight community of friends. In my research on neighborhood restaurants, customers also used references to the fictional Cheers bar to explain why they frequented a particular restaurant. On comment cards, they would write, "It's like Cheers here!" or, when I asked them to describe their favorite place to dine out, they did so in this way: "A place where everybody knows your name, like Cheers."

Not only can film offer us a fresh vantage point on work, film occasionally informs how we make sense of our roles in the new economy. In fact, at one of the restaurants I studied, the movie *As Good As It Gets* came up in multiple interviews with both customers and servers. For example, like the comfort Carol provides to Melvin, one waitress named Jessica had been a routine, habitual, reliable person in her customers' social landscape. Jessica volunteered the following story as an example of her connection to customers:

> There are many people who don't realize what it means to be a waitress, and they probably make as much money or less than I do, but they all feel superior to me. I have customers who say I can never leave. They have told me that if they come in and find out I quit, they are going to be like Melvin in *As Good As It Gets* and go to my house and get me to come back and wait on them! They're just trying to tell me how much they appreciate my service, and I realize that. It's like a little warm fuzzy. (Erickson, 2004b, p. 561)

Jessica's use of the film reveals that servers are drawn into emotional performances through multiple routes—not only managerial strategies, but also their own sense of craft and the collaborative construction of a sense of community. Jessica's customers made use of *As Good As It Gets* as a common text that could help them explain her significance to them. Jessica retells the story to counter the more prevalent attitude of "many people who don't realize what it means to be a waitress." Taken together, these instances demonstrate that popular culture depictions of service workers do occasionally supplement or help explain people's responses to one another within market exchanges.

Popular culture representations of work and social life allow us to hold up fictional interactions at a distance and scrutinize them together: How do popular representations compare to our own experiences in the marketplace, the neighborhood, or the office? What are the social conditions and organizational cultures of the places we work and consume? As a scholar of work, I find films to be one of the many avenues of access to analyzing working life and improving the theoretical insights we use as tools to do our own intellectual and scholarly work. Film is one of many tools we can use as a vehicle for thinking critically about how we perform our roles in the new economy, with particular attention to how our sense of self is affected through the service interactions and the social connections that emerge within the marketplace.

References

Erickson, K. (2004a). Bodies at work: Performing service in American restaurants. *Space and Culture, 7*(1), 76–89.

Erickson, K. (2004b). To invest or detach? Coping strategies and workplace culture in service work. *Symbolic Interaction, 27*(4), 549–572.

Erickson, K., & Pierce, J. (2005). Farewell to the organization man: The feminization of loyalty in high-end and low-end service jobs. *Ethnography, 6*(3), 283–313.

Geto Boys. (1996). Still. On *The Resurrection* [CD]. Houston, TX: Rap-a-lot Records.

Hochschild, A. (1983). *The managed heart: Commercialization of human feeling.* Berkeley: University of California Press.

Leidner, R. (1993). *Fast food, fast talk: Service work and the routinization of everyday life.* Berkeley: University of California Press.

Oxford English Dictionary. (n.d.). Tchotchke. Retrieved December 1, 2007, from http://www.oed.com/

Paules, G. F. (1991). *Dishing it out: Power and resistance among waitresses in a New Jersey restaurant.* Philadelphia: Temple University Press.

Pierce, J. (1995). *Gender trials: Emotional lives in contemporary law firms.* Berkeley: University of California Press.

Toennies, F. (1957). *Community and society.* (C. P. Loomis, Trans.). East Lansing: Michigan State University Press. (Original work published 1912)

Reading 5.2

Reel Families: Family Life in Popular Films

Janet Cosbey

Popular films reveal our ideals and expectations about family life in myriad ways. Some films perpetuate cultural stereotypes about families and the roles that various family members play. Other films, fewer in number, challenge preconceived ideas and offer alternative views of family life in contemporary society. The portrayal of families in film provides sociological material that can be used to explore the family as social group and institution, in much the same way as family stories in novels (Cosbey, 1997). Using the sociological imagination, films about family can be analyzed for the connection between social structure and individual experiences and outcomes.

As filmgoing audiences, do we see the families presented as a reflection of real life, as the actual lived experiences of families at particular points in time and in particular locations? While belief persists that movies reflect reality (hooks, 1996), "reel life" does not act as a mirror for "real life." Movies "screen" and frame social reality, and they reflect ideological images of interaction, relationship, and community (Denzin, 1991). Cultural ideals about families, how they are constituted and expected to act, are deeply ingrained. When we see what looks familiar or normative, we often accept these representations without question. In turn, the Hollywood film industry, in appealing to the largest audiences possible, has given American society widely accepted views of family, reflecting the ideal, if not the real, culture. In this section, the focus is on three dimensions of American family life that are the subject, directly or indirectly, of many Hollywood films and areas of study within family sociology: gendered notions of family roles; racial, ethnic, and social class differences; and family violence.

The discussion of the first two films, *Mona Lisa Smile* (2003) and *Mr. Mom* (1983), focuses on gender in families. Beliefs about family life are rooted in gendered ideas about how a woman/wife/mother and a man/husband/father should act. Sociologists note that gender relations and family life are so intertwined that it is impossible to understand one without the other (Coltrane, 1998). The second set of films, *What's Cooking?* (2000) and *The Pursuit of Happyness* (2006), highlight the ways in which race, ethnicity, and social class position families in the social order and shape family experiences. Last, *The Great Santini* (1979) and *Sleeping With the Enemy* (1991) are films that dramatize the problem of patriarchal power in the family that leads to violence. All six of these films provide a window on family life for sociological analysis.

Gender and Families

Mona Lisa Smile and *Mr. Mom* both provide stories about the breadwinner/homemaker model of family life, what is historically known as the "cult of domesticity."

Mona Lisa Smile takes place in the 1950s and demonstrates the stronghold of the ideological view of family prevalent during that time period. Functionalist theorists such as Parsons and Bales (1955) described this family arrangement in terms of the expressive, caregiving role for women and the instrumental, breadwinner role for men. In this version of family life, women stay home and care for the family, meeting the family members' emotional needs while taking responsibility for the domestic work of making a home. Men, on the other hand, are responsible for earning a living (outside of the home) and protecting their family members from the outside world.

Mr. Mom provides one version of a modern adaptation of family roles and responsibilities. The man in the family plays the expressive role by staying at home and caring for the household and children, while the woman plays the instrumental role working outside the home to support the family economically. Significantly, these gender roles remain unchallenged and unchanged, regardless of who performs them. Both movies offer an opportunity to critically examine the traditional gendered structure of the family, one in historical context, the other in contemporary society.

Mona Lisa Smile (Mike Newell, 2003)

In *Mona Lisa Smile,* Katherine Watson (Julia Roberts), a feminist, forward-thinking, idealistic new professor takes a job teaching art at conservative all-female Wellesley College in the 1950s, hoping to inspire her students to see beyond their expected roles.[1] Katherine leaves her boyfriend behind in California and begins her new life in Massachusetts, ready to change the world, or at least women's place in the world. Katherine approaches her teaching with the earnest belief that she can open the minds of the young female students she meets so that they can challenge the traditional societal expectations of the day and do something more with their lives than become "just a housewife."

As a professor Katherine is confident that she can challenge the women she teaches to think differently about their lives and their futures. She encourages them to think critically and question their life choices. During one scene, she shows her students four advertisements depicting women's lives and asks them to consider the role of women in the future. The students at first are reluctant to take her seriously and are dubious about the questions that she raises. Katherine soon finds the situation at Wellesley not to be what she expected and is ready to pack up and leave. Though she is teaching the best and the brightest students who are always prepared for class, eager to learn and demonstrate their knowledge, she is disappointed to find their sole ambition remains finding a husband. The young women, as daughters of the upper class, are being groomed to take their place as wives and mothers within that social class. As Katherine wryly comments early in the film, it seems as though she is at a finishing school rather than a college, "I thought that I was headed to a place that was going to turn out tomorrow's leaders, not their wives."

[1]According to Roger Ebert (2003), the screenwriters based their script on Hillary Clinton's experiences at Wellesley in the 1960s.

The breadwinner/homemaker model of family life that Katherine finds her students so committed to was part of the historical anomaly of the 1950s. During World War II, women had been recruited into factory work to replace the men who had gone to war. With the changes wrought by industrialization, including the separation of work from home, all women except the most economically privileged were part of the waged economy. The difference in the World War II era was that for the first time in history, women were doing men's work for men's wages. Once the war was over, women lost their jobs or were moved to lower-paying positions to open the labor market for the returning soldiers. At the same time, there was a cultural emphasis on gender-specific roles within the family—and women's place was at home, caring for their husbands and children. Movies during the period reversed the government-produced advertising to entice women into the factories during the war, with images of women happily caring for home and family. Post–World War II movies were a celebration of "Harriet the Happy Homemaker," replacing the World War II images of "Rosie the Riveter" (Edelman, 2008). This version of family life was supported by a strong postwar economy and government policies such as low-interest mortgages to veterans making the American dream of marriage, family, and home available to many working-class as well as middle-class families (Cherlin, 2008).

In the aftermath of World War II, and in the wake of U.S. economic, political, and social transformation, the decade of the 1950s was a "throwback to the Victorian cult of domesticity with its polarized sex roles and almost religious reverence for home and hearth" (Skolnick, 1991, p. 52). The fact that Katherine was not successful with the students at Wellesley is not surprising; the majority of women who attended college in the 1950s had no career plans and dropped out in large numbers to marry (Skolnick, 1991). However, it is important to remember that this was a reflection of the time, rather than a demographic or social trend, and that only members of the upper-middle and upper classes could send their daughters to college. Despite the fact that this period was short-lived, and that most women have engaged in some type of wage labor throughout history, this ideology has been fixed in cultural constructions and memories of family structure. Gendered expectations about family have shaped public policy, as well as individual choices and experiences, across the generations, although in reality it was "the way we never were" (Coontz, 1992). Regardless, the myth has lived on, shaping our beliefs and expectations about family life into the present day.

The depiction of 1950s femininity, etiquette classes, and homemaking skills emphasized in *Mona Lisa Smile* is so far removed from the college experience of women today that we might feel a sense of superiority that we are no longer subject to these outmoded views about women. The beginning of the second wave women's movement is often identified as the critique of the homemaker role, including the isolation and desperation of educated middle-class women who were relegated to keeping house in the suburbs, offered by Betty Friedan in *The Feminine Mystique* (1963). Despite the changes wrought by the women's movement, there are still many remnants of traditional gender roles at work in families today. Gendered ideas about women's and men's work and family roles are evident even in families that espouse support for women working outside of the home and more equitable relationships in the home. There may be more flexibility today for both women and men, but the legacy of the homemaker/breadwinner ideal continues to structure our lives, choices, and potential.

One of ways that women's lives are most directly affected is that, at the same time that women, especially those with young children, are increasingly participating in the paid labor force, they are still expected to come home and work a "second shift" in housework and family care (Hochschild, 2003). Research reveals how much of this work falls to women in families; on average, married women spend about three times as much time on routine housework chores as do married men (Coltrane, 2000). As seen in *Mr. Mom,* it is the structure of these gendered positions in the family that is difficult, if not impossible, to change.

Mr. Mom (Stan Dragoti, 1983)

The gendered division of labor in the public and private realm is the central theme of the movie *Mr. Mom.* With the economic downturn in the automobile industry, Jack Butler (Michael Keaton) loses his executive job. After failed attempts to find employment, he stays home to care for the house and children as "house-husband" while his wife, Caroline (Terri Garr) goes to work as an executive for an advertising agency. It is difficult to accurately estimate how many families fit the stay-at-home father/employed mother configuration today, but of the 23 million married couples with children under 15.4% (1 million) are families with a father who stays home full-time (Fields, 2003). Though this film is a comedic look at this particular family situation, it aptly illustrates the gendered division of labor at home and work, and the legacy of the cult of domesticity.

The mishaps that Jack "Mr. Mom" Butler experiences during his domestic reign illustrate the stark gendered divisions in the homemaker/breadwinner model. The exaggerated tribulations that Jack endures capture the 1950s model of family life with a mom left in charge of the home while the dad was absorbed in his work. He does not know, or need to know, what is involved in maintaining the home and caring for their children; that is the work of the housewife/mother. Thus, when Jack takes over at home, he is a bumbling mess.

Jack soon finds himself overwhelmed with laundry, cooking, cleaning, and other household chores. He is frustrated with the never-ending menial tasks. The majority of these chores require little thought or skill, and on top of that, they must be done constantly. His only social contacts during the day are his children and, though he tries to engage them in conversation and seeks their help with the housework, he is left feeling frustrated and alone. His reactions, humorous in the role-reversal, illustrate the frustration and discontent of the suburban housewife described by Betty Friedan so long ago. Jack shouts at his wife: "My brain is like oatmeal. I yelled at Kenny today for coloring outside the lines! Megan and I are starting to watch the same TV shows, and I'm liking them! I'm losing it."

While Caroline is also initially uncomfortable with the role switch, she grows increasingly enamored with her new job, as Jack becomes increasingly disappointed with his. Caroline is sympathetic to Jack as she confesses that she once felt much the same way. Prior to Jack's job loss, Caroline had strictly adhered to the 1950s model of homemaker, including the belief that she freely chose this way of life because it was best for her family. However, her success at work fosters a new sense of confidence and vitality she did not have in her life as a wife and mother. Interestingly, she gives this up and returns home when Jack is called back to work. In the end, the traditional family is restored, with perhaps greater empathy between Jack and Caroline about their respective roles, but no deeper understanding of how to break free of their gendered destinies.

From very different vantage points, both films highlight gendered arrangements within and assumptions about the family. Ultimately, neither challenges the gendered division of labor. In *Mr. Mom,* Jack and Caroline trade positions in the family for a period of time, but do not question the normative structure of work caring for the home and work for wages outside the home. In *Mona Lisa Smile,* Katherine rejected the normative gender order. She wanted her students to see that they too could have personally fulfilling lives of their own choosing, but she left the school before she could see any of this realized. As she drives away, we are left with the image of her students running after her car, perhaps a glimpse into their longing for the freedom and independence she represents. Yet, the societal expectations imposed by their parents, their peers, and their professors so strongly advocate that a woman's proper place is in the home that they will end up as housewives, as Katherine feared they would.

Race, Ethnicity, and Social Class

The United States continues to become more racially and ethnically diverse as we move forward into the 21st century. In urban areas, one is likely to find ethnic enclaves, what early sociologist Robert Park described as "a mosaic of little worlds" (Park, Burgess, MacKenzie, & Janowitz, 1925/1984). One change that has occurred since Park first observed immigrant settlement patterns in Chicago in the early 20th century is that today's immigrant population is more likely to come from Latin America (53%) and Asia (25%) than from Europe (Migration Policy Institute, 2005).

Another change is that Park predicted that, over time, immigrant groups would assimilate and become part of the "melting pot" where ethnic differences were muted and/or erased. Today, sociologists note that differences between immigrants in terms of human, social, and political capital produce different (segmented) assimilation patterns between families and within families by generation (Booth, Crouter, & Landale, 1997). Another important pattern is biculturalism, where immigrants maintain many aspects of their ancestral culture, such as language and traditions, while becoming increasingly able to function within U.S. society over time. Regardless of these changes, one consistency over time is that nationality, ethnicity, and race intersect in the context of family, shaping experiences and outcomes within and across generations.

Social class also shapes family experiences, values, traditions, and beliefs. Social class impacts almost every facet of family life from determining who and when to marry, having and rearing children, where and how families live, and even the ways in which family members interact with one another. The influence of social class, as well as race and ethnicity, on family life, status, and well-being are the subjects of the films *What's Cooking?* and *The Pursuit of Happyness.*

What's Cooking? (Gurinder Chadha, 2000)

What's Cooking? is the story of one Thanksgiving day in the lives of four families of different racial/ethnic and economic backgrounds living in the same Los Angeles neighborhood. The race and ethnicity (Vietnamese, Latino/a, Jewish, and African American) of the families highlights their unique histories in U.S. society, at the same time that they share common challenges within their families. Two dominant themes in the film are biculturalism and intergenerational conflict.

One of the cornerstones of culture is the food we eat and share. It is Thanksgiving Day in *What's Cooking?* and turkey is the centerpiece of each family's meal. However, biculturalism is apparent in the ethnic foods, décor, and conversation at each family table. In the Nguyen family, the mother, Trinh (Joan Chen), struggles to hold onto Vietnamese customs and incorporate U.S. culture, symbolized by her preparation of their Thanksgiving dinner with traditional Vietnamese food, alongside the Kentucky Fried Chicken (KFC) that is bought to substitute for the turkey she burned. We see food prepared across the kitchens: tamales, spring rolls, candied yams, and Grandma Grace's (Ann Weldon) signature macaroni and cheese. The Avilas cheer for "Señor Turkey" as their cooked bird is pulled from the oven, and Elizabeth (Mercedes Ruhl) announces that there is flan or pie for dessert.

In each of the families, tension, and sometimes outright conflict, becomes apparent between the generations as the events of the day unfold. In immigrant families there is often tension between the immigrating first generation and their offspring, as the second generation incorporates the values and norms of the new culture into their self-identity. The Nguyens worry about their children as they watch them become increasingly American in their dress and manners. The oldest son, Jimmy (Will Yun Lee), who is away at college, tells the family that he cannot return home for Thanksgiving, though in truth he is having dinner with his girlfriend's family, the Avilas. Their teenage daughter, Jenny (Kristy Wu), flaunts her flashy style and her involvement with a white boy. When her grandmother finds condoms in her pocket, Jenny's mother fears the worst for her daughter. Most frightening to the parents is that their teenage son, Gary (Jimmy Pham), may be involved with gangs and violence.

In the upper-middle-class African American Williams household, the father and son are estranged due to their political differences. Dismayed that his son wants to go to a historically black college and major in African American studies, Ronald (Dennis Haysbert) angrily asks, "What could you do with a degree like that?" Michael (Eric George), in turn, believes that his father, a conservative Republican, has "sold out" because he works for a governor who has the reputation as a bigot. Ronald experienced upward mobility and achieved success through education and assimilation into the white dominant culture. Michael, on the other hand, advocates a politics of difference where racial history, identity, and community are the pathway to social, political, and economic power (Kerchis & Young, 1995).

Across the street, in the Jewish household, the Seeligs have warned their daughter Rachel (Kyra Sedgwick) not to "come out" to the relatives who will be joining the family for Thanksgiving dinner. The visit is awkward at best, and when the dinner conversation shifts from small talk to a lively discussion concerning the governor, Rachel points out that he is prejudiced against gays and lesbians. To her mother's dismay, Rachel's Aunt Bea (Estelle Harris) loudly announces, "Rachel is a lesbian—you know, like Ellen."[2]

Within each of these families, ethnicity, race, culture, and class intersect, shaping their lives, opportunities, and experiences. Throughout the film, we find examples of the ways in which stereotypes shape one group's view of and reaction to another. For example, Gina Avila (Isidra Vega) is dating Jimmy Nguyen, and when she brings him home for Thanksgiving dinner, her family asks if he likes Jackie Chan movies, while her brothers make karate gestures toward him throughout the evening. Jimmy is worried about his family meeting Gina because he fears their disapproval, which leads him to lie to his parents and not return home for Thanksgiving. When he finally does bring her home, his family smiles at her and whispers to one another that they like her because she looks Vietnamese.

In their own way, the families are attempting to find their place in American society, with some members of each family resisting mainstream American culture, while others embrace it fully. Each of the families faces a set of challenges that reflect the experiences and concerns that many contemporary American families face regardless of social class or ethnic origin. The stories of the four families are interconnected in surprising ways that parallel the cultural pluralism of American society and show that families, even from radically different backgrounds, have much in common. All of these families deal with issues of child-rearing, gender roles expectations, communication, infidelity, divorce, economic struggles, and generational conflicts. Although similarities that all American families share are emphasized, the unique history and culture of each family, including the realities of racism, ethnocentrism, and xenophobia, are addressed. The challenge of pursuing and achieving the American Dream concerns all of the families who hope that their children will have broader opportunities and realize upward mobility, as expressed by Mrs. Nguyen when she emphatically repeats to her children, "Education is the most important thing!"

The Pursuit of Happyness (Gabriele Muccino, 2006)

The Pursuit of Happyness is based on the true story of Chris Gardner (Will Smith), a financier, who rose from homelessness to become a multimillionaire. The movie is based in San Francisco in the 1980s during a downturn in the U.S. economy. At first glance, this film appears to be retelling the classic "rags to riches" story. Rooted in core American values, the belief that individual effort is a vehicle to upward mobility and economic success is reinforced in cultural stories such as film and literature. Regardless of the opportunity structure, most Americans believe that

---

[2]Ellen DeGeneres is a popular, award-winning stand-up comedienne, television host, and actress. She famously came out as a lesbian on her television show, *Ellen,* in 1997.

you can make your own fortune, and that family poverty and its related problems are the result of personal failure to work hard and take advantage of the equality of opportunity they believe exists in U.S. society (Sawhill & McLanahan, 2006). *The Pursuit of Happyness* dramatizes individual success in achieving the American Dream; however, it also highlights the contrast between the lives of the rich and poor. It illustrates the tenuous grasp most people have on the American dream and the way that family life is shaped and changed by economic (mis)fortune.

When we first meet Chris, his wife is working double shifts to try to make ends meet. She ends up leaving, but unlike most mothers, she relinquishes custody of their son when Chris pleads, "I met my father for the first time when I was 28 years old. When I had children, my children were going to know who their father was." Chris joins the ranks of fathers with full-time custody of their children, approximately 3% of all families today (Kreider, 2007).

In short order, he also joins the ranks of homeless families, as he and his son are evicted from their apartment. Unemployed and unable to find affordable housing, like many homeless parents, Chris makes use of public space such as shelters, bus stations, and public bathrooms to meet their personal needs and to rest for the night. However, unlike most homeless adults, Chris obtains a six-month (unpaid) internship at Dean Witter. In contrast to the other interns who can work late, Chris faces the challenges of working parents (he needs to pick up his son from daycare) and the homeless (he has to rush to the shelter to secure a bed for the night). Despite the complications of his circumstances, he overcomes all odds and wins the coveted full-time job at the end of the internship.

This happy ending resonates with the American Dream. Chris, willing to make high-risk choices that in the short term make him a single homeless father, reaps the rewards of financial success and finds "happyness." In reality it is unlikely that (homeless) families living in poverty will ultimately move into the upper middle class, regardless of their personal qualities and level of motivation. The changing structure of the economy has had a major impact on families, as they struggle with increasing unemployment, housing foreclosures, and declining opportunities. For most poor and near-poor families, the most representative part of Chris's story is the spiral down as the wage-earner in the family leaves, he becomes a single parent, he and his son lose their housing and rely on emergency shelter, and he struggles for survival on a daily basis. In this story, the economic vulnerability of American families is revealed; many families in the United States are one or two paychecks away from poverty, and even homelessness.

Family Violence

The Great Santini and *Sleeping With the Enemy* are films that illustrate some of the causes and consequences of violence in family relationships. Both films exemplify the patriarchal model of domestic violence (Cherlin, 2008). This model explains domestic violence as rooted in men's power and control, which is reinforced through patriarchal laws and customs of society. Historically, women and children were legally considered chattel, or property of men. Men had the power in relationships and violence, or the threat of violence, was one way to maintain their position in the family. Historically, "domestic discipline" of wives was a reality and despite laws prohibiting wife beating, it was not until the latter part of the 20th century that domestic violence and child abuse became public social issues (Straus, Gelles, & Steinmetz, 1980).

The Great Santini (Lewis John Calino, 1979)

The Great Santini is the story of a Marine pilot and his family living in South Carolina in the early 1960s. Bull Meechum (Robert Duvall) had been a daring Marine pilot during World War II and is now a "warrior without a war" struggling to relate to his family and the world around him. Bull sees his family as under his command, in much the same way that he controlled the men in his troops. His relationship with his family is an authoritarian one, based on a belief in strict discipline and obedience as a sign of respect. Despite signs that Bull loves his family, he abuses and controls his wife and children, especially his oldest son, Ben (Michael O'Keefe).

Bull and his wife Lillian (Blythe Danner) are the parents of four children, two sons and two daughters. Whereas Bull, as his name implies, epitomizes hypermasculinity, his wife Lillian is the prototype of traditional femininity. Bull is a Marine pilot who drinks too much, thrives in dangerous combat, and gives direct orders to his subordinates in the military and at home. He is so caught up in his performance of masculinity—demonstrating strength and toughness and maintaining constant control—that he cannot forge a healthy relationship with his family (Kaufman, 2001). He is troubled and bewildered by his failure to have the relationship he longs to have with his family, particularly his son. The characterization of Bull reflects the personal and emotional struggles of men attempting to aspire to cultures' prescribed norms of hypermasculinity, which leave little room for perceived "feminine" traits such nurturing and caretaking. The story of Bull also captures the struggles of families living under the weight of such extreme patriarchal rule, which can lead to dysfunctional relationships, abuse, and violence.

The patriarchy or male dominance theory maintains that societies in which men are encouraged to be powerful and in control create and condone domestic violence. Our cultural norms support "ideal" male/masculine behavior as controlling, dominant, competitive, and aggressive, leaving out caring, nurturing, and concern for the welfare of others—traits thought of as ideally feminine. In societies that foster these gendered norms, men will continue to express their anger through violent and abusive behavior (Johnson, 2005). In *The Great Santini,* we see how this damages the relationship between a father and his son. In *Sleeping With the Enemy,* we see the dynamics of physical and emotional abuse in a marital relationship.

Sleeping With the Enemy (Joseph Rubin, 1991)

The main character in *Sleeping With the Enemy* is Laura Burney (Julia Roberts), a young wife who struggles to free herself from an abusive marriage to wealthy financier Martin (Patrick Bergin). As the movie begins, we are shown a rich, young, white couple who live in a gorgeous beachfront home. It soon becomes clear, however, that behind the image of the ideal marriage, something is terribly amiss. To outsiders it looks as though handsome, successful Martin is a perfect husband, but "behind closed doors," the darker side of their relationship is revealed. Martin is jealous, possessive, controlling, and abusive, with a hair-trigger temper. Laura's life with him is a living hell. Martin proudly parades his attractive wife around at parties for her to be admired by his friends and colleagues, but when they are alone, he abuses her unmercifully psychologically and physically.

Laura is desperate and feels trapped; like many battered women she lacks the economic independence and external social support that are necessary to leave her husband. Laura is isolated, with no family to offer support and no real friends, only acquaintances. Martin is a charming, charismatic, attractive, successful man who, to the outside world, appears to treat his wife well, so she would not likely have been believed if she tried to make the truth of her situation public.

The patriarchal nature of male violence and the gendered dimensions of domestic violence are both illustrated by this film. Martin is extremely jealous and possessive, as if he owned Laura as well as the luxurious beachfront home and all the furnishings and accessories in it that his wealth afforded. He tries to control every aspect of her life and their life together. This control is evident when he lashes out at her for supposed imperfections, such as bathroom towels that are slightly uneven in the way they hang on the towel rack. To control and aggressively criticize such seemingly insignificant household details illustrates the psychological and emotional abuse that can exist alongside physical abuse.

Martin also had economic control in their home. As a traditional housewife, Laura has no income and little to call her own, which makes her escape all the more challenging. The types of violence captured in this film represent what sociologists call "intimate terrorism," violence that is embedded in a larger pattern of power and control that permeates the relationship (Johnson, 2007). In these relationships, men use money, physical brutality, male privilege, tactics of isolation, and threat to keep their female partners under their complete control.

Like most victims of intimate terrorism, Laura is fearful and submissive in her relationship with Martin. Martin's constant bullying makes her doubt herself, and her timidity and low self-esteem are paralyzing. Like most victims of intimate terrorism, Laura plans her escape slowly and gradually, gathering the resources she needs over time. However, unlike most victims, she does so alone, without seeking help from those around her. Dramatically, Laura stages her own death to escape. Though deathly afraid of the water, she takes swimming lessons, and then fakes her death in a drowning accident and literally disappears. She moves thousands of miles away to begin a new life.

While the ending of *Sleeping With the Enemy* is dramatic, it portrays the only option that some battered women feel they have, killing in self-defense (Gillespie, 1990). However, a much more likely scenario is for women to die at the hands of an intimate partner; in the United States, approximately one-third of female murder victims are killed by an intimate partner, compared to 3% of male murder victims (U.S. Department of Justice, 2007). Nonetheless, legal cases involving battered women who killed abusive partners have made public the issue of long-term battering and the self-defensive use of lethal violence as a way to end the violence (Ogle & Jacobs, 2002).

Both *The Great Santini* and *Sleeping With the Enemy* illustrate the patriarchal nature of family violence and the damage that it causes to familial relationships. While *The Great Santini* focuses on a father/son relationship, *Sleeping With the Enemy* focuses on marital violence from the battered woman's perspective. The source of the abuse in both films is the male head-of-household who wields power over and instills fear in his family members. Both films highlight gender inequality in marriage and the family, and point to structural arrangements that contribute to family violence.

Conclusion

We learn many lessons about family life from watching films. Many of the movies produced in Hollywood are unrealistic escapist fare, capitalizing on dramatic situations to sell tickets. However, we still look to films for messages about how we should live our lives. What we learn from the movies provides us with a template for family life, which we often use to measure our own experiences. Sometimes the images and ideals we glean from film can lead to stereotypical and unrealistic expectations about family life. These expectations also can divert us from the real issues that families face in today's society, such as stresses imposed by an increasingly competitive job market, instability in terms of wages and benefits, the possibility and sometimes the reality of losing one's job due to downsizing or other factors, as well as lack of adequate health care benefits and fear of the devastation a serious illness or medical condition could bring.

However, some motion pictures challenge conventional stereotypes and tell gripping stories about realistic, albeit fictional, families. Films can be a powerful and persuasive vehicle for making us look at the world in a new and different way. Movies can inspire us to critically examine the lives we lead and understand ourselves and our relationships within families more clearly. Applying our sociological imagination to these films also enables us to develop a keener sense of awareness and understanding about the lives of others in families unlike our own.

References

Booth, A., Crouter, A., & Landale, N. (Eds.). (1997). *Immigration and the family: Research and policy on U.S. immigrants*. New York: Routledge.

Cherlin, A. J. (2008). *Public and private families*. New York: McGraw-Hill.

Coltrane, S. (1998). *Gender and families*. Thousand Oaks, CA: Pine Forge Press.

Coltrane, S. (2000). Research on household labor: Modeling and measuring the social embeddedness of routine family work. *Journal of Marriage and the Family, 62*, 1208–1233.

Coontz, S. (1992). *The way we never were: American families and the nostalgia trap*. New York: Basic Books.

Cosbey, J. (1997). Using contemporary fiction to teach family issues. *Teaching Sociology, 22*, 227–233.

Denzin, N. (1991). *Hollywood shot by shot: Alcoholism in American cinema.* New York: Walter de Gruyter.

Ebert, R. (2003, December 19). Mona Lisa Smile. *Chicago Sun-Times,* Retrieved May 2, 2009, from http://rogerebert.suntimes.com/apps/pbcs.dll/article?AID=/20031219/REVIEWS/312190304/1023

Edelman, R. (2008, March). *From Rosie the riveter to Harriet the happy homemaker: Women on screen during and after World War II.* Unpublished lecture, New York Council for the Humanities, Pember Library, Granville, NY.

Fields, J. (2003). *America's families and living arrangements: 2003.* Current Population Reports, P20-553. Washington, DC: U.S. Census Bureau.

Friedan, B. (1963). *The feminine mystique.* New York: W. W. Norton.

Gillespie, C. (1990). *Justifiable homicide: Battered women, self-defense, and the law.* Columbus: Ohio State University Press.

Hochschild, A. (2003). *The second shift.* New York: Penguin.

hooks, b. (1996). *Reel to real: Race, sex, and class at the movies.* New York: Routledge.

Johnson, M. P. (2005). Domestic violence? It's not about gender—or is it? *Journal of Marriage and Family, 67,* 1126–1130.

Johnson, M. P. (2007). Domestic violence: The intersection of gender and control. In L. L. O'Toole, J. R. Schiffman, & M. K. Edwards (Eds.), *Gender violence: Interdisciplinary perspectives* (2nd ed., pp. 257–268). New York: New York University Press.

Kaufman, M. (2001). The construction of masculinity and the triad of men's violence. In M. S. Kimmel & M. A. Messner (Eds.), *Men's lives* (pp. 4–17). Boston: Allyn & Bacon.

Kerchis, C. Z., & Young, I. M. (1995). Social movements and the politics of difference. In D. A. Harris (Ed.), *Multiculturalism from the margins: Non-dominant voices on difference and diversity* (pp. 1–28). Westport, CT: Greenwood.

Kreider, R. M. (2007). *Living arrangements of children: 2004* (Current Population Reports, P70–114). Washington, DC: U.S. Census Bureau.

Migration Policy Institute. (2005). *A new century: Immigration and the U.S.* Washington, DC: Author. Retrieved May 2, 2009, from http://www.migrationinformation.org/Profiles/display.cfm?ID=283

Ogle, R. S., & Jacobs, S. (2002). *Self-defense and battered women who kill.* Westport, CT: Praeger.

Park, R. E., Burgess, E. W., MacKenzie, R. D., & Janowitz, M. (1984). *The city: Suggestions for investigation of human behavior in the urban environment.* Chicago: University of Chicago Press. (Originally published 1925.)

Parsons, T., & Bales, R. F. (1955). *Family, socialization, and interaction process.* New York: Free Press.

Sawhill, I., & McLanahan, S. (2006). Introducing the issue. *The future of children: Opportunity in America, 16,* 3–17.

Skolnick, A. (1991). *Embattled paradise: The American family in an age of uncertainty.* New York: Basic Books.

Straus, M. A., Gelles, R. J., & Steinmetz, S. K. (1980). *Behind closed doors: Violence in the American family.* New York: Anchor Books.

U.S. Department of Justice. (2007). *Homicide trends in the U.S.* Retrieved November 18, 2008, from http://www.ojp.usdoj.gov/bjs/homicide/intimates.htm

CHAPTER 6

Global Connections

Max Cohen: Restate my assumptions: One, mathematics is the language of nature. Two, everything around us can be represented and understood through numbers. Three, if you graph the numbers of any system, patterns emerge. Therefore, there are patterns everywhere in nature. Evidence: the cycling of disease epidemics; the wax and wane of caribou populations; sun spot cycles; the rise and fall of the Nile. So, what about the stock market? The universe of numbers that represents the global economy. Millions of hands at work, billions of minds. A vast network, screaming with life. An organism. A natural organism.

Pi (1998)

Kimberly Joyce: If I couldn't be white and I also couldn't be Asian, then my third choice would be African American because I've always wanted to be a gospel singer, and also black men are more forgiving if your butt gets big. Except I'd definitely want light skin and Caucasian features like . . . Vanessa Williams or Halle Berry. And finally, you know, no offense or anything, Randa, but my very last choice would be Arab. I mean, truth be told, you're not in a very enviable position. There's a lot of resentment in this country toward the Middle East and there's a lot of stereotypes floating around which I don't think are true because in the short amount of time I've known you, you haven't tried to bomb anybody and you currently smell okay to me.

Pretty Persuasion (2005)

Max Cohen's observation that the global economy is a "vast network . . . an organism" can be found in sociological theories of globalization. As sociologist Piotr Sztompka (1994) wrote, all people living on the globe constitute a social entity such that "one may speak of a global structure of political, economic, and cultural relations extending beyond any traditional boundaries and binding separate societies into one system" (p. 86). In this chapter, the focus is on global connections, with two readings addressing different dimensions of the "complex connectivity" that is globalization (Tomlinson, 1999, p. 2).

In the first reading, "Scripting an Enemy: Portrayals of Arab Terrorists in American Film," Williams and Linnemann explore Hollywood images of Islamic terrorism before and after the 2001 terrorist attacks on the United States. Using the framework of cultural criminology, they identify the processes of socially constructing Arab Muslims as the Other in ways that have far-reaching implications for national (in)security. However, they point out that this is not a new practice. From the earliest movies featuring sheiks and charlatans to the "bombers, belly dancers, or billionaires" to the most recent presentations of hijackers and terrorists, Arabs have been stereotyped as the Other for nearly a century in Hollywood.

Using four films, two made before 9/11 and two made afterward, the authors trace the scripting of an enemy from the threat of Soviet communism in the *Rambo* films to the Islamic jihad in *The Kingdom* (2007). In *Rambo III* (1988), the mission into Afghanistan is to supply the *mujahideen* with weapons in their battle against the Soviet invading troops. Off-screen, in real life, the CIA reportedly financed an Islamic guerrilla organization with $200,000 per month plus weapons and supplies for their fight against the Soviets (Coll, 2004, p. 4). With the political tides changing, Hollywood followed suit, producing films like *True Lies* (1994) in which anti-American Arabs were the violent enemy to be conquered.

The message intensified after 9/11, though there is also recognition of the complexity of the political, economic, and military linkages between the West and the Middle East, for example in *Syrianna* (2005). However, the viewing audience is reminded of the gulf between the cultures, reinforcing a narrow view of Arabs as oil suppliers or terrorists (Said, 1997). Williams and Linnemann caution us to view films critically, especially when the subject matter is (other) cultures framed as the enemy.

In the second reading, Gonzales turns his attention to the politics of immigration in the film *Dirty Pretty Things* (2002), providing a sociological analysis at the macro and micro levels. He notes that the dual (and often conflicting) structures of the labor market and systems of immigration push immigrants to the margins, where they are rendered invisible to the larger society. The contradictions of the situation are made clear as the immigrants are sought out to fill the unmet needs of a shifting economy, and then have to fight exploitation and violation of their humanity at every turn.

Gonzales encourages application of the sociological imagination to understand immigration in the global economy, and he uses the stories of desperation told in the film to highlight the human costs. Reminding us that human agency is also involved in the process of immigrant adaptation, we see the social networks that allow characters Okwe and Senay to survive the most challenging of circumstances. At the same time, despite the bonds and loyalties established, they cannot overcome the structured inequality they face.

Together the two readings invite us to think about how our lives, opportunities, resources, and worldviews are shaped by our location in the world today. Through film we have a closeup view of the networks of globalization and the various "flows of capital, commodities, people, knowledge, information, ideas, crime, pollution, diseases" across national and geographic boundaries (Tomlinson, 1999, p. 2). Watching this movement, we can think about the institutional domains across cultures (the economy, military, family, religion) and how individuals in different parts of the world are shaped by these domains as well as give shape to them.

References

Coll, S. (2004). *Ghost wars: The secret history of the CIA, Afghanistan, and Bin Laden, from the Soviet invasion to September 10, 2001.* New York: Penguin.

Said, E. W. (1997). *Covering Islam: How the media and the experts determine how we see the rest of the world.* New York: Vintage Books.

Sztompka, P. (1994). *The sociology of social change.* Cambridge, MA: Blackwell.

Tomlinson, J. (1999). *Globalization and culture.* Chicago: University of Chicago Press.

Reading 6.1

Scripting an Enemy

Portraits of Arab Terrorists in American Film

L. Susan Williams and Travis W. Linnemann

Ahmed: When an Arab sees a woman he wants, he takes her.

The Sheik (1921)

It is the classic chicken and egg question: does media shape or reflect culture? We do not profess to resolve the question of a causal relationship between film and culture, but we do offer a glimpse into the ways that particular groups of people are represented in film in the context of history and culture. Specifically, we examine portrayals of Arabs in film before and after the 2001 terrorist attacks on the United States. We use the theoretical framework of cultural criminology to explore the ways that Otherness is socially constructed and enemies are scripted in film.

Cultural Criminology, Otherness, and Film

We incorporate social constructionist concepts to understand the silver screen as a cultural mirror in which we view ourselves, especially in terms of our national identity as Americans and our relationship to people from other nations. Much of this work is taken from cultural criminology, which provides a framework for understanding how "Otherness" is socially constructed. Cohen (2002) argued that British culture changed significantly after World War II. It is not unusual for in-group cohesiveness to be held together by a foreign enemy. After the war's conclusion, certain groups and behaviors (gang activity, punk rockers, and general rowdiness) came to be perceived as a threat to the normative contours of the country. Cohen argued that these ripples of concern, growing into widespread, collective responses, could best be understood within a sociopolitical context. Such movement creates what Cohen called *folk devils*—a term applied to an outsider(s) portrayed as blameworthy for a number of social problems.

The hot pursuit of folk devils intensified into a mass movement that Cohen referred to as *moral panic.* The creation of a new or renewed enemy reaffirms insider status by solidifying social and political boundaries, a mechanism used to quickly (if falsely) quiet cultural strain and ambiguity. Once people are gripped by moral panic, they also become highly susceptible to distorted media messages and conspiracylike threats. Pervasive campaigns associated with moral panic adopt strategies of spreading unsubstantiated news reports, gossip, and urban legends, which can translate into misdirected policies and practices.

Cultural criminology provides a critique resting between "collective behavior organized around imagery, style, and symbolic meaning, and that categorized by legal and political authorities as criminal" (Ferrell, 1995, p. 25). This perspective asserts that mediated images mobilize a network of interrelated cultural, political, and legal forces, and, most often, these are organized to promote the interests of a ruling elite. Since these interests are cloaked in cultural symbols, distributed publicly, and embedded in everyday practices, they are often unrecognizable as elite interests, and therefore remain largely unchallenged. Sometimes, intersections of culture and national events define emergence of public controversies, shaping a specific evolving discourse. One such event was the September 11, 2001, terrorist attacks, and the subsequent intensification of the global war on terror.

Sociologist David Altheide (2006) argued that the 9/11 terrorist attacks provided a catalyst for a particular propaganda campaign. In the aftermath of 9/11, critics claimed that the U.S. government fostered a culture of fear in which citizens willingly yielded long-standing rights and liberties for the sake of safety and security. The public was constantly reminded of imminent threats with the widely publicized "risk of terror" color chart, with warnings ranging from green (low) to orange (high) to red (severe risk of terrorist attacks). This strategy, among others, contributed to a culture of fear, influencing Americans to support offensive strategies in a world where danger lurks, threatening safety and even survival. Mediated images were used extensively, as Altheide documented, at the same time that other images were kept out of public view. For example, a ban on televised images of flag-draped caskets of U.S. soldiers killed in Afghanistan and Iraq muted public reactions to the human cost of the war on terrorism and helped to quash critics of the war.

Arab Images and Stereotypes in Film

Long before 9/11, Hollywood images of Arabs and Muslims were negative.[1] In fact, Muslim Arabs have been stereotyped racially and religiously as the "cultural other" from the beginning of American cinema (Ramji, 2005). Beginning in the 1920s with Rudolph Valentino's roles as "The Sheik," Arabs in film have been represented as "thieves, charlatans, murderers, and brutes" (Simon, 1996). Throughout the 20th century, Arabs in the movies were subject to what some have called the "three B syndrome," being portrayed as "bombers, belly dancers, or billionaires" (Qumsiyeh, 1998).

Over the past couple of decades, the old stereotypes have been replaced: "the sheik and lusty despot have slowly disappeared, leaving hijackers, kidnappers and

[1]As Sheehan (2008) points out, Hollywood lumps together Muslims and Arabs as "one homogenous blob," despite the fact that only one-fifth of the world's Muslims are Arabs (p. xiii).

terrorists. Muslim women have disappeared behind the chador and burka" (Ramji, 2005). In a survey of more than 900 film appearances of Arabs prior to 2001, Jack Sheehan (2001) found only a dozen positive images. In review after review, Sheehan demonstrated the power of negative stereotypes, using an Arab proverb to underscore the influence of recurring images on beliefs and attitudes: "By repetition even the donkey learns" (p. 1).

By using stereotypes to play to the public's sense of what and who is dangerous, cinematic narratives and images powerfully reinforce social norms. Set within an uneasy history of Arab cultural images, the production of anti-Arab images in film and attendant responses intensify as political events heat up. Paralleling the U.S. involvement with Middle East tensions and the Gulf War, box office and DVD sales associated with anti-Arab sentiments spiked whenever Middle East political and conflict events were publicized (Ramji, 2005). For example, the 1990 film *Navy Seals* depicted Islamic terrorists as an extreme threat to the westernized "civilized world." In another example, the film *Not Without My Daughter* (1991)—based on an American woman's attempt to escape her violent Iranian husband's cruelty—ranked highest in ticket sales the week after the Persian Gulf war began. Many argue that such films increase patriotism and, when coupled with a clear definition of "the enemy," substantially increase negative stereotyping and discrimination against identified groups (Oswald, 2005).

We wondered if the events surrounding 9/11 intensified anti-Arab sentiments in film. In a previous work (Williams & Linnemann, 2007), we found 54 blockbuster action films released from 1985 through 2007 with plots or significant dialogue pertaining to terrorism and people of Arab ethnicity. Considering a lag of about two years from the presentation of a popular theme or subject matter to the release of a film, we concluded that the number of terrorism films increased after the events of 9/11 and the subsequent "war on terror." In particular, the theme of Muslim terrorist threat reached an all-time high. We also observed a qualitative change, not only in how Arabs are portrayed in action films (as terrorists), but also in the role of American characters as victims or defenders. In this chapter, we focus on four movies, two films released prior to 9/11, *Rambo III* and *True Lies,* and two films released after 9/11, *Syriana* and *The Kingdom.*

Films Made Before 9/11

Rambo III (Peter MacDonald, 1988)

Rambo III is third in a series focused on Vietnam War veteran John James Rambo (Sylvester Stallone). Viewers initially meet Rambo in *First Blood* (1982) as an unstable homeless drifter suffering from post-traumatic stress disorder after his return from Vietnam. Passing through a sleepy town, Rambo offends a power-hungry sheriff (Brian Dennehy), who tries to prove a point to the disheveled transient. After a confrontation in which the sheriff draws "first blood," Rambo unleashes his training and wrath on the sheriff and his deputies, giving them "a war they won't believe." Rambo is eventually taken into custody by his former commanding officer, Colonel Trautman (Richard Crenna), but not before he inflicts considerable damage. The themes of the film—disillusionment about the Vietnam

War, post-traumatic stress, and homeless veterans—were a reality in 1980s America. Even today, almost half of the approximately 196,000 homeless veterans in the United States are from the Vietnam era (National Coalition for Homeless Veterans, 2008).

The second movie, *Rambo: First Blood Part II* (1985), finds Rambo in an unnamed prison camp, presumably as punishment for his actions in the first movie. Trautman has come to recruit Rambo for a mission to rescue several prisoners of war (POWs) in Vietnam, still alive years after the war ended. Rambo eventually accepts, returning to Vietnam to fight the communist Russians. Again, the movie focuses on the enduring specter of Vietnam, as well as the threat of communist Russia. It is clear that the folk devil threat in this series centers on communism.

Rambo III opens in Thailand. Col. Trautman has come to recruit Rambo for a covert mission assisting the *mujahideen* freedom fighters against the Soviets in Afghanistan.[2] Though Rambo turns down the mission, he relents when Trautman goes alone and is captured. An early scene establishes the tyranny of the Soviet invaders and nobility of the Muslim freedom fighters. In an interrogation by a Russian commander, Trautman describes the great courage and honor of the *mujahideen*:

Trautman: You expect sympathy? You started this damn war, now you'll have to deal with it.

Russian commander: And we'll deal with it. It's just a matter of time before we achieve a complete victory.

Trautman: There won't be a victory. Every day your war machine loses ground to a bunch of poorly armed, poorly equipped freedom fighters. The fact is that you underestimated your competition. If you had studied your history, you'd know that these people have never given up to anyone. They'd rather die than be slaves to an invading army. You can't defeat a people like that. We tried; we already had our Vietnam, now you're going to have yours.

Trautman's description of valiant warriors fighting for freedom stands in contrast to the Russian invaders who cannot win, despite their military advantage. U.S. involvement in Vietnam is compared to Russian involvement in Afghanistan, suggesting that both conflicts represent the struggle of free people against communist oppression. Later in the movie, Rambo meets with *mujahideen* leaders and learns of their struggle against the Soviets:

Masoud (mujahideen leader): You must not judge us before you understand why we are not ready to help. Most of the Afghan people are very strong, and we are determined not to be driven from our land. Our children die of disease, mines, and poison gas, and our women are raped and killed. Last year in the Valley of Ghaman, the next valley, 6,000 Afghans were killed. Pregnant women were cut with bayonets and their

---

[2] *Mujahideen* is a term used to describe Muslim warriors engaged in a *jihad,* or struggle for God. In the late 20th century and early 21st century, the term came to describe various armed fighters who subscribe to militant Islamic ideologies and identify themselves as *mujahideen,* although there is not always an explicit "holy" or "warrior" meaning of the word.

babies were thrown into the fires. This is done so that they will not have to fight the next generation of Afghans. Yet no one sees anything or reads anything in the papers.

The *mujahideen* leader continues, explaining to Rambo the motivation and nobility of the fight in which his people engage, a fact underscored (and never contested) in the movie:

Masoud: What you see here, are the *mujahideen* soldiers, holy warriors. That's why this war is a holy war, and there is no true death for the *mujahideen*, because we have taken our last rights, and we consider ourselves dead already. To us, death for our land and God is an honor. So, my friend, what we must do is to stop this killing of our women and children. If getting this man Trautman free so he can return to the free world and tell what happens here is necessary, then of course we will help. Leave us now, so we can speak among ourselves and find the best way to free this man.

This scene depicts the *mujahideen* as very similar to Rambo, valiant underdogs fighting to save their families and way of life from a foreign/communist threat. This theme of independent heroism, which pervades all three *Rambo* movies, reflects core American values. In the context of this movie, American freedom fighters are aligned with courageous Muslim allies who share the same ideals. Further, these heroes are strong men, muscled specimens beyond what can reasonably be attained by average men. The powerful "hard body" of Rambo symbolizes the remasculinization of American values and politics in the world at large in the post–Vietnam era (Jeffords, 1994).

Across the three *Rambo* movies we witness the political transitions of the 1980s. While anti-Communist sentiments remained stable, *Rambo III* introduced the alignment of American and Arabian forces, fighting the good fight. We know now, of course, that this alliance quickly dissolved with the dissolution of the Soviet Union. In 2002, a special edition DVD of *Rambo III* that included "Afghanistan: Land in Crisis" was released. This feature included commentary from Stallone, the film's producers, and several experts on the social and political history of the Soviet/Afghani conflict. In the feature, Stallone described the release of the movie as "unlucky timing," since it coincided with the end of the Cold War and a "new friendship" with the Soviets. He recalls being very nervous regarding the film's vilification of the Soviet Union in light of new diplomatic inroads, underscoring the power that popular film is seen as exerting on public opinion. As we will see, American moviemakers transferred the folk devil designation once reserved for Soviets onto Arabs.

True Lies (James Cameron, 1994)

The second pre-9/11 movie we examine, *True Lies*, is a comedic action movie that tells the story of Harry Tasker (Arnold Schwarzenegger), a secret American operative, as he tries to balance the mundane details of his cover life while combating international terrorists. The plot of *True Lies* focuses on Tasker's attempts to thwart the efforts of a fictional Muslim extremist organization, Crimson Jihad. The duality of Tasker's life—secret and daring on one hand; an unsatisfying job, difficult children, and marital problems on the other—is an effective device that characterizes contemporary America. Just as the typical American must carry on mundane details of life since 9/11 while dealing with the potential threat of another attack, Tasker manages his day-to-day life with the ever-present threat from Crimson Jihad.

When compared with *Rambo III,* released six years earlier, the representation of Arab Muslims in *True Lies* is dramatically different. This movie uses a variety of cinematic devices to construct the sense of a potent threat to America by Muslim extremists. The Arabs are presented as "sexist, violent brutes and religious fanatics" who are "fiercely anti-American" (Tuman, 2003, p. 126). Perhaps the most obvious symbol is the name of the terrorist organization in this film, the Crimson Jihad. Recall, *jihad* is a term meaning a war for God, or holy war, a designation that blurs the boundary between Islamic extremists and the vast majority of Muslims who live peacefully. By naming antagonists in this movie Crimson Jihad, the connection between Islam and terrorism is reinforced. Again, the plot skillfully manages fear, even through comedic action, further solidifying the us-versus-them dichotomy.

Despite the decided anti-Arab slant, the film's antagonist, Salim Abu Aziz, is allowed to reveal his reasons for attacking America:

Salim Abu Aziz (Crimson Jihad commander): You have killed our women and our children, bombed our cities from afar, like cowards, and you dare to call us terrorists! Now, the oppressed have been given a mighty sword at which to strike back at their enemies. Unless you, America, pull all military forces out of the Persian Gulf area immediately and forever, Crimson Jihad will rain fire on one major U.S. city each week until our demands are met.

Later, and prior to destroying an uninhabited island, the Crimson Jihad gathers at a warehouse to arm the nuclear warhead. In this scene, the moviemakers enlist several powerful images that depict the potency of Arab terrorism as a threat to American lives. In one scene, the Crimson Jihad leader gives a rousing speech in Arabic. The use of Arabic without subtitles is an effective method to emphasize fundamental differences between Arabs and Americans, and the incomprehensibility of Arab people, language, and culture.

During the speech scene, the Crimson Jihad are wearing fatigues and head scarves often attributed to Arab culture, which seems out of place for a group of covert terrorists. Again, this visual display effectively communicates an Other status and the inherent threat of Arab terrorism. At several points during the speech, the soldiers shoot their weapons in the air, conjuring images of anti-American sentiment that one might witness on the evening news. Finally, as the nuclear warhead is armed, it is draped in an American flag and entombed in cement, symbolically sealing the fate of America at the hands of Islamic terrorists.

As with the *Rambo* films, *True Lies* promotes the ideology of American hegemonic masculinity. In these movies, American interests and notions of freedom from tyranny are articulated by violent hypermasculinity. Rambo and Tasker accomplish their goals through brute force and the liberal use of automatic weapons. Once again, we see that action movies incorporate concerns about terrorism within broader cultural values of gender, bolstering the image of America's muscle men as heroes. This is set in contrast to the view of Arab women who are stereotyped in Hollywood films as sexualized objects (the belly dancer) and "bundles in black" (veiled women wearing burqas)—submissive and subordinate to men (Blauvelt, 2008).

Ultimately, our hero foils the plans of the Crimson Jihad, dispatching them with entertaining, if not artistic, skill, while saving his wife, daughter, and the American people. The identity of Arab has been transformed to a polarized Other, a clear enemy to Americans and freedom. Beliefs are also transformed. The audience is left with the not-so-subtle reminder that America will prevail over its (Arab-Muslim) enemies. The stage is set for a full-blown *moral panic,* provided a sufficiently serious trigger event occurs. The events of 9/11 provided just such an occurrence.

Films Made After 9/11

Syriana (Stephen Gaghan, 2005)

The revolving plot of the movie *Syriana* centers on players involved in Middle East diplomacy and the oil trade. The film details how reasons for American involvement in the region are far more complicated than often suggested by mass news media. The script intimates that long-standing U.S. presence in the Middle East is influenced by big business, a continual military agenda, and at times dark, secret alliances. The movie's key character is Bob Barnes (George Clooney), a veteran covert CIA agent sent to the region, an assignment that he hopes will be his last mission.

As the plot unfolds, Bob is betrayed by his foreign contact, an agent named Mussawi (Mark Strong). Though we are never directly told of Mussawi's affiliation, the use of violent imagery draws parallels to real-world events, including the Iraqi conflict. In a harrowing scene, Mussawi interrogates Bob, using various torture devices to obtain information from him:

Mussawi: Bob, what do you know about the torture methods used by the Chinese on the Falun Gong? Huh? Method number one. What's your guess?

As Bob resists giving up information, his interrogator continues:

Mussawi: Water dungeon? Did you guess water dungeon? Number two method? Number two, twisting arm and putting face in feces. Not interested in two? Number three. Number three is called "pulling nails from fingers." What do you think Bob? Number three sound good to you? The purpose is to get the monks or whatever to recant their beliefs. What if I had to get you to recant? That would be pretty difficult right? Because if you have no beliefs to recant, then what? Then you're fucked is what. You're going to give me the names of every person who's taken money from you.

At this point, Mussawi rips off one of Bob's fingernails and continues:

Mussawi: Oh, that is disgusting.

Bob Barnes: Come on Jimmy, you're not one of those Koran thumpers!

Mussawi, obviously agitated by Bob's rejoinder, rips off another nail and begins to beat Bob.

Mussawi: My name is Mussawi. You fucking fuck, . . . [followed by more expletives] this is a war! [more expletives] Give me the fucking names! I'm cutting his fucking head off. I'm going to cut your head off, Bob!

In this dramatic scene, we learn a number of things. First, we learn that Arabs, even those who profess to be allies, cannot be trusted. Second, Arabs are predictably crude and uncivilized. Third, America's enemies are capable and willing to use incredible violence to accomplish their goals. Finally, these actions are connected to Islamic extremists, or "Koran Thumpers." By extension, all Arabs become identified as an Other.

Toward the end of the scene, Mussawi reminds Bob that he is a casualty of war and thus expendable. Just as Mussawi starts to use a knife to behead Bob, he is interrupted. However, the intention is clear. The use of this imagery is a powerful device to help the audience connect a fictional portrayal of Arabs to real-world violence, something they might see on television news. Again, a folk devil emerges.

The story told in *Syriana* involves a complex web of fictional actors, woven together to portray a variety of interests. For the United States, what is at stake is the continued presence of the American military in the region and a lucrative oil deal. For Arabs, it is a continued market for oil and natural gas resources, as well as an enemy to oppose. While transactions between corporations and governments are the central theme, the film takes care to depict Arabs not simply as capitalists (as tends to happen with portrayals of Americans), but as radical fundamentalists who *choose* violence.

In the following scene, a cleric explains to several young boys (who later lead an attack on an American oil tanker) how the West is not only different, but antithetical to their own culture:

Cleric (while praying in a mosque): They will try to disguise the difference to make Muslims who speak about religion appear to be fanatics, or backwards people. They will tell us the dispute is over economic resources or military domination. If we believe that, we play right into their hands with only ourselves to blame. It is not possible to bridge the divide between human nature and modern life through free trade. Impossible. The divine and the worldly are but a single concept. The Koran. No separation of religion and state. The Koran. Instead of kings legislating and slaves obeying. The Koran. The pain of modern life cannot be cured by deregulation, privatization, economic reform, or lower taxes. The pain of living in the modern world will never be solved by a liberal society. Liberal societies have failed. Christian theology has failed. The west has failed.

The dialect of the scene, which is set in a mosque, is spoken entirely in Arabic and presented in subtitles, which is an effective method to illustrate fundamental and hierarchical differences between the two cultures. As the cleric asserts, everything that represents the West has failed, and the only truth that remains is Islam.

While the vilification of Arab culture is subtle, scenes such as the two examined here stand as reminders that not only are Arab and Western cultures vastly different, but they will never be reconciled. Interestingly, as the Muslim cleric purports to denigrate the West, Western ideals are upheld in the film as superior.

Syriana represents a major cinematic production with a complex plot that does not portray a clear enemy/hero dichotomy. Rather, in *Syriana* the line between Arab evil and American righteousness is blurred. However, while American complicity in Middle East violence is suggested, there remains a very real threat represented by Arab culture. Perhaps the most salient suggestion is that Arabs in the Middle East and Americans in the West are separated by far more than distance. There exists a vast cultural separation that may never be drawn near.

The Kingdom (Peter Berg, 2007)

The Kingdom refers to Saudi Arabia and portrays Middle East politics and violence following the fictional bombing of an American compound on Saudi soil. Before the movie begins, an opening timeline provides the audience with background of U.S. involvement in the region and with the Saudi Royal Family, Osama bin Laden, and Saddam Hussein. The timeline reveals events complicated by Middle East oil trade, spanning several presidential administrations.

The opening scene is perhaps the most powerful in conveying the "us" (righteous Americans) versus "them" (violent Arabs) dichotomy. The film opens on an American oil company compound in Riyadh, Saudi Arabia, during a day that resembles any sunny day in Midwest America. Americans are grilling outside, working in their yards, and even playing softball in a distinctly American community transplanted to the desert. However, stark contrasts to a bucolic American cultural scene soon become evident. The compound is heavily guarded by a Saudi police force, complete with barricades, machine gun nests, and tanks. Members of a terror cell wait for an attack to unfold from the rooftop of a distant building.

Without much dialogue, scenes flash from the softball game in Riyadh, to the nearby rooftop where snipers watch, to the United States where the movie's hero, Ronald Fleury (Jamie Foxx), attends show-and-tell at his son's school. Meanwhile, on the rooftop, a small boy draws pictures as another sits on an adult's lap, watching the softball game through binoculars. Then the attack begins. Terrorists, disguised as Saudi security, hijack a truck and drive through the compound, shooting unarmed American men, women, and children. Finally, as another disguised attacker feigns to help the Americans, ushering them his way as they run in panic, he detonates a bomb exclaiming, "There is no God but Allah, and Muhammad is his messenger." Many die, as one man on the rooftop forces his child to watch through binoculars. Here, as elsewhere in the movie, acts of violence are predicated on apparent Islamic fanaticism that is displayed in subtitles. The message clearly connects Islam to ruthless violence.

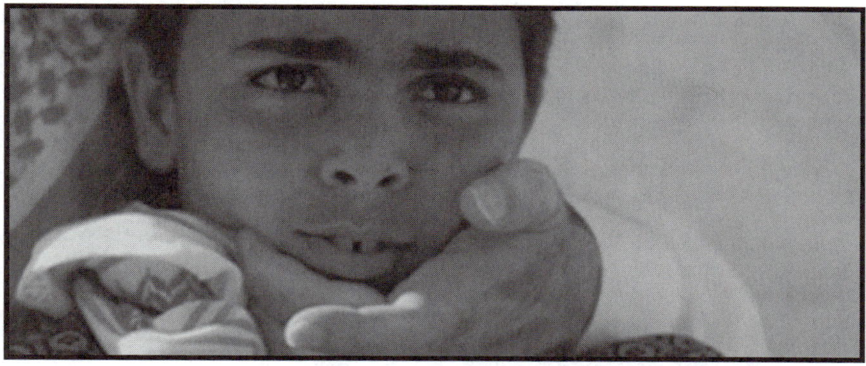

Photo 6.1 The portrayal of "us" versus "them" in *The Kingdom*. In opening scenes, a terrorist forces his child to watch the bombing of an American base in Saudi Arabia, while in America, a loving father attends show-and-tell at his son's school.

Back in America, Fleury is interrupted by news of the attack. He explains to his son, "Daddy's got to work." As Fleury's elite FBI counterterrorism unit assembles in Washington, D.C., other FBI colleagues converge on the site in Riyadh. Surveying the carnage, those at the site note the loss of innocent lives while making preliminary plans with Fleury in D.C. just as a second, even larger bomb explodes. As news of the second attack and the death toll of over 100 reaches Fleury, he briefs his team, advising them that a close colleague has died. One of the FBI agents, Janet Mayes (Jennifer Garner), begins to cry as she hears the news, but stops as Fleury whispers something in her ear. In the initial briefing, we learn more about the region's scenario:

Adam Leavitt (Jason Bateman): Sir, any chance in hell that we get to go over there and use our hands?

Ronald Fleury: You already know the answer; why ask the question?

Adam Leavitt: If there was ever a time to put boots on Saudi sand . . .

Ronald Fleury: There is no way, Adam.

FBI supervisor: The Saudi Royal Family cannot appear as they are losing control. If they lose control, they lose control of the people . . . lose control of the people and lose control of the oil, and that is not going to happen.

Another government official warns that, "American presence in Saudi Arabia is in large part the justification for the bombing to begin with." Further action in the region will only incite more violence in a country that is "one of our few remaining allies in the Middle East." To this, the team underscores the importance of striking back at the attackers, regardless of provocation:

FBI Director Robert Grace: Not to go after criminals because they might harm you is not really a policy of the FBI. You see, we try not to say uncle. We try.

Janet Mayes: Let me put it to you another way. Al-Qaeda lost the first phase of this war. They know it. So, a new zero-sum phase has begun. If you are a westerner or a moderate Arab and you won't join us, we will let loose the truly talented murderers. Men like Abu Hamza. These are operational commanders, men who plan, organize, train, brainwash, and preach extreme violence. These are the men we are fighting.

This dialogue, as with the imagery of the opening scene, frames attackers as elusive and ambiguous actors, difficult to locate among other nonviolent Muslims, while also possessing supernatural abilities to "organize, train, brainwash, and preach extreme violence." This metaphor is not foreign to the heightened sense of fear toward Arab culture following 9/11. This fear is rarely placed in context of highly trained American military, clearly superior to that of Saudis. Such a ploy contributes to a moral panic that justifies extreme offensive tactics.

Later, Fleury uses diplomatic loopholes to gain access to the crime investigation. Here the team is assigned a Saudi counterpart, police Colonel Faris Al Ghazi, an honorable man with many similarities to Fleury. As the film progresses, the bond between Fleury and Al Ghazi grows. We learn that Al Ghazi is a family man who became a police officer to protect his fellow citizens and honor his country. As this relationship grows, so does the duality of "good Arabs" (Al Ghazi) and "bad Arabs" (Hamza, terrorists). However, the noble traits displayed by Al Ghazi are not connected to Arab culture, but are norms shared by Americans and Saudis, namely heroism and love for family and country. The bad Arab stereotype remains intact.

The film culminates with a succession of scenes in which the Saudi/American team uncovers the location and identities of the Riyadh bombers, leading to a dramatic gun battle. During the fight, Al Ghazi is shot twice by a young Saudi boy. Hamza is also killed by Fleury's men, but as he lies dying, a young boy rushes to his side, and he whispers in the boy's ear. When they return home, it is obvious that they are all distraught, not necessarily over the loss of unnamed lives, but over the loss of their new colleague and friend, Al Ghazi. Leavitt asks what Fleury whispered in Mayes's ear at the initial briefing, when she first learned of their friend's death. The scene then flashes to Riyadh, to the young boy and his mother, who asks him of Hamza's whisper. To that, Fleury and the young boy respond:

Fleury: I told her we were going to kill them all.

Boy: He said, "Don't fear them my child. We are going to kill them all."

With this ending, we are again struck with similarities between heroic Americans fighting to preserve our way of life and violent terrorists seeking to destroy us, complicating the issue of good versus evil. A powerful cinematic device goes beyond simple us-versus-them as depicted in *True Lies.* Here, we see that not all Arabs are evil, but their position is culturally relative, just as that of Americans.

At the end of the day, though, *The Kingdom* carries a divisive message that continues to feed fear rather than reason, revealing the inevitability of violence grounded in cultural difference.

Discussion

The following quote, highlighted on a protest poster, offers a social critique that sits uneasily against the backdrop of fear in the post-9/11 American society:

> A person who has a sense that her life is meaningful and her life is in her hands is in fundamental ways more alive than a person who does not. In that sense, on September 11, terrorists used airplanes to kill thousands of people, and politicians and media used the event to kill a little bit of everyone who survived. (Crimethinc Worker's Collective, 2004)

In contrast to media messages and images about the Islamic threat, the occasional reader of the poster is invited to face a different reality, one that asks for a critical eye toward media and the U.S. government. This alternate story may be more uncomfortable in the short run; introspection, on an individual or national level, almost always is. Using the approach of cultural criminology invites us to turn a lens on the power structure that defines groups as outsiders and as a threat, weighing the cost of living in a culture of fear and mistrust.

We conclude by discussing three trends—how terrorism themes in film coincide with national events, the characterization of terrorist (Arab) faces and culture, and the transition of specific horror to a generalized culture of fear.

Terrorist Films and National Events

The portrayal of Arabs and Arabic culture in American films changed to reflect broader sociopolitical contexts in recent U.S. history. In the early 1980s, the image of a Russian enemy served as a convenient articulation of foreign fear—a kind of xenophobia that makes for good film as well as for reinforcement of cultural boundaries. As U.S. foreign policy shifted from involvement with the Soviet Union following the end of the Cold War, the characterization of Arabs as a threat to American interests intensified. Though Hollywood movies have included anti-Arab sentiments throughout moviemaking history, the fall of the Soviet Union, corresponding roughly with the Gulf War in 1990–91, brought a rapid escalation of the demonization of Arabs in American film.

Oversimplification of good versus evil, with a nationalist face on each, is effective propaganda and masks underlying ideologies. The more subtle the message and cinematic device, the more powerful and effective the process of creating the Other. With the splintering of the Soviet block, a threat loosely portrayed as Russian persisted in American film, and political discourse prevailed to construct a generalized sense of insecurity. The Gulf War, the attack on the USS *Cole*, and bombings of the Oklahoma federal building and U.S. embassies reinforced this ideology. An ideology of fear does not require concrete evidence of an enemy actually on our doorstep, only that they might be someday.

Following the events of 9/11, depictions of terrorism in film shifted almost exclusively to faces of Arabic origin. This is not surprising, given the tremendous impact of that day on U.S. society. However, factual accounts document that the threat of Arab terror was hardly a salient construct on the American collectivity prior to 9/11. Not only did these events change many facets of daily life, but also media images permeated America's psyche. America now has one enemy, it seems, and the face is Arab.

Faces of Terror

In the early 1980s, with the ghosts of Vietnam and involvement of the Soviet Union, Russians were the embodiment of evil portrayed on screen. This is easily observable in movies such as the *Rambo* trilogy. America was continually involved in battling communism and was seen as champion of the free world.

Interestingly, the *Rambo* movies allow us to view through a hyperbolic Hollywood lens the transition of one group from heroic ally to vile terrorists. As late as 1988, Arabs were still presented as good people fighting for freedom against Russian tyranny. Following 9/11, Russians virtually disappeared from the cinematic subconscious, despite the fact that Russia still maintains a substantial military force and the world's largest stockpile of nuclear weapons.

During this period of cinematic production, Arabs remained the noble warrior, doing exactly what Americans would expect if our homeland were under attack. Masoud, the *mujahideen* leader, proclaimed: "Most of the Afghan people are very strong, and we are determined not to be driven from our land. . . . What you see here, are the *mujahideen* soldiers, holy warriors."

Following the events of 9/11, the admirable traits of Arab freedom fighters as portrayed by American film virtually disappeared. Gone is the scourge of communism and present is a new global threat from non-Christian fundamentalists, portrayed as endangering our way of life. Such portrayals overgeneralize, leading the naïve observer to believe all followers of Islam are potential terrorists. It also distorts reality. While most Americans believe that Christianity is predominant globally, less than one-third of the world population reports their religious affiliation as Christian (ABC News, 2007). To connect the American idea of religious "minorities" to evil, films such as *Syriana* and *The Kingdom* serve to clearly connect Islamic fundamentalism to terrorism. The silver screen operates as an ideological tool for a culture of fear.

A Culture of Fear

With subsiding support for the war in Iraq (World Public Opinion, 2007), the cinematic focus continues to link Arab culture with terrorism. As a term of popular discourse, terrorism is loaded with "culturally-specific meanings" that are transcoded in film as non-Christian and Arab (Jackson, 2006). The link between politics and cinema becomes clear in the participation of the U.S. Department of Defense in the production of anti-Arab and anti-Islamic Hollywood films. *Iron Eagle* (1986), *Death Before Dishonor* (1987), *Navy Seals* (1990), *Patriot Games* (1992), *True Lies* (1994), *Executive Decision* (1996), *Rules of Engagement* (2000), and *Black Hawk Down* (2001) were all films made with the assistance of the Department of Defense (Blauvelt, 2008). In *Rules of Engagement,* written by former Secretary of the Navy James Webb, justification for killing civilians is provided in the story line when it is discovered that even children were armed and trying to kill U.S. soldiers.

We end this discussion by asking the so-what question: Why does any of this matter? What are the implications of a culture of fear perpetuated and enhanced—perhaps even created—by a celluloid image? As Barry Glassner (1999) reminded us, we are all too often afraid of the wrong things. We focus almost exclusively on a country and a people—or our idea of them—when in actuality they pose little threat to our way of life.

In such an increasingly borderless globe, group differences may become less defining. We do not downplay the significance, and the horror, of 9/11. We do invite readers to question and critique media representations of cultures, especially as they parallel political and economic realities in an increasingly global network of interdependent social worlds.

References

ABC News. (2007). *Most Americans say they're Christian: Varies greatly from world at large.* Retrieved January 25, 2008, from http://abcnews.go.com/sections/us/DailyNews/beliefnet_poll_010718.html

Altheide, D. (2006). *Terrorism and the politics of fear.* Lanham, MD: Alta Mira Press.

Blauvelt, C. (2008). Aladdin, Al-Qaeda, and Arabs in U.S. film and TV. *Jump Cut: A Review of Contemporary Media, 50*(spring). Retrieved November 30, 2008, from http://www.ejumpcut.org/currentissue/reelBadArabs/text.html

Cohen, S. (2002). *Folk devils and moral panics* (3rd ed.). New York: Routledge.

Crimethinc Worker's Collective. (2004). *Forget terrorism: The hijacking of reality.* Retrieved May 9, 2009, from http://www.crimethinc.com/texts/pastfeatures/forgetterrorism.php

Ferrell, J. (1995). Culture, crime, and cultural criminology. *Journal of Criminal Justice and Popular Culture, 3,* 25–42.

Glassner, B. (1999). *The culture of fear: Why Americans are afraid of the wrong things.* New York: Basic Books.

Jackson, R. (2006). *Religion, politics and terrorism: A critical analysis of narrative of "Islamic terrorism."* Center for International Politics Working Paper Series. Retrieved November 20, 2008, from http://www.socialsciences.manchester.ac.uk/disciplines/politics/researchgroups/cip/publications/documents/Jackson_000.pdf

Jeffords, S. (1994). *Hard bodies: Hollywood masculinity in the Reagan era.* Piscataway, NJ: Rutgers University Press.

National Coalition for Homeless Veterans. (2008). *Facts and media.* Retrieved January 24, 2008, http://www.nchv.org/background.cfm

Oswald, D. L. (2005). Understanding anti-Arab reactions post 9–11: The role of threats, social categories, and personal ideologies. *Journal of Applied Social Psychology, 35*(9), 1775–1799.

Qumsiyeh, M. B. (1998, January). 100 Years of anti-Arab and anti-Muslim stereotyping. *The Prism*. Retrieved November 30, 2008, from http://www.ibiblio.org/prism/jan98/anti_arab.html

Ramji, R. (2005). From *Navy Seals* to *The Siege*: Getting to know the Muslim terrorist, Hollywood style. *The Journal of Religion and Film, 9*(2). Retrieved April 28, 2009, from http://www.unomaha.edu/jrf/Vol9No2/RamjiIslam.htm

Sheehan, J. (2001). *Reel bad Arabs: How Hollywood vilifies a people.* Portland, OR: Interlink.

Sheehan, J. (2008). *Guilty: Hollywood's verdict on Arabs after 9/11.* Northampton, MA: Olive Branch Press.

Simon, S. J. (1996). Arabs in Hollywood: An undeserved image. *Latent Image: A Student Journal of Film Criticism* (spring). Retrieved November 30, 2008, from http://pages.emerson.edu/organizations/fas/latent_image/issues/1996–04/arabs.htm

Tuman, J. S. (2003). *Communicating terror: The rhetorical dimensions of terrorism.* Thousand Oaks, CA: Sage.

Williams, L. S., & Linnemann, T. W. (2007). *From Rambo to World Trade Center: The demonization of Arabs in American cinema.* Unpublished manuscript.

World Public Opinion. (n.d.). *Global poll: Majority wants troops out of Iraq within a year.* Retrieved December 1, 2007, from http://www.worldpublicopinion.org/

Young, J. (2007). *The vertigo of late modernity.* Thousand Oaks, CA: Sage.

Reading 6.2

Dirty Pretty Things

Migration, the State, and the Contexts of Survival in the Global City

Roberto G. Gonzales

Okwe: We are the people you don't see. We are the ones who drive your cabs; we clean your rooms . . .

Dirty Pretty Things (2002)

International migration and immigrant adaptation have long been important subfields of sociology. Among its many strands, the sociology of immigration provides sociological frameworks to understand the important questions of why people migrate; how immigrants experience life, work, and the adaptation process in the host country; and how they, in turn, shape society, influence labor markets, and affect government expenditures. This important subfield also helps us to understand the ways in which social structures facilitate the assimilation and acculturation processes for certain immigrant populations, while serving to marginalize others.

The quickening pace of globalization and the fluid movement of capital, ideas, and culture have prompted social scientists to question the salience of the nation-state and argue that national borders have weakened (Mann, 1990; Tambini, 2001). However, the complexities of contemporary international migration provide evidence to counter such claims. Particularly, during the past three or four decades, there has been an intensification of labor migration and recruitment from developing countries to developed countries. Within these rapid changes, contradictions have become evident, as immigration law has failed to keep pace with labor needs created by the new economies. As a result, immigrants find themselves caught in a complex web of interests. One important and unfortunate consequence of this mismatch between immigration policies and the needs of capital is the growth of large, vulnerable migrant populations in developed countries to serve the needs of local and national economies. Without access to important benefits and protections, however, few of them are given the means to successfully integrate into society and the economy.

Dirty Pretty Things and the Immigrant Metropolis

Through a consideration of film and video, teachers and students alike have been able to engage important topics, such as immigration, beyond the textbook, thereby providing an additional dimension from which to flesh out important and sometimes abstract concepts. *Dirty Pretty Things* (2002) offers a means of exploring the consequences of complex immigration processes through its depiction of contemporary immigrants in a global city. It provides both micro- and macrolevel perspectives of immigration, as well as a tightly woven story about immigrant survival in the face of a labor market and an immigration system that oftentimes leaves immigrants out on the margins of society and renders invisible these countless, faceless low-wage immigrant workers.

Dirty Pretty Things offers a critical lens through which we might examine and critique the various mechanisms that impact the lives of international migrants. While its setting is contemporary London, the film's narrative allows us to draw comparisons with U.S. cities such as New York, Chicago, or Los Angeles. On the macrolevel, it offers a glimpse into a city that has seen a dramatic transformation in its racial and ethnic landscape, largely due to demographic and economic shifts. As immigrants have been sought out to fill the needs of the economy, the film demonstrates the ways in which these arrangements produce vulnerability and exploitation.

As the immigrant characters in the film are looked on to provide needed cheap labor, they are often rendered marginal within society and subject to the punitive arm of the state's immigration raids and deportation processes.

On the microlevel, *Dirty Pretty Things* captures what some of the day-to-day struggles involved in immigrant life might look and feel like. Using the sociological imagination (Mills, 1959), we see the structure of the global economy, and the individual-level experiences of immigrants today. Through the film, we come to know the characters and relate to the processes of adaptation, which often involve sacrifice, loss of social status, and desperation. While some manage to hold on to their dreams and maintain their dignity and integrity, these limits are tested at every turn.

Globalization and Increasing Inequality

Dirty Pretty Things brings together macro- and microlevel analyses in a carefully crafted case study of contemporary immigration, and the not always fluid or easy incorporation processes immigrants face in what sociologist Saskia Sassen (2001) and others call global cities. According to the literature, cities such as Amsterdam, London, Frankfurt, Hong Kong, Los Angeles, New York, Paris, Sydney, Tokyo, and Zurich, have emerged as command centers for the global economy and, as such, have undergone massive and parallel changes. Within these cities is a growing stratum of jobs employing high-income professionals and low-wage jobs held by immigrants. Inherent in this arrangement is increased inequality. While the connection immigrants and low-wage service work have to the global economy dominated by finance and specialized services may not be directly obvious, Sassen argues that immigrants fulfill a number of important functions that are essential to keep the society and economy running smoothly.

In fact, these global cities incorporate large numbers of immigrants in activities that service strategic sectors of the economy. Sassen argues that these new economies require a flexible workforce to service the needs of the professional and managerial classes. These professionals require services that facilitate an intensive time commitment to work and have grown accustomed to certain amenities that are dependent on a low-wage service workforce. A corresponding and related growth in these cities' informal economies increases the level of flexibility in a broad range of activities and contributes to a growing presence of casual labor markets and a growing casualization of the employment relation. Immigrants are in a position that renders them a likely workforce, and they are absorbed into these low-wage jobs, meeting these various needs as an invisible "serving class."

Seemingly taking its cue from this literature, *Dirty Pretty Things* chooses to focus on this often underdiscussed and invisible side of the contemporary city. London, as seen through these lenses, is bifurcated between the haves and the have-nots and a city where many of its ethnic minorities and foreign-born struggle to make ends meet. Represented as a multiethnic city, where most of the immigrants toil in low-wage service and informal sectors, it is also a relentless city that many of these marginalized immigrants want to leave.

We see the city through the lives of its immigrant characters. Therefore, we see the world of low-wage and unauthorized immigrants. From this perspective, London is as much a city of low-wage workers and ethnic enclaves as it is a corporate world-class city. Dense ethnic communities, like Chinatown, serve the dual purposes of keeping poor immigrants out of public sight when they go home, while also allowing immigrants to hide some of their own secrets, as Okwe (Chiwetel Ejiofor) does when he attempts to hide Senay (Audrey Tautou) from immigration authorities. We also see cab stands, run by North African immigrants to serve the more affluent, and outdoor markets staffed by a collection of immigrants from across the globe.

The hotel, however, provides the most salient example and critique of the cleavages of inequality that are produced by globalization, as it is a site of both production and consumption, of service and leisure. While tourists and global professionals see the hotel as a place of leisure amid city touring or business, it is where many others make their living, servicing the needs of the guests. We see this through the film's main characters, as well as when scanning the faces of the hotel maids who line up for work every day. These two populations live in radically different worlds, yet come together under these arrangements. The hotel also brings together most of the film's central characters, and we see the ways in which their lives are impacted by the dual processes of migration and globalization. This dichotomy is expressed well by Sneaky (Sergi López) in his statement to Senay, "Come on. You've cleaned their shit for so long. Now you can be one of them."

Contexts of Reception

While the literature on global cities covers the ways in which globalization has manifested within cities and among immigrant populations, it still leaves many questions unanswered, namely why certain immigrants achieve some success in the integration process and others do not fare quite as well. Indeed, contemporary immigration challenges much of the conventional wisdom on immigrant assimilation, as today's immigrants are absorbed differently into the immigration system and in the labor market than were international migrants at the turn of the 20th century. Much of the contemporary scholarship on immigrant adaptation has focused on the interplay between individual level characteristics and structural considerations in examining how and why immigrant families fare differently. Sociologists Alejandro Portes and Rubén G. Rumbaut (2001) argue that modes of incorporation in the adaptation process of immigrants are shaped by various contexts of reception. Immigrants bring skills in the form of education, job experience, and language knowledge. The currency of these skills, however, is determined by the larger contexts that receive their respective groups. Portes and Rumbaut point out that among other contexts, the policies of the receiving government and the conditions of the host labor market are instrumental in structuring opportunities.

The Categorizations of the State

While global processes create the need for immigrant workers, the laws of the state shape the ways in which they are incorporated into the economy and society. As Aleinikoff (2001) contends, the state creates immigration laws that shape immigrant integration, and therefore plays a significant role in opportunities for work, rights, and social benefits. Moreover, it regulates who stands inside or outside the law and who qualifies to participate in society, as it dictates who has access to resources. Sociologist Cecilia Menjívar (2000, 2006) builds on this notion by adding that citizenship as legal status—by way of granting rights and responsibilities to the individual in the state—plays the dual role of determining immigrants' membership in society, while also conditioning their understanding of their place within that society.

The categories under which immigrants are designated within a country's particular immigration system determine the extent to which they are entitled by law to participate in society. Certain designations allow immigrants to work, vote, receive benefits, and travel outside of the country. However, not all immigrants have access to such privileges. Portes and Rumbaut (2001) outline three important government responses to immigrants—exclusion, passive acceptance, and active encouragement—and their benefits and limitations.

Immigrants who enter the country without the proper authorization encounter exclusion by the host country and end up in the underground economy. These unauthorized migrants are excluded from most forms of participation, as they cannot legally work, vote, drive in most states, receive benefits, nor travel outside of the country. These immigrants have little to no rights in the host country and, as such, are outside of the law. At any time, these immigrants can be jailed and deported. They often live and work in the shadows of society and are susceptible to employers and others who prey on their vulnerability.

On the other hand, immigrants who are granted legal status enjoy a range of entitlements in the host country. They can legally participate in the society and labor market and are eligible to receive state entitlements. Beyond these rights, however, the government does not endow special benefits to the vast majority of these immigrants. Portes and Rumbaut refer to this type of government response as a passive acceptance. However, some immigrants receive an active reception by the government. This group is made up of refugees and a small number of asylum seekers, who are provided with government assistance for resettlement. This type of government support is important because it offers these newcomers access to an array of resources that do not exist for other immigrants, including job-skill training programs, adaptation programs, access to special loans, and monetary assistance. For those who possess high human capital, governmental assistance can translate into an opportunity for rapid upward mobility.

Having an understanding of the ways in which the state shapes immigrant integration is helpful in understanding the different experiences of the immigrant characters in *Dirty Pretty Things.* The particular legal designations of the state's system mark immigrants as either insiders or outsiders. Kitty Calavita (1998, 2005) argues that immigration laws, rather than regulating immigration, control and marginalize immigrants. That is, immigrants are conditioned to recognize and accept their place in society and to be fearful of the watchful eye of the state. They understand the ways in which the laws restrict them and behave accordingly (Chavez, 1991, 1998; Coutin, 2000, 2002; Hagan, 1994). Susan Coutin (1996, 1998) concurs, asserting that these laws in effect criminalize migrant behavior. However, Coutin (2002) points out that migrants do exercise agency as they create what she calls legitimate spaces for work, political, and social life.

Dirty Pretty Things provides a useful representation of immigrants living in contemporary London, through a detailed examination of its five central characters—Okwe, an unauthorized immigrant from Lagos who works clandestinely as a cab driver and hotel porter; Senay, a Turkish asylum seeker who, out of necessity, violates the conditions of her visa by working as a maid in the hotel and accepting rent money; Ivan (Zlatko Buric), the Russian immigrant doorman at the hotel; Mr. Juan, or Sneaky, the Spanish head porter at the hotel, who meddles in the underground economy by operating an illegal organ selling business, and Guo Yi (Benedict Wong), an ethnically Chinese refugee (presumably from Southeast Asia) who works as a mortician.

These characters fit into the range of Portes and Rumbaut's taxonomy of governmental reception of immigrants: exclusion, passive acceptance, and active encouragement. Through these lenses, we can see the ways in which their respective statuses shape their experiences and their available options.

Okwe, the film's central character, is an unauthorized immigrant. As such, he is excluded from access to any supports in London. He works without the authorization of the state, collecting fares as a cab driver and receiving his wages as a hotel porter in cash payments. Although he has plenty of human capital by virtue of his medical training, his immigrant status locks him out of the formal economy and opportunities to participate in regular society. He is constantly reminded of the limitations of his status throughout the film as he lives his day-to-day life in the shadows, hidden in dense ethnic communities, and constantly on the hideout from immigration officials. Even his good friend, Guo Yi, reinforces this reality when he cautions him: "You're an illegal, Okwe. You don't have a position here. You have nothing. You are nothing." We see that Okwe internalizes this position when he lectures Senay, "For you and I, there is only survival. It is time you woke up from your stupid dream."

Similarly excluded, Senay finds herself living in the shadows and working clandestinely. Her immigrant status differs from that of Okwe as she is actually legally in London, seeking asylum. However, the conditions of her petition are that she cannot receive money in the form of wages or rent payments. Unable to survive under the existing law, Senay works illegally at the hotel. Like Okwe, she finds that she cannot make ends meet with one low-wage job, so she offers her friend and coworker, Okwe, her couch for rent. Once she violates the conditions of her status, she finds herself outside the law and propelled into the punitive arm of the state. Senay dreams of a better life in New York, but she is continually forced to confront the limitations of her status. "Always we must hide," she tells Okwe in frustration.

The remaining characters are seemingly legal immigrants of one particular status or another. While we do not know as much about Ivan (his character is not central in the story line), we can assume that he is a legal immigrant, albeit without access to higher paying jobs. He works as the doorman of the hotel. As such, he is subjected to the sometimes harsh cycles of London's weather and does what he can to earn extra money on the job. We see and hear of Ivan's schemes to hustle unknowing hotel guests and his accepting money to keep quiet about the goings on at the hotel.

Sneaky's situation is slightly more complicated. A Spanish immigrant (also presumably legal) employed as the head porter of the hotel, Sneaky supplements his income by moonlighting in the underground economy. Like many of the immigrants in the film, Sneaky has realized that he cannot make ends meet by merely working at the hotel. His position at the hotel and his legal status earn him a good position within the underground economy. Unlike Okwe who must hide, Sneaky is able to use legal status as a "middle-man minority" (Light & Bonacich, 1988) between well-established professionals such as doctors and underground counterfeiters to serve an endless supply of desperate immigrants in need of his services. Sneaky appears to also have more money and material possessions than the others. He drives a Mercedes, drinks champagne, and wears expensive clothes. When he says, "my whole business is based on happiness," he has put into simple terms the complexity of the underground economy and the different interests involved. Moreover, this admission also speaks to the ways in which Sneaky has internalized the ideals of capitalism and a desire to assimilate.

Positioned similarly as legal immigrants, Ivan and Sneaky face certain limitations, presumably due to their inability to get better jobs. As a result, both are entrepreneurial, doing what they can to hustle extra money. Sneaky, however, seems to have a better grasp of navigating both worlds in order to be successful in the informal economy.

Finally, Guo Yi, Okwe's closest friend, is a refugee of Chinese origin, perhaps from Southeast Asia. Of all the characters, Guo Yi has probably the best position within the formal economy, as a mortician. While we do not know what governmental benefits Guo Yi might receive, we know he has a good and stable job and a car. We do not see Guo Yi struggle like the other characters. He has a position within society and has a decent job to match his education and skill levels. As such, Guo Yi is also in a position to assist his friends, namely Okwe and Senay. He provides ongoing advice to Okwe and assists him in finding a place to hide Senay. He also provides Okwe and Senay a much needed ride to the airport.

The Conditions of Work

While the government context is important in determining where immigrants begin their lives in the host country and the resources available to them, their labor market experience is arguably as important in determining social and economic mobility. Unlike immigrants at the turn of the 20th century who were almost uniformly labor migrants being absorbed into expanding industrial economies, contemporary immigrants across the globe are received both as professionals with advanced degrees and as labor migrants with little formal education or skills. Portes and Rumbaut (2001) underscore the importance of such human level variables in determining where and how immigrants enter the labor market. This theorizing is consistent with the findings of segmented labor market scholars, who examine the structural and institutional constraints of work (Averitt, 1968; Massey et al., 1998; Piore, 1979).

Such scholars argue that the labor market stratifies firms and workers into primary and secondary sectors. Whereas the primary sector meets basic demands of the economy, the secondary sector meets fluctuating or seasonal demands and relies primarily on lower paid, flexible, labor intensive jobs primarily occupied by immigrants. Highly skilled immigrants in fields such as health care, engineering, and computer science begin their lives in the new country with professional jobs that pay well. However, certain groups of immigrants and racial minorities are systematically excluded from particular employment opportunities, while slotted into the lower levels of the labor force and therefore unable to gain access to better jobs that provide higher wages, job security, and opportunities to advance. These workers are said to earn lower wages than domestic workers, even when human capital is held constant.

The immigrants seen in *Dirty Pretty Things* are overly represented within the secondary sector of the labor market. With the exception of Guo Yi, they all occupy the bottom rungs of the labor force. Cab drivers, maids, and janitors are all nonwhite and speak English with accents. We do not fully know to what extent these barriers are due to low levels of education, English language fluency, country of origin, racial or ethnic group, or exclusion from the regular workforce because of immigrant status. However, these are indeed among the range of barriers that direct immigrants into secondary sector jobs. Among the film's characters, we know that Okwe is trained as a doctor and forced into the world of low-wage work because of his immigrant status. As such, he experiences a loss of status and is left to piece together low-wage jobs to make ends meet. His coworkers at the hotel and the taxi stand, and the immigrants captured by the camera's eye within the hospitals and on the streets, make up a less educated and skilled low-wage work force.

Scholarly research has demonstrated that most of these immigrant workers, without advanced levels of education, have a difficult time breaking into primary sector jobs. These immigrant characters are invisible to the vast majority of society, yet they comprise increasing numbers in secondary and informal labor markets. As such, they work for low wages and take the dirty jobs that most of the native-born shun. To survive in the city, they have to hustle, take more than one job, work illegally, and participate in the informal economy. Okwe works around the clock, with very little sleep. We do not know whether he is saving money for his family in Lagos, but we do know that his two jobs do not pay him well. This is corroborated by the evidence from other characters. Senay cannot survive in London without work, so she works illegally. She also cannot make ends meet, so she rents her couch to Okwe. Ivan does what he can to make extra money, taking advantage of every opportunity at the hotel to make an extra dollar.

Like Okwe, Guo Yi possesses a high degree of human capital. He speaks the Queen's English and has been educated in the medical field. We do not know whether he was a doctor in his home country, but we do know that he has enough formal education to qualify for a job as a mortician. Refugee status and the accompanying privileges allowed him to translate his previous education and job training into a professional job that pays enough money to afford a car and other amenities. Guo Yi's character allows us to see the effects of both government reception and labor market participation. His training and background match that of Okwe, but as a certified refugee, Guo Yi is able to reap the benefits of his education and training in the host country, whereas Okwe cannot. Similarly, Guo Yi is a legal immigrant in London and entitled to many of the benefits as other immigrants like Ivan and Sneaky. However, his education clearly allows him to bypass the lower strata of the labor market. While we do not know a lot of the specifics of the immigrants in the background of the film, Guo Yi's position in London's society and economy stands in direct contrast with those immigrants—the maids in the hotel, the janitors in the hospital, and the cab drivers—who possess low human capital, experience negative governmental reception, and experience discrimination in the labor market because of their racial or ethnic backgrounds.

Viewers do not get much insight into the domestic lives of the immigrant characters. Most of the film's activity takes place at the hotel or other places of work. However, we do see Senay's apartment, where Okwe rents the couch. This arrangement between Senay and Okwe tells us at least a couple of things. First, and especially for Okwe, home is only for sleeping (and even that is very limited). Between his two jobs, he works around the clock, with little time for a break (we even see him falling asleep at the hotel and taking herbs to help him stay awake). Second, the apartment is not big enough for both Okwe and Senay. They share one key, taking turns in the apartment. When Okwe is there, Senay is at work and vice versa. The apartment is small and quaint. While it is not run-down, it does not have many amenities and the plumbing is bad.

Human Agency and the Ethnic Community

This essay would be incomplete if I did not address the human agency that is involved in the process of immigrant adaptation, survival, and, for some, success. Despite extreme obstacles in the migration and incorporation processes, immigrants and their families manage to lead lives of dignity, to send children to college, and to improve their circumstances over time. A great many, however, are structurally locked into impoverished lives, poor and desperate communities, and jobs with very few opportunities for advancement. Even within these circumstances, opportunity is born out of struggle, as these same conditions have spawned labor movements, mass protest, and ongoing struggle for better lives. There is evidence of such agency throughout this film: Senay's pursuit of her dreams takes her over each barrier she faces; Okwe succeeds in guarding his immigrant status while preserving his morality; and Ivan finds ways to take advantage of the system and enjoy himself in the process.

The best articulation of agency in *Dirty Pretty Things* is in its depiction of the ethnic networks of immigrants and the advantage those bonds garner. In Portes and Rumbaut's modes of incorporation thesis, the ethnic community makes up the third category of importance in the adaptation process. Many scholars argue that immigrant coethnic networks cushion the impact of cultural change and provide immigrants with a wider set of resources from which to find jobs, housing, economic assistance, and support (Gibson, 1988; Waters, 1999; Zhou & Bankston, 1998). At the same time, some scholars argue that among networks of poor immigrants, there are insufficient resources to leverage advantage to individual members (Menjívar, 2000).

Dirty Pretty Things captures this tension. Certainly, the film depicts numerous examples of the ways in which low-wage and unauthorized immigrants exercise agency in assisting each other day to day: Okwe and Senay provide each other mutual support in their initial apartment sharing relationship; Okwe and Ivan help Senay when immigration officials show up at the hotel looking for her; Guo Yi offers Okwe a place to sleep when Okwe and Senay are forced to leave her apartment, as well as provides information on an apartment in Chinatown for Senay; Okwe assists the North African immigrants who sell their organs for passports; Senay teaches Okwe that it not useless to hold onto dreams; Okwe saves Senay from Sneaky; and they all band together in the end to outsmart Sneaky, beat the system, and get Okwe and Senay out of London.

Yet, the bonds between the film's characters are not enough to improve each other's working or living conditions. While Okwe and Senay manage to escape London (we might suspect that a similar fate awaits Senay in New York, and perhaps even for Okwe if he cannot practice medicine), the film does not offer an alternative to a life of low-wage work and fear for its immigrant characters who remain. Most of them do not have connections adequate enough to provide entrée into better paying jobs, for example. Okwe is powerless to help Senay find a good job when she is forced to leave the hotel, and Guo Yi does not have the ability to assist Okwe in his attempt to reestablish his status in the medical field. "All I bring you is bad luck," Okwe tells Senay, noting his limitations in helping her or himself move out of their structurally locked positions.

While it is clear that this is *the* point the film wants to make, by doing so it fails to provide any alternative for its immigrant characters. Human agency is reduced to the ability or inability to escape London or just to ease the day-to-day pain. This is reminiscent of the popular wave of urban films of the early 1990s—notably, *Boyz n the Hood* (1991) and *Menace II Society* (1993)—wherein the protagonists discover that the only path toward upward mobility is to leave the violence of the neighborhood.

Just as these films did not the offer alternatives of remaining in the community and working toward making it better, *Dirty Pretty Things* does not help viewers to envision a world where immigrants can organize collectively for civil, human, and worker rights.

Work, the State, and the Construction of Vulnerable Migrants

Shortcomings aside, *Dirty Pretty Things* gives viewers a no-holds-barred look at the implications of contemporary immigration policies and economic practices. Taken together, the harsh realities of the state and the limited opportunities of the labor market create lives of struggle and marginality for low-skilled immigrants. In fact, for many of these immigrants, the dual forces of the immigration system and the labor market work in tandem to create a low-wage labor force that is responsive to the needs of capital and vulnerable to a whole host of people—employers, immigration officials, and the dangerous world of the underground. It is perhaps here, where *Dirty Pretty Things* launches its harshest critique and where it makes its most important contribution to the immigration debate. As such, it provides a means through which to make micro and macro connections, identify some of the key mechanisms that shape the fates of the immigrant characters, and gain insight into the consequences of the ways in which contemporary cities absorb and marginalize unskilled and unauthorized immigrants, producing vulnerability and industries of opportunists who take advantage of their limited and limiting circumstances.

Like the vulnerable immigrants in Leo Chavez's seminal work *Shadowed Lives* (1998), the characters in the film live *shadowed lives.* They must be careful not to draw attention to themselves, as it could lead them to jail, to deportation, or without the means to earn a living. They live in fear, often hidden in the shadows. They take bad jobs out of desperation and do their best not to interact with government agencies or officials because doing so could arouse suspicion and jeopardize their employment or result in deportation.

Okwe and Senay constantly look over their shoulders. As a woman, Senay may be even more vulnerable than Okwe. The conditions of her asylum request seem to be out of touch with the reality of contemporary urban life and survival. She has very little choice in London, other than to try to earn money. However, doing so puts her on the wrong side of the law and out of status. In other words, by violating the conditions of her asylum request, Senay loses the legal rights to be in the country and thus becomes like Okwe, unauthorized. The seemingly tenable decision to work has dire consequences. When the immigration officials come for Senay, she is forced to go into hiding and change jobs and apartments. When the veil of security is lifted, Senay has to confront the totality of her vulnerability, as she has little choice other than to take an undesirable job at a sweatshop. When immigration officials show up at Senay's workplace, all of the workers run out. This is reminiscent of a similar scene in the film *El Norte* (1983) and suggests that all of the workers in the sweatshop are working without proper authorization and susceptible to immigration raids.

Senay's vulnerability becomes more extreme when immigration officials single her out to her employer. As a result, she is left even more vulnerable, as not only her labor is exploited but also her body. Her identification, not only as an unauthorized worker but as one who is being sought out by immigration authorities, gives the sweatshop owner a means to exploit her. This puts Senay in a dangerous situation, as her employer uses this information to blackmail her into performing sexual acts in exchange for his silence. As an immigrant out-of-status and as a woman, Senay's character illuminates the precariousness of life for unauthorized women. Without the protections of the state, many immigrant women are vulnerable to any number of employers, government officials, and men who hold power over them.

The film goes to great lengths to portray the manner in which immigrant vulnerability can manifest itself in extreme desperation. The central story line of the film underscores this point as its immigrant characters are revealed to be so desperate to gain legal status with the entitlements that come along with it, they are willing to sell their organs. This is reminiscent of the plot in *Sympathy for Mr. Vengeance* (2002), in which Ryu, the film's protagonist, turns to the black market to sell his own organs in order to pay for his sister's surgery. The horror of these desperate acts serves as a strong metaphor for these characters' utterly desperate living situations, as they willingly give up literal and physical parts of themselves in order to survive. This stark example is a chilling reminder of the dangers involved in the migration experience, as many international migrants risk their lives to cross into developed countries and have to continue to make life-preserving decisions once in the receiving country. This calls to mind images of bodies stacked up on makeshift rafts, humans locked in trunks and trailers, deaths from starvation and dehydration, and lives of indentured servitude that have come to be a part of the migration narrative.

Further, the theme of migrant vulnerability in the film speaks to the complicated relationships between sending and receiving countries. That is, the ways in which limited opportunities in sending countries and the enticement of available jobs at higher wages, prompt and push people toward migration. While we do not have the opportunity to view life in any of the characters' homelands (neither do the vast majority of citizens who come into contact with migrants), we are left to conclude that it must be dire enough that people would risk their lives and uproot themselves to travel across borders into strange lands.

Dirty Pretty Things compels us to ponder the link between the cumulative effects of migration and a subpopulation so desperate it is willing to trade their vital organs for money to live. Upon arrival, with very few options, many migrants find that while wages may be higher, the material conditions of immigrant life, coupled with the fear and anxiety produced by systems of surveillance and enforcement of the state, produce poor, desperate, and scared migrant workers. The characters live in the shadows of the receiving society, under poor working and living conditions, risking violence and imprisonment.

This vulnerability and extreme desperation make these immigrants susceptible to the unscrupulous acts of others. This reality illuminates the world in which contemporary immigrants live and work. Fearful of deportation and jail, immigrants become systematic prey to employers looking for cheap and pliant labor, they become less than their original selves, and they sacrifice their own values and even their own bodies to survive. Given these contexts, sacrifice is not only conceivable but a necessary requirement for survival.

Okwe and Senay, too, come to understand the contours of this world. Throughout the film we see Okwe doing his best to keep a low profile and stay out of the sight of authorities. But when he discovers a human heart in the hotel toilet, he finds himself caught in a dilemma between his need for survival and his sense of morality. In his confrontation with Sneaky, Okwe is quickly forced to make a decision. As Sneaky proceeds to call the police, so that Okwe can report what he has found, Okwe must assess his choices and his own vulnerability. He is left to hang up the phone, choosing to protect himself over doing what he believes is right. Sneaky's warning serves as an important reminder: "If you're so concerned, go to the police. Get yourself deported."

However, through his contact with desperate immigrant organ donors and the frustration of his own limitations, Okwe begins to come to terms with the complexities of such decisions. This growing awareness is brought to the fore when Okwe realizes that the one he loves, Senay, has given up her virginity and promised to sell an organ for a passport in order to escape. Okwe's dilemma between morality and survival comes together, as do the pieces of the film's complicated puzzle, as neither Okwe nor the viewer can overlook Senay's loss of innocence (and the right to control her own body) as a direct consequence of her quest for survival in a cruel, cruel city.

Photo 6.2 Migrant vulnerability in *Dirty Pretty Things.* Okwe (Chiwetel Ejiofor) uses his medical skills to save a fellow immigrant after a botched surgery to harvest organs.

Conclusion

For sociologists and moviegoers alike, *Dirty Pretty Things* offers a glimpse into the world in which low-wage immigrants toil, struggle, and pursue their dreams. As a case study, it provides a snapshot of the contemporary global city. Behind large finance centers and corporations, beyond the glitz of the city, we find the foundational infrastructure, the women and men who keep the global economy running. Through this medium, the filmmakers can delve deeply into the questions of how the macro is manifested in the micro; that is, how do complex systems of global labor impact the lives of the most vulnerable, the low-wage and unprotected migrants.

Engaging with film alongside academic scholarship enhances our understanding of the theoretical while revealing to us the ways in which we might recognize sociology in mediums such as popular film. When used appropriately and creatively, film can animate important conceptual frameworks and provide a human face and engaging story line. In turn, our all-too-often causal viewing of films can be fortified by the development of critical lenses and, most important, a sociological imagination.

References

Aleinikoff, T. A. (2001). Policing boundaries: Migration, citizenship, and the state. In G. Gerstle & J. Mollenkopf (Eds.), *E pluribus unum? Contemporary and historical perspectives on immigrant political incorporation* (pp. 267–291). New York: Russell Sage Foundation.

Averitt, R. T. (1968). *The dual economy: The dynamics of American industry structure.* New York: Norton.

Calavita, K. (1998). Immigration, law, and marginalization in a global economy: Notes from Spain. *Law and Society Review, 32*(3), 529–566.

Calavita, K. (2005). *Immigrants at the margins: Law, race, and exclusion in southern Europe.* New York: Cambridge University Press.

Chavez, L. R. (1991). Outside the imagined community: Undocumented settlers and experiences of incorporation. *American Ethnologist, 18*(2), 257–278.

Chavez, L. R. (1998). *Shadowed lives: Undocumented immigrants in American society.* Fort Worth, TX: Harcourt Brace College Publishers.

Coutin, S. B. (1996). Differences within accounts of U.S. immigration law. *PoLAR: Political and Legal Anthropology Review, 19*(1), 11–19.

Coutin, S. B. (1998). From refugees to immigrants: The legalization strategies of Salvadoran immigrants and activists. *International Migration Review, 32*(4), 901–925.

Coutin, S. B. (2000). *Legalizing moves: Salvadoran immigrants' struggle for U.S. residency.* Ann Arbor: University of Michigan Press.

Coutin, S. B. (2002). Questionable transactions as grounds for legalization: Immigration, illegality, and law. *Crime Law and Social Change, 37*(1), 19–36.

Gibson, M. (1988). *Accommodation without assimilation: Punjabi Sikh immigrants in an American high school.* The Anthropology of Contemporary Issues Series. Ithaca, NY: Cornell University Press.

Hagan, J. M. (1994). *Deciding to be legal: A Maya community in Houston.* Philadelphia: Temple University Press.

Light, I., & Bonacich, E. (1988). *Immigrant entrepreneurs.* Berkeley: University of California Press.

Mann, M. (1990). *The rise and decline of the nation state.* Oxford, UK: Blackwell.

Massey, D. S., Arango, A., Hugo, G., Kouaouci, A., Pellegrino, A., & Taylor, J. E. (1998). *Worlds in motion: International migration at the end of the millennium.* Oxford: Oxford University Press.

Menjívar, C. (2000). *Fragmented ties: Salvadoran immigrant networks in America.* Berkeley: University of California Press.

Menjívar, C. (2006). Liminal legality: Salvadoran and Guatemalan immigrants' lives in the United States. *American Journal of Sociology, 111*(4), 999–1037.

Mills, C. W. (1959). *The sociological imagination.* London: Oxford University Press.

Piore, M. (1979). *Birds of passage: Migrant labor in industrial societies.* New York: Cambridge University Press.

Portes, A., & Rumbaut, R. G. (2001). *Legacies: The story of the immigrant second generation.* Berkeley and New York: University of California Press and Russell Sage Foundation.

Sassen, S. (2001). *The global city: New York, London, Tokyo* (updated 2nd ed.). Princeton, NJ: Princeton University Press.

Tambini, D. (2001). Post-national citizenship. *Ethnic and Racial Studies, 24*(2), 195–217.

Waters, M. C. (1999). *Black identities: West Indian immigrant dreams and American realities.* New York: Russell Sage Press.

Zhou, M., & Bankston, III, C. L. (1998). *Growing up American: How Vietnamese children adapt to life in the United States.* New York: Russell Sage Foundation.

CHAPTER 7

Social Change and the Environment

Donna: Have you ever read about the life of Mahatma Gandhi, Ronnie? Or Henry David Thoreau on *Civil Disobedience*? That essay changed my life—it taught me that people have a right to stand up and speak out when an injustice is being done—that it's their obligation, their duty as human beings. Gandhi believed that one person with the truth was a majority—could win. Even women are fighting for their rights, Ronnie!

Born on the Fourth of July (1989)

Neville (talking to Anna about Bob Marley): He had this idea. It was kind of a virologist idea. He believed that you could cure racism and hate . . . literally cure it, by injecting music and love into people's lives. When he was scheduled to perform at a peace rally, a gunman came to his house and shot him down. Two days later he walked out on that stage and sang. When they asked him why, he said, "The people, who were trying to make this world worse . . . are not taking a day off. How can I? Light up the darkness."

I Am Legend (2007)

In the film *I Am Legend* (2007), it is 2012 and scientists have genetically reengineered a virus that is the cure for cancer. Three years later, the virus has mutated and seemingly killed most of humanity and turned the survivors into monster-like creatures who can only survive in the dark. An earlier film version of the story, *The Omega Man* (1971), is set in 1977 two years after a virus caused by biological warfare between the People's Republic of China and the Soviet Union has killed the majority of earth's population, leaving a few mutant survivors. Both films project into the future, exploring how science run amok can destroy humankind and make the planet unsafe for living creatures. Robert Neville (Will Smith), explaining why he keeps searching against all odds for a cure, brings together key themes addressed in this chapter: war, violence, working for justice and peace, science and the environment, and the possibility of creating a better world in the future.

In the first reading, Kathryn Feltey explores nonviolence in film as a strategy for challenging systems of domination, injustice, and inequality, and as a way of life. Violence is endemic in film and society; it is seen as the way to gain and maintain power. Internationally, an orientation to war is the organizing principle of relations, such that a significant amount of national resources are allocated to the production and distribution of arms and the training and maintenance of military forces. Further, as a society we invest our energy into expecting and preparing for the worst; what we imagine as possible in the future guides our decisions and actions in the present. Accordingly, "People who cannot imagine peace will not know how to work for it" (Boulding, 1998).

Feltey asks us to use nonviolence as presented in film as a starting point for imagining a world in which *Satyagraha* and *ahimsa* are the organizing principles for nations and communities. Using *Gandhi* (1982) and *The Long Walk Home* (1990), nonviolence as a model for social change is examined in the movement for home rule in India and the Montgomery bus boycott in the movement against racial apartheid in the United States. The methods used are designed to challenge and change the social order, and to shift the paradigm of power from might to right. As Donna (Kyra Sedgwick) tells Ron (Tom Cruise) in *Born on the Fourth of July* (1989), Gandhi believed that "one person with the truth . . . could win."

The film *Witness* (1985) provides a contrast between a contemporary crime-ridden urban Philadelphia and the bucolic peaceful setting of an Amish community not far away in miles, but a century away in lifestyle. Detective John Book (Harrison Ford), hiding out on an Amish farm, tells his partner, "Where I'm at is maybe 1890. . . . Make that 1790." Not only is the standard of living "primitive" by modern standards without the conveniences of technology, but the community is completely devoid of diversity across all dimensions. In sociological terms, this is a folk society, a *gemeinschaft*, where kinship ties, shared values, and a simple division of labor are central to the social order. As Durkheim wrote in the 19th century, there is a high degree of cohesion and integration in a society based on mechanical solidarity. In Amish communities, the social order is reinforced by the Ordnung; nonviolence is central to this way of life.

In the second reading, Christopher Podeschi uses science fiction films as "future myths," stories about the possible future direction of society. These myths are based on visions of society and its relationship with nature. Podeschi explores culture as a "site of struggle" where resistance to the exploitative relationship between society and nature is manifest in a call for the valuing of nature through sustainable economies and technology, as opposed to reproductive discourses where nature is a resource to be exploited for human need through technology.

Examining the most popular science fiction films of the second half of the 20th century, Podeschi finds that most are based on reproductive discourses that present a "technologically saturated" future where environmental consequences are ignored. Exceptions to this approach are found in films that give warning about the dangers of nuclear warfare, such as *Planet of the Apes* (1968) and *Testament* (1983), and technology that involves some sort of reproduction of humanity, as in *The Empire Strikes Back* (1980) and *The Matrix* (1988). In keeping with the reproductive discourse, nature and animals are portrayed as dangerous, needing to be conquered or tamed, and of value only in so far as they can be colonized to serve humankind.

Together these two readings challenge us to think about social change and the future of society. What does society look like when people are asked to imagine a future they are not afraid to enter? Research with groups of people in countries around the globe reveals that their imagined futures share some common elements: communities are more rural than urban; lifelong education is valued; there is no spectator-leisure industry; nation-states become less significant; peacekeeping brigades replace military armies; and living in harmony with the environment is a priority (Bakker, 1993). In a similar vein, sociologists in future studies have identified a set of shared global values that promote the future health of all societies: individual responsibility; treating others as we wish them to treat us; respect for life; economic and social justice; nature-friendly ways of life; honesty; moderation; freedom (expressed in ways that do not harm others); and tolerance for diversity (Bell, 2004). Ultimately, our survival on the planet depends on the answer to a question posed by climatologist Jack Hall (Dennis Quaid) in *The Day After Tomorrow* (2004): "Will we be able to learn from our mistakes?"

References

Bakker, J. I. (1993). *Toward a just civilization: A Gandhian perspective on human rights and development.* Toronto: Canadian Scholars Press.

Bell, W. (2004). Humanity's common values: Seeking a positive future. *The Futurist* (September-October), 30–36.

Boulding, E. (1998). Peace culture: The problem of managing human difference. *Cross Currents, 48*(4). Retrieved May 8, 2009, from http://www.crosscurrents.org/boulding.htm

READING 7.1

THE ONLY POSSIBLE SOLUTION?

The Challenge of Nonviolence to the Hegemony of Violence in Film

Kathryn Feltey

Violence in film has received extensive attention from academics, journalists, political pundits, and the general population for some time now. As a viewing audience, we have grown accustomed to increasingly violent images in the media, with widespread acceptance of violence as the "normal" response and remedy to a wide range of threats, conditions, and problems. Not only do audiences accept the violence that is presented, but there has been an increase in popular demand for more graphic depictions of violence (Slocum, 2000). With the advantages of technology and big budgets, filmmakers have complied, giving audiences more "explicit, even exaggerated, portrayals of aggression and its grisly, splattering aftermath in shootings, slashings, explosions, and crashes" (Iadicola & Shupe, 2003, p. 55). The sensibility seems to be, as an Interpol agent exclaims in the film *Nighthawks* (1981), "To combat violence, you need greater violence. To defeat a violent people you need to be trained to react in a given situation, with ruthless, cold-blooded violence as well."

An interesting debate in the literature on violence in film focuses on the role of American cinema as either challenging or supporting dominant values and norms, while many argue that Hollywood serves as an agent of both social control and change (Slocum, 2000). Violence in this debate is interrogated in terms of how it is presented, who is using it, and to what end. Whether we see violence as legitimate and justified depends on the answers to these questions. One way that violence in film increases solidarity, and therefore contributes to the social order, is by identifying a common enemy for the viewing audience. The face of the enemy has changed throughout the history of cinema, influenced by the historical and cultural context (see Williams and Linnemann, Chapter 6). For example, films from the World War II era featured German and Japanese enemies, while Vietnam and post-Vietnam films such as the *Rambo* films focused on defeat of the Viet Cong and the communist threat. Violence in film has also served to highlight competing forces for social change, in films such as *Matewan* (1987), in which West Virginia coal miners, struggling to form a union, battle against the company and armed state agents.

A central concern about violence in film is whether the violence is contextualized for the viewing audience or presented gratuitously. Films that use violence in a ritualistic display of stereotypical masculinity (e.g., *Die Hard: With a Vengeance,* 1995) or for its shock value, as in hyperreal violent films (e.g., *Sin City,* 2005), ignore the causes and effects of violence on people and the communities in which they live. Symbolic violence, on the other hand, couples emotions and images of violence, requiring the viewing audience to grapple with the meaning of violence in the social contexts where it occurs. Films using symbolic violence include historical explorations, what Ed Guerrero (Chapter 3) calls "historical agonies," such as *Schindler's List* (1993), *Rosewood* (1997), and *Hotel Rwanda* (2004).

Despite the interrogation of film violence and concerns about decontextualized gratitutous violence, the use of violence or the portrayal of human struggles and conflict as requiring violent methods is, for the most part, left unchallenged. Even when the story is one in which social change is the goal, violence is not seen as (part of) the problem, only the misuse of violence by unhinged individuals or unjust rulers or regimes. In the right hands, violence in film becomes a celebration of righteous victory, the triumph of good over evil, and justice served. This is also true at the level of interpersonal conflict. For example, in films about domestic violence, the victimized wife is most victorious when she can use violence more effectively than her abusive husband, as in *Sleeping With the Enemy* (1991) (see Cosbey, Chapter 5) and *Enough* (2002) (see Sutherland, Chapter 4).

Widespread acceptance of violence as the only way to protect an individual or a nation from external threat is legitimized across social institutions, including the media (Iadicola & Shupe, 2003). The ubiquitous presence of violence in film promotes a worldview of violence as not only necessary, but as the only effective method for resolving conflict, settling differences, protecting interests and resources, and gaining and maintaining power. As Lieutenant Jean Rasczak explains in the film *Starship Troopers* (1997), "Naked force has settled more issues in history than any other factor. The contrary opinion 'violence never solves anything' is wishful thinking at its worst. People who forget that always pay. . . . They pay with their lives and their freedom."

What can the "wishful thinking" of nonviolence, as presented in film, teach us about alternative methods for challenging injustice and creating social change? Interestingly, the few films that fit this definition are documentaries and feature films based on historical events (true stories) involving the use of nonviolence. While countless fictional and fantasy films include aggression, violence, force, and warfare, very few use nonviolence as a central theme, either as a way of creating social change or as a way of life.

The paucity of alternatives to violence in film can be understood by thinking about film text as part of cultural hegemonic processes (Cooper, 1999). Hollywood's contribution to these processes can be seen in the ways that movies have promoted dominant American myths and values, such as individualism, power as dominance over, racial superiority, heteronormativity, and material success. Violence to protect and promote the American way of life is an integral part of these processes. Ultimately, hegemonic meanings and values are experienced as practices that define reality, and reinforce that definition so completely, that it becomes difficult for many to resist, much less challenge or change what is defined as real (Williams, 2001).

In this reading, we will examine films that challenge the hegemony of violence by making nonviolence, as a method of resistance, a source of political power, and/or way of life, central to the story. For each film we will consider the following: (1) how nonviolence is presented—that is, what is the story of nonviolence in the context of the film; (2) whether the nonviolence story challenges or supports dominant values and norms in society; and (3) how alternatives to violence, as portrayed in film, might provide us with models for creating meaningful social change.

The Films

The first film we will consider is a biopic of the life of the one person who is most associated with nonviolence historically, Mohandas Karamchand Gandhi (Mahatma Gandhi). Making *Gandhi* (1982) was a 20-year dream of the film's director and producer, Richard Attenborough. Funding for the film was a problem; the project was rejected by every studio in the film industry. Finally, it was made with private money and only purchased for distribution after it was completed (Briley, 1996). Nonviolence changing the political and economic structure of a country across the ocean was apparently not seen as a potential moneymaker in the United States. Of course, the critical and popular acclaim of the film was tremendous. *Gandhi* was widely distributed and won a total of eight Academy Awards, including Best Picture, Actor, Director, Art Direction, Cinematography, Costume Design, Editing, and Original Screenplay. It was also nominated for Best Makeup, Original Score, and Sound.

The second film, *The Long Walk Home* (1990), is a fictional story set in Montgomery, Alabama, in 1955 during the 381-day bus boycott inspired by Martin Luther King, Jr., and his teachings on nonviolent protest. The film centers on two women, Odessa Cotter (Whoopi Goldberg) and Miriam Thompson (Sissy Spacek). Odessa has been working as a domestic in the Thompson household for nine years. Miriam is living the white, middle-class, mid-century life of a housewife and mother. The relationship between Miriam and Odessa, structured by class, race, and the regional politics of the U.S. South, provides insight into the political impact of the boycott within and between their respective communities.

The last film, *Witness* (1985), differs from the first two in that it is a work of fiction, without reference to particular historical events, although it takes place in the late 20th century in Pennsylvania. The film is a thriller/drama that explores two cultures, the Amish and English (non-Amish). Nonviolence as a value and way of life in the Amish community stands in stark contrast to the violent world of urban America. Sociologically, we gain insight into these cultures through the eyes of the outsider who visits. Rachel Lapp (Kelly McGillis) and her son Samuel (Lukas Haas) are immediately exposed to violent crime when they arrive in Philadelphia and Samuel witnesses a murder, while John Book (Harrison Ford) experiences deep culture shock when, after being shot and seriously wounded, he hides out with the Amish for a time to escape the men who are pursuing him.

In all three films, nonviolence is central to the story. However, it is important to note that it is not the absence of violence that qualifies these films as nonviolent. Rather it is the centrality of nonviolence not just as a method for changing the outcome of a particular situation, but as a complete, holistic way of life that structures the ways that people create and sustain their social worlds.

The Story of Nonviolence in Film

The story of nonviolence, as told through the selected films, is linked to and rooted in spiritual and religious beliefs. In *The Long Walk Home* and *Witness,* Christian beliefs are the foundation for choosing nonviolence and creating a nonviolent way of life. In *The Long Walk Home,* Odessa and her family find solace and comfort in their Christian beliefs and church community. Once the bus boycott has begun, Odessa joins in by refusing to take the bus to work, walking the nine miles to the Thompson home.[1] We see Odessa arrive home exhausted from long days at work and the long walk home, struggle to put her church shoes on swollen and bleeding feet, and attend services where Martin Luther King, Jr., gives stirring sermons of inspiration and hope to the weary boycotters. We never see King, but hear his stirring words of nonviolence, Christian faith, and fighting for justice at the Holt Street church:

> We are here, we are here this evening because we're tired now. And I want to say, that we are not here advocating violence. We have never done that. I want it to be known throughout Montgomery and throughout this nation that we are Christian people. . . . The only weapon that we have in our hands this evening is the weapon of protest. . . . And we are not wrong, we are not wrong in what we are doing. If we are wrong, the Supreme Court of this nation is wrong. If we are wrong, the Constitution of the United States is wrong. If we are wrong, God Almighty is wrong. If we are wrong, Jesus of Nazareth was merely a utopian dreamer that never came down to earth. If we are wrong, justice is a lie: love has no meaning. And we are determined here in Montgomery to work and fight until justice runs down like water and righteousness like a mighty stream.

In *Witness,* the Amish practice nonresistance, based on the belief that Christians should "turn the other cheek" when confronted by an enemy, and only return good to those who do them harm. At one point, Eli Lapp (Jan Rubes) tries to explain to his grandson, Samuel, why guns and violence are wrong, according to Amish beliefs: "This gun of the hand is for the taking of human life. We believe it is wrong to take a life. That is only for God. Many times wars have come and people have said to us: you must fight, you must kill, it is the only way to preserve the good. But Samuel, there's never only one way. Remember that." Confused by what he has seen in Philadelphia, Samuel tells his grandfather he would only kill the "bad men" who can be identified by what they do: "I can see what they do. I have seen it." Here is where we learn from Eli about the separateness of the Amish community from the larger society, as he tells his grandson that "having seen, you become one of them. . . . What you take into your hands, you take into your heart. Wherefore, come out from among them, and be ye separate, saith the Lord, and touch not the unclean thing."

---

[1]While much of the historical focus on the boycott has been on the leadership, the success of the movement has been attributed to "the nameless cooks and maids who walked endless miles for a year to bring about the breach in the walls of segregation" (Burks, 1990, p. 82).

Gandhi's approach to nonviolence is shaped by Hinduism, but he regards all religions as contributing to a message of unity and nonviolence. As he tells the group gathered with him during his fast to bring an end to the rioting between Hindus and Muslims, "Each night before I sleep, I read a few words from the Gita and the Koran and the Bible." He goes on to "share these thoughts of God" with his companions: "I will begin with the Bible where the words of the Lord are, "Love thy neighbor as thyself" . . . and then our beloved Gita, which says, "The world is a garment worn by God, thy neighbor is in truth thyself" . . . and finally the Holy Koran, "We shall remove all hatred from our hearts and recline on couches face to face, a band of brothers."

Based on these beliefs about the oneness of humanity, he developed the concept of *Satyagraha,* which he also called truth-force or soul-force, based in part on the principle of *ahimsa* (doing no harm in thought, word, or deed to any living being). While Gandhi recognized that the existence of a society necessitates *himsa* (the destruction of life) in small and subtle ways, he promoted the goal of minimizing harm at the individual and collective level (Lucien, 1984). This includes the way that people choose to live, their use of available resources, and the dependence of their lifestyle on the exploitation of others. We see in the film, for example, a stark contrast between the simple life that Gandhi lives in community with others and the Indians who are in government who live like the British, with all of the status symbols of material success. In one scene, when Gandhi first returns to India, a reception in his honor is held at the home of a successful Indian lawyer, during which we see:

> A splendid peacock, its tail fanned in brilliant display, lords it on a velvet lawn. A woman in a sumptuous silk sari is trying to feed it crumbs. Behind her, Gandhi's reception is in full spate—silver trays, tables covered in fine linen, Indian servants, a swimming pool, a small fountain, the grounds filled with Indian millionaires and dignitaries gathered with their wives to meet the new hero from South Africa. (Briley, 1982)

We soon discover that Gandhi's vision of India under home rule is the India of "seven hundred thousand 'villages' not a few hundred lawyers in Delhi and Bombay." He explains that the leadership of the home rule movement will never be successful without this understanding: "Until we stand in the fields with the millions who toil each day under the hot sun, we will not represent India—nor will we ever be able to challenge the British as one nation."

Unity is central to Gandhi's nonviolence—and this extends across national boundaries, as well as social divisions within a society by caste, religion, and gender. All of us, according to Gandhi, are human beings, no matter which side we are on, as he explains to photographer Margaret Bourke-White (Candice Bergen) when she visits him in prison, "Every enemy is a human being—even the worst of them. And he believes he is right and you are a beast. And if you beat him over the head you will only convince him. But you suffer, to show him that he is wrong, your sacrifice creates an atmosphere of understanding—if not with him, then in the hearts of the rest of the community on whom he depends."

The idea that passive nonresistance is a powerful response to violent conflict is a consistent message throughout the movie *Gandhi.* Early in his career, when Gandhi was a lawyer in South Africa, he was confronted with institutional and individual racism against Indians. Speaking with his Indian compatriots about the legal discrimination they experience, he urges them to join the nonviolent battle ahead, saying:

> I am asking you to fight! To fight against their anger, not to provoke it. We will not strike a blow, but we will receive them. And through our pain we will make them see their injustice, and it will hurt—as all fighting hurts. But we cannot lose. We cannot. They may torture my body, break my bones, even kill me. Then, they will have my dead body—not my obedience.

He stresses that fighting to expose injustice means being "willing to die a soldier's death" while never striking a blow.

The Challenge of Nonviolence in Film

Nonviolence in *Gandhi* and *The Long Walk Home* is the method and philosophy of the larger social movements occurring in India and the United States. In *Gandhi,* we see the struggle for Indian independence from British rule. Promoting the idea that the British can only rule if India allows herself to be ruled, Gandhi encourages Indians, Muslim and Hindu alike, to resist British rule through noncooperation. He stresses that reclaiming India does not require defining the British as an enemy to be conquered. Instead, Gandhi focuses on the humanity of all parties involved, saying, "I want to change their minds—not kill them for weaknesses we all possess."

Changing minds is not just a matter of debate, but is achieved through action. In the film, we see campaigns organized to communicate, symbolically, politically, and economically, that Indians will not participate in systems that oppress and exploit them. Gandhi's approach was systematic: first he gave careful consideration to the problem at hand, then he would begin communicating with the person(s) involved. Next he made the communication public, and last he informed the person(s) that he would resort to public campaigns to force the issue. According to Gandhi's grandson, he always did this with great politeness and with no intentions of inconveniencing the opposition (Meyer, 2005).

For example, in the film, Gandhi meets with British authorities after the Amritsar massacre of 1919, in which General Reginald Dyer (Edward Fox) ordered his troops to fire on a meeting of unarmed Indian civilians—men, women, and children—at Jallianwalla Bagh, an enclosed courtyard, resulting in 400 deaths and 1,500 injured. In the shocked aftermath of this brutal attack, we see a meeting between the British (represented by the viceroy, two generals, a naval officer, two senior civil servants, and a senior police officer) and Indian leadership. The scene opens with the viceroy stating, "You must understand, gentlemen, that His Majesty's government—and the British people—repudiate both the massacre and the philosophy that prompted it."

Gandhi's response links the actions and philosophy that the British "repudiate" to the business-as-usual domination of one country by another: "We think it is time you recognized that you are masters in someone else's home. Despite the best intentions of the best of you, you must, in the nature of things, humiliate us to control us. General Dyer is but an extreme example of the principle. It is time you left." A flummoxed officer rhetorically asks, "You don't think we're just going to walk out of India?"

Gandhi, demonstrating the last course of action, announces that Indians will refuse to cooperate with British rule. He politely responds to the officer, "Yes. In the end, you will walk out, because 100,000 Englishmen simply cannot control 350 million Indians if those Indians refuse to cooperate. And that is what we intend to achieve: peaceful, nonviolent, noncooperation—'til you, yourselves, see the wisdom of leaving, Your Excellency."

Nonviolence as the organizing principle was a challenge to the British colonial authority; it was difficult to justify wholesale violence against a population that refused to use violence defensively, much less offensively. To try to contain the dissent and undermine the solidarity occurring among the Indian population, the British used military and police force, the legal system, and manipulation of Muslim-Hindu conflicts. In one scene, Gandhi is tried for sedition for advocating the overthrow of the British government. When asked by the Advocate General at the trial whether he denied the charges, Gandhi responds, "Not at all." Turning to the judge, he continues, "And I will save the Court's time, M'Lord, by stating under oath that to this day I believe nonco-operation with evil is a duty. And that British rule of India is evil. . . . And if you truly believe in the system of law you administer in my country, you must inflict on me the severest penalty possible." Gandhi's challenge is not limited to the judge in this one situation, but to the system of law established by the British to protect their (unjust) rule of India. By pushing the British to treat him as a criminal for his advocacy of home rule and sentence him to prison, the injustice of their domination and control over India is further revealed, to the British themselves and to the rest of the world.

The strategy of civil disobedience that Gandhi employed was central in the civil rights movement under the leadership of Dr. Martin Luther King, Jr. King was influenced by the teachings of Gandhi, as was Rosa Parks' whose arrest for refusing to give up her seat to a white bus rider sparked the Montgomery bus boycott. In the film *The Long Walk Home,* we only hear about Rosa Parks, arrest through discussion in the black community about staying off the buses. Historically what we know is that Rosa Parks, an activist with the NAACP, had attended a workshop at the Highlander Folk School where she learned about Gandhi's nonviolent methods of protest just a few months before her arrest (Fisk, 2000).

In *The Long Walk Home,* we enter into the Jim Crow South of the mid 20th century, where racial apartheid is reproduced through legally reinforced social customs and practices. In one scene, Miriam drops off Mary Catherine (Lexi Randall) and her friends at a park to play, under Odessa's supervision, while she goes to the beauty parlor. Odessa and the children are summarily escorted from the park by a young police officer, who yells at Odessa in front of the children that the park is for whites only. He is later forced to apologize to the children (and Odessa) when Miriam complains to

Photo 7.1 The Jim Crow South as depicted in *The Long Walk Home.* Odessa (Whoopie Goldberg) and the children wait for an apology from the police officer who threw them out of the "whites only" park earlier that day.

the police chief about her daughter being thrown out of the park. Acknowledging that the children were with their maid, she defensively exclaims, "It's not like she was paradin' her own children around the park, for heaven's sake!"

In another example, we see black maids going to work on the bus, following the prescribed ritual of entering the front of the bus, dropping their coins in the fare box, exiting the bus and reentering through the rear door to find their seat at the back of the bus. Applying a sociological perspective, we understand that this ritual symbolically reinforces the legal system of racial apartheid, where blacks are defined and treated as different from and inferior to whites. It is a daily reminder of everyone's place in the racial/class hierarchy of the society. The disruption of this practice, enacted by several women prior to Rosa Parks' famous refusal to move,[2]

[2] Early in 1955, 15-year-old Claudette Colvin refused to get up when the white section filled and the bus driver told the black riders to move back. The police were called; upon their arrival Claudette said, "I done paid my dime, I ain't got no reason to move." E. D. Nixon, former president of both the state and local NAACP chapters, prepared to take on Colvin's case until he learned she was pregnant and thought she would be discredited by the white press. He turned down the cases of two more women before he learned that Rosa Parks had been arrested. Rosa agreed to let her friend from the NAACP turn her case into a cause for the movement (Jones, 1995, p. 262).

reveals that systems of inequality are dependent on the acceptance and acquiescence of the oppressed. When the subordinated group refuses to obey unjust laws, the dominant group has no choice but to reinforce the law (throwing Odessa out of the park, arresting Rosa Parks on the bus) or change the law.

The bus boycott was a nonviolent challenge to the economic and political racial inequality on which the social order rested. For Miriam, it is an inconvenience since Odessa is frequently late to work, and exhausted as a result of her long walk. Miriam offers to drive her to and from work several days a week, but keeps this from her husband Norman (Dwight Schultz) who has joined the White Citizen's Council and is working to break the boycott and the burgeoning civil rights movement. When he finds out, he shouts angrily, "Here I am trying to hold my head up as a white man in this town, and you're carting a nigger maid." He forbids Miriam to drive, telling her that she needs to leave the decision making to him as the head of the household, the one who knows best. He goes on to explain that there can never be a real relationship between the races, "We don't know her, can't ever know her." On the other side of town, Odessa's husband is unhappy when she adds Miriam to their evening prayer, warning that "She don't know us, and she don't want to know us."

Yet, Miriam's beliefs about race and race relations are changing, and she is beginning to see Odessa not as a member of a racial group, or even as her maid, but as a woman who, like herself, is a wife and mother. At one point, she tells Odessa that she does "the real mothering," even to the extent of taking care of Mary Catherine when she had the chicken pox, "and you hadn't even had them yourself." She goes on to wonder, "would I have done that for your daughter?" We, the viewing audience know the answer, as a southern white woman she never would have been in the position of caring for Odessa's children, who are growing up in a different Montgomery than the one Miriam's children know.

Odessa's children are directly affected by the bus boycott. Odessa has instructed them not to ride the bus under any circumstances, but her daughter, anxious to see her boyfriend, sneaks away from the house and onto one of the empty buses. She soon realizes she is in trouble when two white boys begin harassing her, and she runs from the bus only to realize they are in pursuit. Her brother, who has followed her, intervenes and ends up taking the beating intended for her. A follower of King's teachings, he refuses to strike back; as he repeatedly struggles to his feet, we see his hands at his sides gripped into fists, gradually releasing to an open palm. He has the strength of his convictions and is victorious in overcoming the desire to meet violence with violence. At church that night, he listens joyfully to the sermon, and joins in singing "Marching to Zion" with his worried, but proud, parents and remorseful sister.

The bus boycott continues, and King's house is firebombed; Miriam drives alone down his street to see the damage. We learn about the highly organized alternative transportation system established during the bus boycott when Miriam becomes a regular driver for the carpool. Odessa explains to her that the police have been ticketing and arresting drivers and passengers, and tells her that she could donate money if she wants to help, rather than putting herself at risk. Miriam is determined to do something, not just give money (which she notes is really Norman's anyway) to help bring the boycott to a successful end. Odessa explains to Miriam that the goal of the movement is much more far-reaching, and that the buses are just a beginning in the fight for justice and racial equality. She warns that "When it's all said and done, people are going to look at you, Miss Thompson. And they gonna say you were part of this."

The challenge of the boycott to the white supremacist system is made clear in the final scenes of the movie, which take place in the downtown parking lot that serves as the carpool center. Led by Norman's brother, members of the White Citizen's Council show up threatening the drivers and smashing out car windows. Intending to shut down the taxi service, they confront the work-weary domestics waiting for a ride home, shouting in one voice, "Walk, nigger, walk!" The women join hands, and begin singing, timidly at first and then in stronger voice, a gospel hymn. Miriam stands literally between two worlds, the black women who work as maids in the homes of white women like herself standing in solidarity, and the white men who rule in their homes and the community, by violence if necessary. One of the women reaches out her hand, and Miriam joins the line of women, with Mary Catherine by her side. The men, including Norman, are confused and frightened by the power of this nonviolent stand against their hatred and racism; in the end, they turn and walk away.

In *Witness,* there is no direct nonviolent challenge to dominant rulers or groups in society. The Amish live separately within the larger society, in communities based on shared values and beliefs, including rejection of modern technology and deep commitment to nonviolence. However, the Amish way of life provides an alternative to the dominant culture, demonstrating that violence is not necessary to create and sustain community. In the film, the centrality of violence and aggression in urban Philadelphia stands in stark contrast to the quiet, peaceful rhythm of life in the Amish community.

These two cultures clash early in the film when 8-year-old Samuel observes a homicide in the train station in Philadelphia. Questioned by detective John Book, he reveals that he saw the man who committed the murder; as it turns out a police officer, like the victim. As Samuel and his mother Rachel wait for what will happen next, they observe Book and the chief of homicide in an angry exchange. Samuel, frightened, asks, "Momma, are they angry with us?" She reassures him, "No, no. It's just the English way." Samuel has been introduced in ways large (murder) and small (an angry verbal exchange) to the world outside of the Amish community. However, Rachel makes clear that they want nothing to do with this crime or the community in which it occurred. When Book tells her that they are not free to go since Samuel is a material witness to a homicide, she tells him, "You do not understand, we have nothing to do with your laws!"

The tables are soon turned when Book is forced to hide at the Lapp farm in the Amish community after being shot by the police officer who committed the murder. He is immersed in a world without modern conveniences, where his way of life and his work are disdained. He defensively tells Rachel, "I am a cop. That's what I know and that's what I do." Rachel responds, "What you do is take vengeance, which is sin against heaven!" Reinforcing the divide between their worlds, Book observes, "That is your way, not mine." She corrects him, "It is God's way."

Book passes as a member of the Amish community, dressing like the other men, appearing "plain," which Rachel explains to him means not being vain or taking pride in one's appearance. The juxtaposition between his Amish appearance and who he is as a denizen of the modern world violates the expectations of the townspeople and tourists who visit to see the Amish and their way of life. In one encounter, a tourist asks to take his picture, and Book threatens, "Lady, you take my picture with that thing and I'm gonna rip your brassiere off and strangle you with it! You got that?"

In another scene, Book approaches several young men harassing an Amish man and his family. Rachel tells him that this happens from time to time, and that he should do nothing in response, urging him to "turn the other cheek." Book is furious and, confronting the young men, he explodes, knocking one unconscious and breaking the nose of another. A local man says to the gathering crowd, "Never seen anything like that in all my years!" The explanation given is that he is a cousin visiting from Ohio. The local man observes, "Well, them Ohio Amish sure must be different . . . around here the brethren don't have that kind of fight in them." He shouts after them, "This ain't good for the tourist trade, you know! You tell that to your Ohio cousin!"

The real challenge and power of nonviolence is revealed in the final scenes of the movie when Philadelphia police officers who are involved in the cover-up of the murder show up at the Lapp farm to kill John Book. In a gripping scene, Samuel is told to run to a neighboring farm to safety, but hearing gunshots he returns home, grabs the bell rope, and begins ringing to let the community know they need help. The community responds, making their way across the fields to the Lapp farm where they stand as witnesses to the violence occurring, bringing an end to the siege. In the end, we see Book leaving the farm, returning to his world, with Eli calling in farewell, "You be careful, John Book! Out among them English!"

Nonviolent Models for Social Change in Film

In this reading nonviolence as a method and goal of social change, as well as a way of life, is explored in film. In American society, violence has been naturalized in the sense that we see it as a normal and even necessary part of human life. However, the films analyzed here provide a different perspective on violence, challenging the dominant paradigm and perhaps suggesting alternative routes to creating social change and/or living nonviolently.

First, all three films direct our attention to the interconnectedness of personal experience and social structure, what in sociology we refer to as "the sociological imagination." In all of the films, the social structure and culture shape the lives of the individuals we meet, at the same time that they are social agents influencing, reinforcing, and/or changing the social worlds in which they live. In *Gandhi,* for example, we see Muslims and Hindus working together for Indian home rule, and British military and Indian police who brutalize those defying British rule in India. In *The Long Walk Home,* we see Miriam become an ally in the Montgomery bus boycott and the civil rights movement, while her husband joins a racist organization to break the boycott and undermine the movement. In *Witness,* Rachel and Book can see past their differences and even love one another, at the same time that their differences, rooted in cultural beliefs and practices, are irreconcilable.

Second, two of the films, *Gandhi* and *The Long Walk Home,* provide historical accounts of nonviolence as conceptualized and practiced in social movement activism at particular times and places in the 20th century. In both we see the power of the state, enacted by the police, military, government agencies, and citizen organizations, used to block nonviolent resistance and social change. The protagonists in both films engage in acts of great courage that cost them dearly. Gandhi and his followers are beaten, imprisoned, and killed. In Montgomery, those involved in the boycott are threatened, attacked, and arrested. We watch as Miriam, faced with the choice of complicity or acting outside of her race and class interests, loses her husband, her position and ties in the community, and her way of life. Odessa's son is beaten for protecting his sister from attack.

At the same time, we see individuals transformed by their activism and belief in the possibility of social change. We see groups of people brought together in common cause, breaking down barriers, undermining prejudices. For example, in the 240-mile march to the sea in the 1930 salt campaign in *Gandhi,* "we see an extraordinary variety of participants: old, young, students, peasants, ladies in saris and jewels, Muslims, Hindus, Sikhs, Christian nuns, Untouchables, merchants, some vigorous and determined, others disheveled, tired and determined" (Briley, 1982).

Third, in all three films, nonviolence works for those who choose it as a way of life and vehicle for social change. In *Gandhi* and *The Long Walk Home,* nonviolent social protest challenges and changes the society in which it occurs. The British leave India; Jim Crow laws are dismantled and racial inequality is (and continues to be) challenged in the United States. In contrast, in *Witness* nonviolence is part of the Ordnung, the unwritten social order that guides behavior in the community. In this *gemeinschaft*/folk society, homogeneity and conformity provide the context for shared values, including nonviolence. Eli warns Rachel that she will be shunned if she goes against the Ordnung: "Rachel, you bring this man to our house. With his gun of the hand. You bring fear to this house. Fear of English with guns coming after . . . Rachel, good Rachel, you must not go too far!" Living separately from the dominant culture allows the Amish to protect their way of life and to ensure that all live according to the Ordnung.

Last, these films provide a way of thinking about how we choose to live in society at this point in history. Do we accept injustice, inequality, and violence as the "American way" or do we seek to challenge and change structures that harm those who are not among the privileged in society? To choose nonviolence, do we need to separate from the dominant culture, rejecting values and beliefs rooted in and supportive of consumerism, individualism, accumulation, and violence? How do we go about envisioning the type of world we want to live in and working to create that world? Perhaps movies such as *Gandhi, The Long Walk Home,* and *Witness* provide a beginning for this exploration, as nonviolent alternatives to hegemony in film.

References

Briley, J. (1982). *Gandhi: The screenplay.* Retrieved July 1, 2008, from http://www.gandhiserve.org/video/gandhi_ screenplay.html

Briley, J. (1996). On "Gandhi" and "Cry Freedom": Two pivotal scripts in my life. *Creative Screenwriting, 3,* 3–12.

Burks, M. F. (1990). Trailblazers: Women in the Montgomery bus boycott. In V. L. Crawford, J. A. Rouse, & B. Woods (Eds.), *Women in the civil rights movement: Trailblazers and torchbearers, 1941–1965* (pp. 71–84). Bloomington: Indiana University Press.

Cooper, B. (1999). Hegemony and Hollywood: A critique of cinematic distortions of women and color and their stories. *American Communication Journal, 2*(2). Retrieved November 22, 2008, from http://acjournal.org/holdings/v012/Iss2/articles/brendacooper/ index.html

Fisk, L. J. (2000). Shaping visionaries: Nurturing peace through education. In L. J. Fisk & J. L. Schellenberg (Eds.), *Patterns of conflict, paths to peace* (pp. 159–193). Peterborough, ON: Broadview Press.

Iadicola, P., & Shupe, A. (2003). *Violence, inequality, and human freedom.* Lanham, MD: Rowman & Littlefield.

Jones, J. (1995). The Long Walk Home. In M. C. Carnes (Ed.), *Past imperfect: History according to the movies* (pp. 262–265). New York: Henry Holt.

Lucien, B. (1984). Nonviolence and Satyagraha in Attenborough's *Gandhi. Journal of Humanistic Psychology, 24*(3), 130–141.

Meyer, T. (2005). Creating a culture of nonviolence: A conversation with Arun Gandhi. *Nonviolent Communication.* Retrieved June 25, 2008, from http://www.nonviolentcommunication.com/press/article_PDF/tMyer/Nonviolence_Arun_Gandhi_TMeyer.pdf

Slocum, D. J. (2000). Film violence and the institutionalization of the cinema. *Social Research, 67,* 649–681.

Williams, R. (2001). Base and superstructure in Marxist cultural theory. In J. Higgins (Ed.), *The Raymond Williams reader* (pp. 158–178). Malden, MA: Blackwell.

Reading 7.2

From Earth to Cosmos

Environmental Sociology and Images of the Future in Science Fiction Film

Christopher W. Podeschi

We need to take something for granted at the outset: the relationship between society and nature is problematic. While nature constrains societies (e.g., arid conditions make farming difficult), our concern here is with problems caused by humans, like deforestation and biodiversity loss, the impact of suburban sprawl on water quality, or what will continue to be a big focus for years: global warming.[1] These are serious problems not just for what they do to nature, but because human societies rely on the environment as a home, source of sustenance, and for various other "ecosystem services." Harm to the natural world can ultimately mean harm to people and society.

Responsibility for understanding these problems so that we can improve conditions in the future falls on the shoulders of citizens, politicians, and scholars alike. Among scholars, these problems need attention from disciplines other than the natural sciences like ecology, biology, or geology. Sociology must also play a central role. In this section, I introduce *environmental sociology* and present some of the scholarship in this area. Then, I examine images of the future in science fiction films as a way of looking at the "culture of nature" or how nature and society's relationship with nature are "social constructions." My aims are to demonstrate both the value of close examination of film content to environmental sociology and the contribution sociology can make to future sustainability through deepening understanding of important social and cultural processes.

Environmental Sociology

Talking about environmental problems as caused by humans is actually a good place to start introducing environmental sociology. With regard to climate change, the term *anthropogenic* is used to emphasize that it is caused by human activities (e.g., CO_2 emissions from fossil fuel use). But the term *human activity* is too vague for sociologists. Sociologists are interested in the social and cultural contexts that guide human behavior and history. For example, *humans* built and moved to the suburbs that now

[1] It must be emphasized here that there is essentially no controversy among climate scientists the world over about whether global warming is occurring naturally or is caused by human activities: they believe it is caused by human activities. For background information, see *An Inconvenient Truth* (2006), directed by Davis Guggenheim, and "Climate Change 2007: Synthesis Report: Summary for Policymakers" by the Intergovernmental Panel on Climate Change (2007).

create water-quality problems when it rains,[2] but there's a *sociological* story to tell as well. At present, "growth machines" in many locales compete with other cities and towns to attract industries and people (see Logan & Molotch, 1987). During the Great Depression, the federal government contributed to urban sprawl by backing mortgages for new housing to stimulate the economy. Then, pushed by business interests, the government built the highways that made it easier to live outside of cities (Andrews, 1999). Further, many white Americans sought refuge from urban social conditions and the influx of racial and ethnic minorities to the city (see Szasz, 2007). In the end, it is clear that sprawl-induced water quality problems have a specific social history. Generic talk about "human activity" is inadequate because environmental problems are *social* problems.

Environmental sociologists also emphasize that society and nature are not really separate entities. Consider the consumption and disposal practices of human beings, for example. People in wealthy nations get their sustenance from grocery store shelves and send their waste from porcelain bowls through sewer pipes to treatment plants. Layers of social structure (the capitalist food production and distribution system), culture (dining and waste disposal habits), and technology (modern agriculture and waste treatment systems) have distanced people from nature, but the flow of material and energy that ties human bodies, and thus society, to continuous interchange with the earth *necessarily* remains intact. And this is just one illustration of the myriad ways in which society and nature are inextricably linked (see Carolan, 2005; Dunlap, 2002; Murphy, 2004).

Perhaps surprisingly, sociology long ignored the integration of society and the environment. In an early attempt to define environmental sociology, William Catton and Riley Dunlap (1978) asserted that by ignoring society-environment dynamics and ecological limits to economic growth, sociology was guilty of falling under a "human exceptionalist" or "exemptionalist" paradigm. They argued for a "New Ecological Paradigm" and for sociology to explicitly examine society-environment dynamics.

Recent scholarship by Dana Fisher (2006) provides an excellent example of this "integrationist" perspective.[3] Fisher is interested in why the United States has been reticent on the issue of climate change policy when compared to other nations. This stance is curious given the consensus among climate scientists that global warming is real, caused primarily by the burning of fossil fuels, and promises serious future consequences. Recent reports by the Intergovernmental Panel on Climate Change (the body that shared the 2007 Nobel Peace Prize with Al Gore) confirm this (see Intergovernmental Panel on Climate Change, 2007).[4] Fisher

---

[2] When it rains, some water soaks into the ground and some runs off into streams and rivers. The water runs off especially well from "impervious surfaces" like roads, parking lots, and buildings, and along the way it picks up pollutants, like oil from roads and bacteria from pet waste in yards (see Libes, 2003).

[3] For other examples of integrationist environmental sociology, see Freudenburg, Frickel, and Gramling (1995), Freudenburg and Gramling (1993), and Murphy (2002, 2004).

[4] Recent UN climate talks in Bali, Indonesia, confirm U.S. reticence, with American negotiators remaining "obstructionist until the final hour of the two-week convention and had changed their stance only after public rebukes that included boos and hisses from other delegates" (Fuller & Revkin, 2007). The talks ended with hope that more progress can be made during subsequent negotiations once a new U.S. president takes office in 2009.

notes that while environmental sociologists have looked at *social* explanations for the U.S. position, they haven't yet looked at the role of "the material." Aiming to correct this, she relies on the notion of "conjoint constitution"—that is, the idea that social phenomena like policymaking can be understood as the result of the interaction of natural and social factors (Freudenburg, Frickel, & Gramling, 1995). Fisher looks at recent U.S. politics and finds the level of coal extraction in states to be an excellent predictor of both Senate votes on climate policy and state-level climate policies. Specifically, the presence of coal, a resource used to generate electricity and a huge contributor to climate change, predicts opposition to policies that might ameliorate climate change. She asserts that "including aspects of America's natural resource endowment helps to explain more fully climate change policy in the United States" (2006, p. 488).

While the integrationist approach is an important corrective, many environmental sociologists nonetheless feel it's important to not lose sight of the value of conventional sociology (i.e., sociology that remains focused on social and cultural factors rather than examining society-nature interaction directly). The "Treadmill of Production" model developed by Allan Schnaiberg serves as an excellent example of a conventional approach to environmental sociology (Schnaiberg, 1980; Schnaiberg & Gould, 1994).[5] For a nation like the United States, rather than becoming gradually more environmentally friendly, the Treadmill model envisions continued economic growth and ongoing pressure to prevent or weaken environmental regulations. Continued and increased environmental disorganization can thus be expected. The structure of modern capitalist societies is the cause, and so this path is seen as more or less inevitable absent mass political action to undermine capitalism.

Two phenomena are of central importance. One is what Schnaiberg (1980) calls the "growth coalition." Put simply, the major players in the economy, capitalists, government, and the working class, are unified in their commitment to continued economic growth. The working class has an interest in a higher standard of living and job security, both of which are seen as more likely if the economy is expanding. The government's role is to make sure the economy is profitable for capitalists *and* that the general population is content (see also O'Connor, 1973). These are known as the "accumulation" and "legitimation" functions, respectively. The former is accomplished through investments like highways on which goods can be transported, supporting university research of new technologies, and training future workers. Legitimation takes various forms, but examples include social programs like welfare to the poor or workplace safety regulations. The government becomes committed to growth because expansion provides the revenues via taxation that make it possible to "satisfy the demands of both constituencies" (Schnaiberg, 1980, p. 211).

---

[5] For other examples of environmental sociology that is both primarily conventional and structural, see Foster (1999), Rudel (2005), Warner and Molotch (2000), and York, Rosa, and Dietz (2003). Note that often this sort of work includes measured dependent variables that are nonsocial (e.g., deforestation rate), but these are placed in this category for looking primarily at conventional social factors and processes as driving forces rather than examining "feedback" from the environment.

The other key treadmill phenomenon is *competition* between capitalist firms. Competition compels capitalists to increase profits by limiting the cost of production. From this comes opposition to government regulation of business. But competition leads to growth because it compels capitalists to expand their enterprises through increased or innovative production (Schnaiberg, 1980). This allows a firm to undercut the competition's price and sell more product, increasing profits and possibly gaining investors. In the short run, the innovating firm wins out, but soon competition will invest in innovation and everyone is back to square one in terms of profitability. The difference, however, is that the capacity for production has expanded. And this goes on in every industry. Among other things, growth of output means more resources and energy are used and more pollution and trash are generated. Even with "green" technology, gains in efficiency are offset by growth (Schnaiberg & Gould, 1994).

While these production factors are central for understanding the push for continued growth, the problem of consumption arises as well. Companies making more products need more buyers. A number of strategies are available, such as selling in overseas markets, but marketing and the availability of credit (e.g., consumer credit cards) are most notable for the role they have played in creating demand and, in turn, the high and environmentally unsustainable levels of consumption present in the United States (Schnaiberg, 1980). Status concerns and the feeling of "relative deprivation" cultivated by marketing and by living in a stratified society like the United States are worth considering as well (see Schor, 1999).

Nature, the environment, society-environment relations, and environmental problems are both real phenomena *and* cultural objects (i.e., things imbued with meanings that impact perception and action). Environmental sociologists therefore turn attention to culture as well, and much of this work is "conventional" rather than "integrationist." Survey research looking at people's worldviews and attitudes with regard to the environment is one focus (for example, see Dietz, Kalof, & Stern, 2002; Dunlap, Van Liere, Mertig, & Jones, 2000; Jones & Carter, 1994). There is also a literature on the "social construction" of environmental problems—that is, the social process through which environmental issues get recognized and defined as problems.[6] Environmental problems like global warming are frequently not apparent or clearly problematic in people's everyday lives. As a result, recognition requires effort, called *claimsmaking,* on the part of activists and often scientists (for examples, see Hannigan, 1995; Mazur & Lee, 1993; Ungar, 1998a). The success of claims depends on politics, media attention, and even on the "character" of the claims themselves, since some claims are made in ways that are more likely to draw people's attention and convince them the problem is serious (Snow & Benford, 1988; Snow, Rochford, Worden, & Benford, 1986).

At the same time, *counterclaims* are made by those who would undermine or prevent recognition of an issue. Usually, successful counterclaims *prevent* public recognition of a problem, but with global warming things have followed a different course. McCright and Dunlap (2003) point out that global warming had been

---

[6] This is a problem for all contentious issues, not just environmental problems. See Spector and Kitsuse (1987).

successfully "constructed." Late 1990s poll results indicated that the majority of the American public was concerned and ready to support efforts to fix the problem. But since then, scientific consensus about the reality, human causes, and seriousness of the problem has continued to grow while public support in the United States has waned. Calling this the "delegitimation of global warming as a social problem" (p. 350), McCright and Dunlap show how organizations in the American conservative movement, an "elite-driven network of private foundations, policy-planning think tanks, and individual intellectuals" (p. 352), caused this by seizing on and promoting the views of just five scientists skeptical about climate change. With their financial clout and political influence, these organizations were able to get the climate skeptics' views heard by Congress and the media. Despite holding an exceedingly rare position, their views were presented alongside the view of the majority of climate scientists. These efforts effectively planted doubt in public opinion. Despite the overwhelming consensus among climate scientists, recent polls show that a remarkable number of Americans believe climate scientists disagree about climate change (Nisbet & Myers, 2007). According to McCright and Dunlap (2003), this public confusion "translates into political inaction and policy gridlock—disproportionately favoring powerful interests attempting to construct the nonproblematicity of environmental conditions" (p. 366).[7]

The Culture of Nature in Science Fiction Films

Environmental sociologists also look at the *social construction of nature*. Rather than looking at the social construction (or delegitimation) of environmental problems, the focus here is on meanings of nature itself and on the meaning of society's relationship with nature. Scholars have used survey research to explore these meanings, but they have also studied how nature is represented in cultural artifacts like prime time television shows, documentaries, advertisements, magazines, or children's textbooks (Lerner & Kalof, 1999; Papson, 1992; Shanahan & McComas, 1999; Ungar, 1998b). In my research, I have examined the meaning of nature and of society's relationship to nature in general audience magazines (Podeschi, 2007) and science fiction films (Podeschi, 2002). In both cases I used a historical approach to see if the modern environmental movement has impacted the presentation of nature in popular culture.

I use science fiction films as *future myths*; the opposite of origin myths, future myths are stories about where a society is headed or might be headed. Even though they can be outlandish, these visions are undoubtedly significant to society as hopeful projections, extrapolations, or warnings. In these future scenarios, whether intended or not, the producers provide a vision of the future value of nature and of society's relationship with nature.

[7] Research by Freudenburg, Gramling, and Davidson (2008) complements the work of McCright and Dunlap (2003) by highlighting how antienvironmental campaigns use "SCAMS" or "Scientific Certainty Argumentation Methods," preying on the reliance on statistical probabilism in scientific research.

So the question becomes: What is the *nature* of future myths? Theoretically, I see culture as serving the powerful through the distribution of ideas and information. However, power is tempered, imperfect, and incomplete. Plenty of ideas are circulating that counter the interests of powerful groups. Further, hegemony requires sustained effort and there are no guarantees that it will last. In short, culture should be seen as a place where power matters, but also as a site of struggle. Thus, we might expect science fiction films, in some ways and to some extent, to *transcode* (Ryan & Kellner, 1988) the thinking found in the environmental movement that challenges the existing and exploitative relationship between society and nature.

Using the most popular science fiction films from 1950 through 1999, I devised a scheme for interpreting the nature- and environmentally-relevant material in the films (the films are listed at the end of this reading). This scheme was grounded in efforts by sociologists to understand the key orienting discourses in environmentalism and the society at large (see Brulle, 1996; Olsen, Lodwick, & Dunlap, 1992). I called the environmentalist discourse *resistant* because it demanded changes in the existing society-nature relationship. As an ecological vision, environmentalist discourse calls for things like sustainable economics and technology and for seeing value in nature. Historically, the primary cultural stance toward nature in the United States has been called Manifest Destiny, the *dominant social paradigm* or the *technological social paradigm* (see Brulle, 1996; Brulle, 2000; Olsen et al., 1992), a discourse I call *reproductive* because the way society operates is left unchallenged. Sentiments here include an *anthropocentric* (human-centered) focus on nature as a resource to be used for human needs and wants *and* a trust in the potential of technology to provide great benefits and solve future problems with minimal attention to potential environmental consequences. In the resulting analysis, I explored two dimensions in science fiction film: the relationship of society with nature and the value of nature.

The Relationship of Society With Nature

Almost all of the futuristic films show society "colonizing the cosmos." The scale of the colonization varies. Futures like that in *Total Recall* (1990) only take us as far as the planets of our own solar system—humans have colonized Mars. In other films, like *Alien* (1979), *Aliens* (1986), *Forbidden Planet* (1956), *Planet of the Apes* (1968), and both the *Star Trek* (1979–2009) and *Star Wars* (1977–2005) series, the scale of colonization is grand. In *Alien* and *Aliens,* people have gone so far that there are "frontier" and "core" systems, and in the *Star Trek* films there are galactic political boundaries. Rather than small outposts on distant planets, the films show societal-level colonization and resource extraction. In *Star Wars,* the desert-planet Tatooine is significantly colonized in a manner reminiscent of earthly colonization of lands and peoples. Here humans appear to be intruding colonists, shown as "modern" and "western" relative to the desert people who tellingly fight them in guerilla-fashion. Extensive control is also depicted by the "taming" of a planet and a harvesting of its resources. The Genesis technology from *Star Trek II: The Wrath of Khan* (1982) and the terraformers in *Aliens* are both technologies that tame wild environments on the "frontier" of space and make them useful to humans.

Similarly, the resources of other planets are mined in *Total Recall, Alien,* and *The Empire Strikes Back* (1980). Regardless of the details, this imagery is significant for resonating well with reproductive discourse. One can argue that these films envision a new manifest destiny, with space as the new wilderness frontier (rather than the North American West). Depictions of a planet-hopping future support reproductive discourse by implying that societies need not worry about resource scarcity or sustainability in the here and now.

Technology is important when considering society's relationship with nature as well. Technology involves both transforming and controlling nature to suit social interests and is a means through which societies interact with and impact nature.[8] And technology is of course a major presence in science fiction films. Historically, American culture has trusted and valued technology, viewing it as a source of progress and a benefit to humanity (see Olsen et al., 1992). This is the reproductive discourse. Some environmentally concerned thinkers continue in this tradition, confident future industrial society can develop new, environmentally sound technologies (see Hawken, Lovins, & Lovins, 2000; McDonough & Braungart, 2002). The more conventional resistant discourse, however, in both mainstream and radical environmentalism, questions the value of technology, technological developments, and heavily technological societies given the risks and hazards created by industrial capitalism, chemical-intensive agriculture, nuclear power, and the like (see Brulle, 2000). In this discourse, global warming is seen as resulting from societies' reliance on machines.

For the most part, visions of the future in science fiction films resonate with reproductive discourse by serving as uncritical celebrations of the potential of technology. Overall, future myths envision a technology-saturated future, but most telling is the fact that more than three-fourths of the films I analyzed present a future of extraordinarily powerful technologies. The list is considerable, but a few examples stand out, especially those that have significant power over nature. In many films, space travel is a regular event, but in the *Star Trek* films, spaceships use "warp" drives in which space itself is folded and then restored with "space matrix restoration coils." The film *Forbidden Planet* takes viewers to Altair IV where the technology of a long-extinct alien society is intact and materializes thoughts. Visitors to the planet who learn to use the technology have to be careful, however, because their subconscious desires can create monsters beyond their control. While this amazing device is clearly dangerous, and not a direct projection of future human technology, viewers are told in the end that it is the kind of thing "we" will eventually achieve. Also startlingly powerful, in *Star Trek II,* we are introduced to the Genesis device (mentioned above), which can turn a lifeless environment into a lush paradise in a matter of moments. The device is like a missile and when detonated, "matter is reorganized" at the subatomic level "with life generating results." Rounding out the list is an example that seems tame relative to what we find in some films: in *Back to the Future Part II* (1980), the National Weather Service controls the Earth's weather with clocklike precision. Doc Brown (Christopher Lloyd), one of the main characters, at one point complains how the "post office isn't as efficient as the weather service."

---

[8] Murphy (2004) in fact calls technology "recombinant nature."

Science fiction films promote reproductive discourse; risk and/or environmental consequences stemming from technologies are largely ignored. As a whole, the technologies depicted in the films rarely malfunction and risks are the exception rather than the rule. Telling in this regard is the fact that films from the 1970s and later depict nuclear technology as "naturalized" into the future social landscape. *Forbidden Planet, Star Wars, The Black Hole* (1980), *Back to the Future Part II, Star Trek II, Aliens, Star Trek IV: The Voyage Home* (1986), *Total Recall,* and *Demolition Man* (1993) all present nuclear power as an unproblematic aspect of the imagined future society. In *Back to the Future Part II,* there are even personal reactors that look like food processors. Such validation of this inherently risky technology is clearly controversial from a resistant environmentalist perspective. Further, in futuristic films when technological risk is acknowledged, the source of the trouble is never the technology itself. In *Aliens,* for example, a military force is sent to rescue an off-world colony from hostile alien creatures, but the aliens have built their den over "a big fusion reactor." The technology itself is not a threat, it's a naturalized part of the social landscape, but they do have to be careful not to shoot the reactor when fighting the aliens. Additionally, when risks are depicted, this depiction has to be read in context, and the broader context is frequently one of amazing technology that prevents the film from taking a truly critical or resistant stance. The *Star Trek* series, for example, includes devices that "beam" people from place to place by turning their matter into a signal and then reassembling them at their destination. In *Star Trek: The Motion Picture* (1979), the first film in the series, two people die from a malfunction of the beaming device, but this can't be read as the film taking a resistant stance toward technology in general because the broader context is one of technological splendor.

A remarkable number of films even use a "bait and switch" approach with regard to fears about technology: risk or serious concerns about a technology are acknowledged, but then ultimately contradicted by the plot. One has to wonder about the potential for this technique to cultivate faith in technology. In *Alien,* Ash (Ian Holm), the "artificial person," endangers his human crewmates by following programmed orders that have him work to ensure an extraordinarily violent alien makes it back to earth. This starts as sharp critique of technology: Ash is an important member of the crew, but in contrast to humans, he is valueless and unreflective and so not to be trusted. In *Aliens,* however, we meet and come to love Bishop (Lance Henrikson), the perfected artificial person reprogrammed to protect human life above all else. Three Cold War era films about nuclear technology use bait and switch as well: *Destination Moon* (1950), *The Beast From 20,000 Fathoms* (1953), and *Voyage to the Bottom of the Sea* (1961). *Destination Moon* focuses on a group of men who want to launch a nuclear-powered rocket to the moon despite fearful opposition from the public and the government. They launch anyway, and though they have some trouble getting home, the trip is a success. Concerns about technology are proven unfounded and viewers are given the message that they should trust technological innovations because their unfounded fears will delay the course of progress.

Star Trek II includes a particularly strong example of the bait and switch approach. The Genesis device, with the power to create life and transform environments, is initially strongly criticized. After learning about the technology, Bones (DeForest Kelley), the protagonist doctor, is outraged because of its power and potential for accident or aggressive misuse. Another key protagonist, Mr. Spock (Leonard Nimoy), concurs and points out that if detonated where life already exists, it would destroy that life by replacing its subatomic matrix. Bones is further angered: "We're talking about universal armageddon!" The rest of the film, however, operates to undermine this critique of nature-transforming technology. Genesis is touted for its potential to alleviate "cosmic food and population problems."

The technology is used twice with clear demonstration of its creative potential. The first experiment with the device takes place in a huge underground cavern. The result is a cavern filled with breathtaking natural beauty and fertile, bountiful nature. In the second use, the antagonist Khan (Ricardo Montalban) attempts to destroy the protagonists and their spaceship, the Enterprise, by detonating a stolen Genesis device. The Enterprise escapes the attack and the detonation creates a beautiful and lush planet. Khan's evil is transformed into goodness and beauty. These positive depictions are backed up by the characters themselves. Mr. Spock dies at a critical moment, and in his eulogy Captain Kirk (William Shatner) diffuses sorrow for Spock's death by noting that it took place in the "shadow of new life" (the new planet). Near the end of the film, we are also provided shots of key characters gazing upon the beauty of the new creation. One can certainly read this as a film-length argument for trusting technology to fix environmental problems like resource scarcity. As with colonization of space, resistant discourse about creating sustainable societies is undermined.

On the whole, the films studied resonate with reproductive discourse for valorizing technology or raising and then countering fears about technological risks. There is, however, content more squarely critical of technology. Nuclear technology receives strong negative attention in the 1950s and 1960s. *On the Beach* (1960) and the *Planet of the Apes* films (1968–1972) are essentially psychological horror films about the effects or aftermath of nuclear war. *Them!* (1954) and *The Incredible Shrinking Man* (1957) articulate fears of fallout from nuclear weapons testing.

Another technology that is presented as negative or potentially harmful is technology that involves some sort of union with or simulation of humanity. Specifically, cyborgs, bionics, implants, and robots or artificial intelligence receive sustained critique as fundamentally flawed in futuristic films from 1969 forward. Implanted surveillance devices or other technological connections to the body are repulsed in *Total Recall, Demolition Man,* and *The Matrix* (1999). Cyborg slaves from *The Black Hole* and Darth Vader from the *Star Wars* series provide a critical vision of the union of technology and human bodies. The films *2001: A Space Odyssey* (1968) and *The Matrix* focus on artificial intelligence or simulated humanity gone awry. In these cases, computers become too much like human beings, (i.e., the simulations are too good, and they gain consciousness and agency and threaten or enslave humanity).

The *Star Wars* films actually reverse the bait and switch approach described above and can be read as critical of technology as well, but in a more general manner. These films are set in a highly technological future where the protagonists (the rebels) and antagonists (the empire) rely on powerful and amazing technologies.

On the surface, the films validate technology, but parallel to this is a latent opposition between good and evil serving as a critique of technology. First, the forces for good, the rebels, are grounded in nature. The rebel princess (Carrie Fisher) is from a lush blue planet resembling Earth and Yoda (voice of Frank Oz), the master of the good Jedi Knights, lives on a swamp planet teeming with life. By sharp contrast, the evil empire is grounded almost entirely in built technological environments like mind-bogglingly huge spaceships. Most telling, however, is the Death Star. This planet-sized sphere is the empire's "home" and so functions like an artificial planet. It is also a weapon powerful enough to destroy real planets, and it does destroy the Earthlike planet in the first film. So the films give us real nature destroyed by evil, simulated nature. In addition to this contrast, the lead antagonist, Darth Vader (voice of James Earl Jones), is a cyborg nightmare. Viewers are meant to impute that his body is damaged from a life of violence, and the damaged parts have been replaced with bionics. He is even forced to wear a mask and cannot breathe on his own. At one point, a key protagonist says of Vader, "He's more machine now than man; twisted and evil," implying that the union with technology explains his brutality. In the end, even though both good and evil rely on technology in these films, this latent linkage of good with nature and evil with technology can be read as a warning for future societies.

On the Value of Nature

Turning to the status of nature relative to society in these projected futures, the question is whether the films project a cultural opposition between nature and society, with society more highly valued. On the one hand, futures containing continued domination of nature through colonization and technology can be seen as assigning nature a secondary status relative to society. But here we look at the direct "charge" given to animals and environments in these projected futures.

From a distance, environments are frequently positively charged as beautiful landscapes. But when environments and inanimate nature are more than background and salient in the narrative, they are usually hostile, harsh, or simply devalued. On its own, this clearly stands in contrast to "resistant" environmentalist discourses for devaluing nature and for failing to depict it in an ecological fashion, but it also reinforces that sense that future societies will colonize and control a new "wilderness."

In terms of devaluation of nature relative to society/humanity, *Star Trek II* defines an entire planet in terms of its potential usefulness to humanity. When the protagonists arrive at Regula I, Mr. Spock describes it as "a great rock in space" with "unremarkable ores." In other words, it is seen as good for nothing. *Forbidden Planet* uses a different approach. The inventor of a remarkably powerful robot demonstrates the safety of his creation by commanding it to destroy both a tree and the key protagonist. It vaporizes the tree, but refuses to harm the "rational being." Nonrational life beware!

The *Star Wars* series depicts natural settings as harsh and unpleasant. In the first film, we are introduced to a desert planet, Tatooine; in the second, an ice planet,

Hoth; and finally a swamp planet, Degobah, in the third. All three are depicted as terribly unpleasant places for extreme heat and sterility, extreme cold, and muck, respectively. The swamp planet is perhaps most hostile for it is not only dark and covered with murky water, but filled with organisms that spark fear and disgust, like lizards and giant snakes. The protagonists from the future who find themselves in these places make their displeasure clear. For example, Luke Skywalker, the main protagonist, calls Degobah a "slimy mud hole." Tempering Degobah's negative depiction some is the fact that Yoda, one of the arch-protagonists and symbols of goodness in the films, lives on Degobah and appreciates the place, perhaps because its fecundity fuels his magic power (he's a Jedi Master of the Force). Otherwise, the only nonbuilt and friendly environment in the *Star Wars* trilogy is the forest moon Endor in *The Return of the Jedi* (1983).

The films' depiction of animals parallels that of environments. Counter to resistant discourse, animals are not presented as inherently valuable in ecological terms, but tend to be hostile, disgusting, valued if domesticated, or merely lesser than the "civilized." The films project Arnold Arluke and Clinton Sanders' (1996) "sociozoologic scale" into the future. Good animals are docile pets and tools while bad animals refuse to stay in their proper place, entering society as freaks, vermin that contaminate or disgust, and as demons seeking to kill people.

In the *Star Wars* films the protagonists not only have to contend with harsh environments, but a series of "demons" waiting there to consume them. In *Star Wars* and *The Empire Strikes Back,* tentacled creatures lurk in murky water. In *Return of the Jedi* and *Empire Strikes Back* the protagonists barely escape from enormous worms, one of which could have swallowed their spaceship. The films *Alien* and *Aliens* focus on one type of creature, a part demon, part vermin "sea monster in space" that the planet-hopping humans have encountered (begging the question, who is the alien?). These "wild" animals are demons because they are unbelievably strong and fast, and can easily kill and devour humans. Amazing survivors, the creatures also can sustain themselves and blend physically into any type of environment. In *Alien,* the body of the creatures becomes metallic and imitates the "technology texture" of the ship, a curious and perhaps critical statement about evolution of life in future technological environments. The aliens are vermin because of their similarity with Earth pests. They can reproduce parasitically (in a human host in *Alien*) or by eggs (*Aliens* features a large and disgusting den). They are also reminiscent of snakes, spiders, and insects. The fact that they *infest* places is a particularly clear similarity to vermin. In *Alien,* one creature occupies and travels primarily via the internal structure of the spaceship in which the film takes place. In *Aliens* a whole host occupies the internal structure of a large building complex. In both cases, the films draw a clear parallel to rodent and insect infestation with which audiences are familiar. The infestations inspire dread because the protagonists never know where the aliens may be.

Two other trends counter resistant discourse. First of all, the films analyzed occasionally present us with valued domesticated animals in contrast to the wild and dangerous ones. For example, docile, gentle, and horselike "ton-tons" are opposed to an abominable snowman that captures the hero in *The Empire Strikes Back.* In *Alien,* Jonsie, the beloved spaceship cat, is set in stark opposition to the alien. As domesticated animals are *in the fold* of civilization, such contrasts reinforce a division between society and nature (see Arluke & Sanders, 1996). Secondly, apart from being portrayed as hostile, films denigrate animals as simply of lesser value than humanity. Domesticated animals are no exception; they are valued in contrast to wild creatures, but they are still humans' pets or tools. For one of the best examples of this, we can turn again to the ton-tons from *The Empire Strikes Back.* In the beginning of this film, Han Solo (Harrison Ford) goes out into the hostile environment of the ice planet Hoth riding a ton-ton in search of the hero, Luke Skywalker (Mark Hamill). He does this knowing that his ton-ton is unlikely to survive plummeting temperatures as night falls. The ton-ton does indeed die, but Han saves Luke, notably by using the recently deceased animal's insides for warmth. Not only is the ton-ton domesticated in the service of humans, the ton-ton can be sacrificed in favor of human life.

To close, it must be noted that the devaluation of nature relative to society does not go wholly unchecked. In *Star Trek IV: The Voyage Home,* Mr. Spock tells Captain Kirk that "only human arrogance" would assume humans are the only intelligent form of life on Earth. In *Aliens,* appreciation for alien biology is expressed in scientific terms by one of the key protagonists. The films *Planet of the Apes* and *Beneath the Planet of the Apes* (1970) take a different approach, devaluing humanity by providing sharp critiques of human society as destructive. Such content cannot be ignored, but these are counterexamples against the primary thrust of the films. More common are dangerous or valueless environments and creatures.

In reality, of course, nature can be uncomfortable, scary, and dangerous. However, as the primary emphasis in the films, rather than a balanced or ecological view, this fits and reinforces reproductive discourse on the value of nature. Historically in the United States, nature has been seen as wild and dangerous or as something without value that needs to be controlled for human benefit (see Brulle, 2000; Nash, 1982). By contrast, resistant discourses in the history of both the radical and more mainstream arms of the environmental movement see nature, wilderness, and wildlife as inherently valuable and/or worth preserving, both for themselves and for humanity (see Brulle, 1996, 2000).

Conclusion

This reading introduced and examined basic concepts from environmental sociology. We then saw how environmental sociologists look at the potential for natural resources to impact social and political processes, how socioeconomic structures generate environmentally problematic economic growth, and how social actors politicize scientific perspectives on environmental problems and influence public opinion. Last, we saw how science fiction films can be used to study the culture of nature or the way that nature itself is socially constructed. This work is useful not just for what it reveals about the meaning of nature, but for what it reveals about imagery with regard to nature in the future.

I hesitate to paint my findings with too broad a brush. What's presented here already glosses over a fair amount of specificity worth examining and debating, but what stands out is the way that the films for the most part fail to resonate with resistant or environmentalist discourse. Instead, with some exceptions, they fit reproductive discourse by presenting future technological society in optimistic terms and reinforcing a sharp nature-society dualism by depicting nature as an obstacle or a thing of lesser value than humanity/society. This is the case despite the rise of the modern environmental movement in the early 1960s *and* despite survey results indicating that the culture has moved away from the technological or dominant social paradigm (see Olsen et al., 1992).

Films reflect the time periods in which they are made. For example, from the late 1960s forward, technologies that unite with or simulate humanity are strongly resisted (e.g., Darth Vader; Hal from *2001*), perhaps articulating some anxiety about the integration of computers into society. This generally fits resistant discourse concerns about technology, but is notably distant from *environmental* concerns. In contrast, nuclear technology's potential for environmental harm is debated in science fiction films made during the 1950s and 1960s when fears about nuclear weapons testing and the Cold War were significant. But after this, as the films focused on cyborg technologies, they also "naturalized" nuclear power in the depicted futures. This is particularly interesting given the visibility of the antinuclear and environmental movements and historical events such as Three Mile Island in 1979 and Chernobyl in 1986 (see Rothman, 2000).

Ultimately, the future presented in science fiction film is not one of technological caution, of sustainable living on earth, or of harmony with wild places and creatures, but is instead one of ubiquitous and powerful technology, of colonization of space and other planets, and of conflict and difficulty with hostile environments and creatures. Does this indicate a lack of responsiveness to the environmental movement in popular culture? Maybe it is simply the case that commercial film producers are not interested in gritty science fiction films that question the future or the potential of technology, fearing fewer people will pay for tickets to pessimistic films.

References

Andrews, R. (1999). *Managing the environment, managing ourselves: A history of American environmental policy*. New Haven, CT: Yale University Press.

Arluke, A., & Sanders, C. R. (1996). *Regarding animals*. Philadelphia: Temple University Press.

Brulle, R. J. (1996). Environmental discourse and social movement organizations: A historical and rhetorical perspective on the development of U.S. environmental organizations. *Sociological Inquiry, 66*, 58–83.

Brulle, R. J. (2000). *Agency, democracy, and nature: The U.S. environmental movement from a critical theory perspective*. Cambridge: MIT Press.

Carolan, M. (2005). Realism without reductionism: Toward an ecologically embedded sociology. *Human Ecology Review, 12*(1), 1–20.

Catton, W. R., & Dunlap, R. E. (1978). Environmental sociology: A new paradigm. *The American Sociologist, 13*(February), 41–49.

Dietz, T., Kalof, L., & Stern, P. C. (2002). Gender, values and environmentalism. *Social Science Quarterly, 83*(1), 353–364.

Dunlap, R. E. (2002). Paradigms, theories and environmental sociology. In R. E. Dunlap, F. H. Buttel, P. Dickens, & A. Gijswijt (Eds.), *Sociological theory and the environment: Classical foundations, contemporary insights* (pp. 329–350). New York: Rowman & Littlefield.

Dunlap, R. E., Van Liere, K. D., Mertig, A., & Jones, R. E. (2000). Measuring endorsement of the new ecological paradigm: A revised NEP scale. *Journal of Social Issues, 56*(3), 425–442.

Fisher, D. (2006). Bringing the material back in: Understanding the U.S. position on climate change. *Sociological Forum, 21*(3), 467–494.

Foster, J. B. (1999). *The vulnerable planet: A short economic history of the environment.* New York: Monthly Review Press.

Freudenburg, W. R., Frickel, S., & Gramling, R. (1995). Beyond the nature/society divide: Learning to think about a mountain. *Sociological Forum, 10*(3), 361–392.

Freudenburg, W. R., & Gramling, R. (1993). Socioenvironmental factors and development policy: Understanding opposition and support for offshore oil. *Sociological Forum, 8*(3), 341–364.

Freudenburg, W. R., Gramling, R., & Davidson, D. (2008). Scientific certainty argumentation methods (SCAMS): Science and the politics of doubt. *Sociological Inquiry, 78*(1), 2–38.

Fuller, T., & Revkin, A. C. (2007, December 16). Climate plan looks beyond Bush's tenure. *The New York Times,* p. 1.

Hannigan, J. (1995). *Environmental sociology: A social constructionist perspective.* New York: Routledge.

Hawken, P., Lovins, A., & Lovins, H. (2000). *Natural capitalism: Creating the next industrial revolution.* New York: Little, Brown.

Intergovernmental Panel on Climate Change. (2007). *Climate change 2007: Synthesis report: Summary for policymakers.* New York: The United Nations. Retrieved December 20, 2007, from http://www.ipcc.ch/pdf/assessment-report/ar4/syr/ar4_syr_ spm.pdf

Jones, R. E., & Carter, L. F. (1994). Concern for the environment among black Americans: An assessment of common assumptions. *Social Science Quarterly, 75*(3), 560–579.

Lerner, J., & Kalof, L. (1999). The animal text: Message and meaning in television advertisements. *Sociological Quarterly, 40*(4), 565–586.

Libes, S. (2003). *Why we should all be "Waccamaw waterwatchers."* Distinguished Teacher-Scholar Lecture Series. Conway, SC: Coastal Carolina University.

Logan, J. R., & Molotch, H. (1987). *Urban fortunes: The political economy of place.* Berkeley: University of California Press.

Mazur, A., & Lee, J. (1993). Sounding the global alarm: Environmental issues in the US national news. *Social Studies of Science, 23*(4), 681–720.

McCright, A. M., & Dunlap, R. E. (2003). Defeating Kyoto: The conservative movement's impact on U.S. climate change policy. *Social Problems, 50*(3), 348–373.

McDonough, W., & Braungart, M. (2002). *Cradle to cradle: Remaking the way we make things.* New York: North Point Press.

Murphy, R. (2002). The internalization of autonomous nature into society. *The Sociological Review, 50*(3), 313–333.

Murphy, R. (2004, August). *Technological disasters, natural disasters, environmental disasters: Toward the integration of social constructionism and critical realism.* Paper presented at the Annual Meeting of the American Sociological Association, San Francisco, CA.

Nash, R. (1982). *Wilderness and the American mind.* New Haven, CT: Yale University Press.

Nisbet, M. C., & Myers, T. (2007). Twenty years of public opinion about global warming. *Public Opinion Quarterly, 71*(3), 444–470.

O'Connor, J. (1973). *The fiscal crisis of the state.* New York: St. Martin's Press.

Olsen, M., Lodwick, D., & Dunlap, R. (1992). *Viewing the world ecologically.* Boulder, CO: Westview.

Papson, S. (1992). "Cross the fin line of terror": Shark week on the Discovery Channel. *Journal of American Culture, 15,* 67–81.

Podeschi, C. (2002). The nature of future myths: Environmental discourse in science fiction films, 1950–1999. *Sociological Spectrum, 22*(3), 251–297.

Podeschi, C. (2007). The culture of nature and the rise of modern environmentalism: The view through general audience magazines, 1945–1980. *Sociological Spectrum, 27*(3), 299–331.

Rothman, H. (2000). *Saving the planet: The American response to the environment in the twentieth century.* Chicago: Ivan R. Dee.

Rudel, T. K. (2005). *Tropical forests: Regional paths of destruction and regeneration in the late 20th century.* New York: Columbia University Press.

Ryan, M., & Kellner, D. (1988). *Camera politica: The politics and ideology of contemporary Hollywood film.* Bloomington: Indiana University Press.

Schnaiberg, A. (1980). *The environment: From surplus to scarcity.* New York: Oxford.

Schnaiberg, A., & Gould, K. A. (1994). *Environment and society: The enduring conflict.* New York: St. Martin's Press.

Schor, J. B. (1999). *The overspent American: Why we want what we don't need.* New York: Harper Paperbacks.

Shanahan, J., & McComas, K. (1999). *Nature stories: Depictions of the environment and their effects.* Cresskill, NJ: Hampton Press.

Snow, D. A., & Benford, R. D. (1988). Ideology, frame resonance, and participant mobilization. *International Social Movement Research, 1,* 197–217.

Snow, D. A., Rochford, Jr., E. B., Worden, S. K., & Benford, R. D. (1986). Frame alignment processes, micromobilization, and movement participation. *American Sociological Review, 51,* 464–481.

Spector, M., & Kitsuse, J. I. (1987). *Constructing social problems.* Hawthorne, NY: Aldine de Gruyter.

Szasz, A. (2007). *Shopping our way to safety: How we changed from protecting the environment to protecting ourselves.* Minneapolis: University of Minnesota Press.

Ungar, S. (1998a). Bringing the issue back in: Comparing the marketability of the ozone hole and global warming. *Social Problems, 45*(4), 510–527.

Ungar, S. (1998b). Recycling and the dampening of concern: Comparing the roles of large and small actors in shaping the environmental discourse. *Canadian Review of Sociology and Anthropology, 35*(2), 253–276.

Warner, K., & Molotch, H. (2000). *Building rules: How local controls shape community environments and economies.* Boulder, CO: Westview.

York, R., Rosa, E., & Dietz, T. (2003). Footprints on the earth: The environmental consequences of modernity. *American Sociological Review, 68*(2), 279–300.

Appendix: Sample of Science Fiction Films

I selected two top box-office grossing films in each half-decade, then the next highest grossing film from any point in the decade was added. If more than one film from the same series was selected, these were only counted once toward the total of five for that decade.

1950s: *Destination Moon* (1950), *The Beast From 20,000 Fathoms* (1953), *Them!* (1954), *Forbidden Planet* (1956), *The Incredible Shrinking Man* (1957)

1960s: *On the Beach* (1960), *Voyage to the Bottom of the Sea* (1961), *Fantastic Voyage* (1966), *Planet of the Apes* (1968), *2001: A Space Odyssey* (1968)

1970s: *Beneath the Planet of the Apes* (1970), *A Clockwork Orange* (1972), *Star Wars* (1977), *Alien* (1979), *Star Trek: The Motion Picture* (1979)

1980s: *The Black Hole (1980), Star Wars: The Empire Strikes Back* (1980), *Star Trek II: The Wrath of Khan* (1982), *Star Wars: Return of the Jedi* (1983), *Aliens* (1986), *Star Trek IV: The Voyage Home* (1986), *Back to the Future Part II* (1989)

1990s: *Total Recall* (1990), *Demolition Man* (1993), *Jurassic Park* (1993), *Waterworld* (1995), *The Matrix* (1999)

Film Index for Teaching Sociology

Age and Aging

About Schmidt (2002)

American Beauty (1999)

Away From Her (2006)

The Big Chill (1983)

Bubbeh Lee & Me (1996)

The Bucket List (2007)

Cocoon (1985)

Driving Miss Daisy (1989)

The Evening Star (1996)

For Better or Worse (1995)

A Gathering of Old Men (1987)

Grumpier Old Men (1995)

Grumpy Old Men (1993)

Harold and Maude (1971)

The Hours (2002)

How to Make an American Quilt (1995)

I Never Sang for My Father (1970)

Iris (2001)

Jack (1996)

Mrs Dalloway (1997)

Mrs Palfrey at the Claremont (2005)

Nobody's Fool (1994)

On Golden Pond (1981)

A Rumor of Angels (2000)

Salut Victor (1989)

Shirley Valentine (1989)

Strangers in Good Company (1990)

The Sunshine Boys (1975)

Terms of Endearment (1983)

Trip to Bountiful (1985)

Tuesdays With Morrie (1999)

Unforgiven (1992)

The Whales of August (1987)

Documentaries

Aging in America: The Years Ahead (2003)

Beauty Before Age (1997)

Eager for Your Kisses: Love and Sex at 95 (2006)

42 Up (1999)

Number Our Days (1976)

Shameless: The Art of Disability (2006)

Still Doing It: The Intimate Lives of Women Over 65 (2004)

Still Kicking: Six Artistic Women of Project Arts & Longevity (2006)

Whisper: The Women (1989)

Assimilation, Immigration, Pluralism, Colonialism

America, America (1963)

Bhaji on the Beach (1994)

Cabeza de Vaca (1991)

Chocolat (1989)

Coming to America (1988)

Como Era Gostoso o Meu Francês (How Tasty Was My Little Frenchman) (1971)

Crossing Over (2009)

Dirty Pretty Things (2002)

Double Happiness (1995)

El Norte (1983)

The Emerald Forest (1985)

Hotel Rwanda (2004)

In America (2002)

The Man Who Cried (2000)

The Mission (1986)

Mister Johnson (1991)

Moscow on the Hudson (1984)

My Big Fat Greek Wedding (2002)

My Family (1995)

Quilombo (1984)

Rabbit-Proof Fence (2002)

Documentaries

Los Trabajadores/The Workers (2003)

College Life

American Pie (1999)

American Pie 2 (2001)

Animal House (1978)

Back to School (1986)

Boys and Girls (2000)

Blue Chips (1994)

Breaking Away (1979)

Bring It on Again (2004)

Campus Man (1987)

College (2008)

College Confidential (1960)

DOA: Dead or Alive (2006)

Educating Rita (1983)

Fast Break (1979)

Good Will Hunting (1997)

The Graduate (1967)

Higher Learning (1995)

Horsefeathers (1928)

The House Bunny (2008)

The Human Stain (2003)

Love Story (1970)

Mona Lisa Smile (2003)

The Nutty Professor (1963, 1996)

The Program (1993)

Old School (2003)

Oleanna (1994)

Real Genius (1985)

Revenge of the Nerds (1984)

Revenge of the Nerds II (1987)

Rudy (1993)

School Daze (1988)

The Skulls (2000)

Spring Break (1983)

The Sure Thing (1985)

Van Wilder (2002)

Where the Boys Are (1960)

Where the Boys Are '84 (1984)

With Honors (1994)

Wonder Boys (2000)

Corporate and Organizational Deviance

The China Syndrome (1979)

A Civil Action (1998)

Class Action (1991)

The Corporation (2003)

Duplicity (2009)

Erin Brockovich (2000)

Fun With Dick and Jane (2005)

Glengarry Glen Ross (1992)

The Hudsucker Proxy (1994)

The Insider (1999)

Lord of War (2005)

Michael Clayton (2007)

The Rainmaker (1997)

Runaway Jury (2003)

She Hate Me (2004)

Thank You for Smoking (2005)

Documentaries

Enron: The Smartest Guys in the Room (2005)

Roger & Me (1989)

Crime, Criminology, Criminal Justice

Absence of Malice (1981)

. . . And Justice for All (1979)

Brubaker (1980)

Casino (1995)

Catch Me If You Can (2002)

Chinatown (1974)

Clockers (1995)

Crash (2004)

Dead Man Walking (1995)

Double Indemnity (1944)

Eye for an Eye (1996)

Goodfellas (1990)

Monster (2003)

Pathology (2008)

Serpico (1973)

Shawshank Redemption (1994)

Training Day (2001)

12 Angry Men (1957, 1997)

Documentaries

Bowling for Columbine (2002)

Brother's Keeper (1992)

The Farm: Angola, USA (1998)

Girlhood (2003)

A Hard Straight (2004)

Lockdown: Inside America's Prisons (2006)

Paradise Lost: The Child Murders at Robin Hood Hills (1996)

Paradise Lost 2: Revelations (2000)

Rape in a Small Town: The Florence Holway Story (2004)

Un Coupable Idéal (Murder on a Sunday Morning) (2001)

What I Want My Words to Do to You: Voices From Inside a Women's Maximum Security Prison (2003)

Culture

After Hours (1985)

American History X (1998)

Borat: Cultural Learnings of America for Make Benefit Glorious Nation of Kazakhstan (2006)

Crazy Like a Fox (2004)

Daughters of the Dust (1991)

The End of Violence (1997)

The Gods Must Be Crazy (1980)

Local Hero (1983)

Lost in Translation (2003)

My Big Fat Greek Wedding (2002)

The New World (2005)

Not Without My Daughter (1991)

Out of the Holes of the Rocks (2008)

Party Monster (2003)

Saving Face (2004)

Slumdog Millionaire (2008)

Smoke Signals (1998)

Songcatcher (2000)

Witness (1985)

Documentaries

Made in China (2007)

Paris Is Burning (1990)

Split: A Divided America (2008)

Voices in Exile (2005)

When We Were Kings (1996)

Deviance

Acts of Worship (2001)

Angus (1995)

Bonnie and Clyde (1967)

Borat: Cultural Learnings of America for Make Benefit Glorious Nation of Kazakhstan (2006)

Cidade de Deus (City of God) (2002)

Crimes and Misdemeanors (1989)

A Home at the End of the World (2004)

Kids (1995)

Party Monster (2003)

Requiem for a Dream (2000)

Taxi Driver (1976)

Documentaries

The Life of Kevin Carter (2004)

Street Life: Inside America's Gangs (1999)

Ethnic Stratification

Angels and Insects (1996)

Braveheart (1995)

Cry, the Beloved Country (1995)

Gandhi (1982)

High Hopes (1989)

Losing Isaiah (1995)

Los Santos Inocentes (Holy Innocents) (1984)

The Man Who Cried (2000)

Metropolitan (1990)

My Beautiful Laundrette (1985)

Nothing but a Man (1964)

A Passage to India (1984)

Racing With the Moon (1984)

Raining Stones (1993)

Salaam Bombay! (1988)

Sammy and Rosie Get Laid (1987)

Sense and Sensibility (1995)

Six Degrees of Separation (1993)

Trading Places (1983)

White Man's Burden (1995)

Who's Who (1978)

Family

Addams Family Values (1993)

Alice Doesn't Live Here Anymore (1974)

American Beauty (1999)

Antwone Fisher (2002)

Baby Mama (2008)

A Cool, Dry Place (1998)

Daughters of Dust (1991)

A Day Without a Mexican (2004)

Erin Brockovich (2000)

Europa, Europa (1990)

Familia Rodante (Family Rodante) (2004)

The Family Stone (2005)

Far From Heaven (2002)

Germinal (1993)

Guess Who (2005)

Guess Who's Coming to Dinner (1967)

Home for the Holidays (1995)

The Incredibles (2004)

Jerry Maguire (1996)

The Joy Luck Club (1993)

Knocked Up (2007)

Kramer vs. Kramer (1979)

Let's Get Married (1960)

Little Miss Sunshine (2006)

Losing Isaiah (1995)

Magnolia (1999)

Marvin's Room (1996)

Meet the Browns (2008)

Monsoon Wedding (2001)

Monster's Ball (2001)

Moolaadé (2004)

Mrs. Doubtfire (1993)

My Family (1995)

The Namesake (2006)

Real Women Have Curves (2002)

The Royal Tenenbaums (2001)

Rudy (1993)

Running With Scissors (2006)

A Simple Twist of Fate (1994)

Sling Blade (1996)

Soul Food (1997)

The Squid and the Whale (2005)

The Stone Boy (1984)

The Virgin Suicides (2000)

White Man's Burden (1995)

White Oleander (2002)

Why Did I Get Married? (2007)

Documentaries

Capturing the Friedmans (2003)

Chicks in White Satin (1994)

The Farmer's Wife (1998)

Love and Diane (2002)

The Motherhood Manifesto (2007)

Family Violence

Bastard Out of Carolina (1996)

Beauty and the Beast (1991)

Black and Blue (1999)

The Burning Bed (1984)

Carousel (1956)

A Cry for Help: The Tracey Thurman Story (1989)

Deadly Matrimony (1992)

Enough (2002)

Fried Green Tomatoes (1991)

Sleeping With the Devil (1997)

Sleeping With the Enemy (1991)

Waitress (2007)

What's Love Got to Do With It? (1993)

Documentaries

Abused Women Who Fight Back: The Framingham Eight (1994)

Defending Our Lives (1994)

Hostages at Home (1994)

Safe: Inside a Battered Women's Shelter (2001)

Shifting the Paradigm: From Control to Respect (1999)

Gender—Doing Gender

Blonde Venus (1932)

The Break Up (2006)

Bus Stop (1956)

College (2008)

Fast & Furious (2009)

Flawless (1999)

Funny Girl (1968)

The House Bunny (2008)

Pirates of the Caribbean: At World's End (2007)

Pirates of the Caribbean: The Curse of the Black Pearl (2003)

Pirates of the Caribbean: Dead Man's Chest (2006)

Pulp Fiction (1994)

Semi-Pro (2008)

Sex and the City (2008)

She's the Man (2006)

Tootsie (1982)

White Chicks (2004)

Yentl (1983)

See also: all the James Bond films

Globalization

Babel (2006)

Blood Diamond (2006)

The Constant Gardener (2005)

A Day Without a Mexican (2004)

Dirty Pretty Things (2002)

El Norte (1983)

Documentaries

The Corporation (2003)

Darwin's Nightmare (2004)

Fast Food Nation (2006)

King Corn (2007)

Mardi Gras: Made in China (2005)

Zoned for Slavery (1995)

Health and Health Care

As Good As It Gets (1997)

At First Sight (1999)

The Doctor (1991)

Gattaca (1997)

John Q (2002)

Miss Evers' Boys (1997)

Outbreak (1995)

Passion Fish (1992)

Philadelphia (1993)

Documentaries

"The American Experience: The Pill" (2003)

Big Bucks, Big Pharma: Marketing Disease & Pushing Drugs (2006)

Deadly Deception: The Mark Hacking Story (2004)

A Healthy Baby Girl (1997)

La Operación (1982)

"Trade Secrets: A Moyers Report" (2001)

Health Care—Experiencing Illness and/or Disabilities

Boys on the Side (1995)

The Bucket List (2007)

I Am Sam (2001)

Living & Dying (2007)

My Left Foot: The Story of Christy Brown (1989)

One True Thing (1998)

On Golden Pond (1981)

The Other Sister (1999)

Regarding Henry (1991)

Terms of Endearment (1983)

What's Eating Gilbert Grape (1993)

Documentaries

"On Our Own Terms: Moyers on Dying in America"

Health Care—Mental Health and Substance Abuse

American Gangster (2007)

As Good As It Gets (1997)

A Beautiful Mind (2001)

Benny & Joon (1993)

Blow (2001)

Clean and Sober (1988)

50 First Dates (2004)

Frances (1982)

Girl, Interrupted (1999)

Harvey (1950)

I Am Sam (2001)

Leaving Las Vegas (1995)

Little Miss Sunshine (2006)

One Flew Over the Cuckoo's Nest (1975)

Postcards From the Edge (1990)

Rachel Getting Married (2008)

Requiem for a Dream (2000)

Running With Scissors (2006)

28 Days (2000)

When a Man Loves a Woman (1994)

Documentaries

Addiction: Why Can't They Just Stop (2007)

The Secret Passage: A Journey of Black Women and Depression (2001)

Health Care—Professionals and Institutions

Awakenings (1990)

Critical Care (1997)

Gross Anatomy (1989)

The Hospital (1971)

House Calls (1978)

One Flew Over the Cuckoo's Nest (1975)

Patch Adams (1998)

Vital Signs (1990)

Documentaries

Sicko (2007)

Health Care—Reproductive Health

Cider House Rules (1999)

Citizen Ruth (1996)

4 Luni, 3 Saptamâni si 2 Zile (4 Months, 3 Weeks, and 2 Days) (2007)

A Handmaid's Tale (1990)

If These Walls Could Talk (1996)

Juno (2007)

Knocked Up (2007)

Rambling Rose (1991)

Vera Drake (2004)

Waitress (2007)

Documentaries

Lake of Fire (2006)

La Operación (1982)

High School

Breakfast Club (1985)

Bring It On (2000)

Can't Buy Me Love (1987)

Can't Hardly Wait (1998)

College (2008)

Dazed and Confused (1993)

Election (1999)

Fame (1980)

Fast Times at Ridgemont High (1982)

Flirting (1992)

Friday Night Lights (2004)

Get Real (1999)

Heathers (1989)

High School Musical (2006)

Just One of the Guys (1985)

My Bodyguard (1980)

Napoleon Dynamite (2004)

Pretty in Pink (1986)

Risky Business (1983)

Rushmore (1999)

She's All That (1999)

Sixteen Candles (1984)

Stand and Deliver (1988)

The Virgin Suicides (2000)

Documentaries

Hoop Dreams (1994)

Men and Masculinity

Affliction (1997)

American Beauty (1999)

American Pie (1999)

The Benchwarmers (2006)

Breakin' All the Rules (2004)

City Slickers (1991)

College (2008)

The Cowboys (1972)

Death of a Salesman (1985)

Fight Club (1999)

40 Days and 40 Nights (2002)

40-Year-Old Virgin (2005)

Full Monty (1997)

The Godfather (1972)

The Godfather: Part II (1974)

The Godfather: Part III (1990)

The Hangover (2009)

He Got Game (1998)

I Love You, Man (2009)

Inside Man (2006)

Knocked Up (2007)

Monster's Ball (2001)

Rio Bravo (1959)

The Searchers (1956)

Superbad (2007)

Take the Lead (2006)

3:10 to Yuma (2007)

Wall Street (1987)

White Chicks (2004)

The Wrestler (2008)

Documentaries

Boys Will Be Men (2002)

Hip-Hop: Beyond Beats and Rhymes (2006)

Hoop Dreams (1994)

Men's Lives (1974)

Michael Kimmel on Gender (2008)

Paris Is Burning (1990)

Tough Guise: Violence, Media & the Crisis in Masculinity (1999)

Wrestling With Manhood (2002)

Mothers and Mothering

Almost Famous (2000)

Baby Boom (1987)

Chocolat (2000)

The Good Mother (1988)

The Grifters (1990)

In America (2002)

The Joy Luck Club (1993)

Little Man Tate (1991)

Mommie Dearest (1981)

Mother (1996)

Not Without My Daughter (1991)

Ordinary People (1980)

The Positively True Adventures of the Alleged Texas Cheerleader-Murdering Mom (1993)

Serial Mom (1994)

Steel Magnolias (1989)

Stepmom (1998)

Surviving My Mother (2007)

Terms of Endearment (1983)

Todo Sobre Mi Madre (All About My Mother) (1999)

What's Eating Gilbert Grape (1993)

Documentaries

Ima Hozeret Habayta (A Working Mom) (2008)

The Motherhood Manifesto (2007)

The Mother's House (2005)

Refrigerator Mothers (2003)

Politics and Political Economy

All the President's Men (1976)

Bobby (2006)

Bob Roberts (1992)

Bulworth (1998)

The Candidate (1972)

Che: Part One (2008)

Che: Part Two (2008)

Dr. Strangelove or: How I Learned to Stop Worrying and Love the Bomb (1964)

Frost/Nixon (2008)

Good Night, and Good Luck (2005)

La Muerte de un Burócrata (Death of a Bureaucrat) (1966)

The Manchurian Candidate (1962, 2004)

Mr. Smith Goes to Washington (1939)

Nixon (1995)

Primary Colors (1998)

Reds (1981)

Three Days of the Condor (1975)

W (2008)

Wag the Dog (1997)

Documentaries

The Big One (1997)

"Surviving the Good Times: A Moyers Report" (2000)

The War Room (1993)

Poverty and Homelessness

Cidade de Deus (City of God) (2002)

Dark Days (2000)

The Fisher King (1991)

God Bless the Child (1988)

Joyeux Calvaire (1996)

Kicking It (2008)

The Pursuit of Happyness (2006)

With Honors (1994)

Documentaries

"America's War on Poverty" (1995)

Life Below the Line: The World Poverty Crisis (2007)

1 More Hit (2007)

Poverty Outlaw (1997)

Prejudice and Discrimination

Bad Day at Black Rock (1954)

Black Like Me (1964)

Boys Don't Cry (1999)

The Defiant Ones (1958)

The Diary of Anne Frank (1959)

Do the Right Thing (1989)

Far From Heaven (2002)

Home of the Brave (1949)

Il Giardino dei Finzi-Contini (The Garden of the Finzi-Continis) (1970)

Intruder in the Dust (1949)

Jungle Fever (1991)

Map of the Human Heart (1993)

No Way Out (1950)

Nothing but a Man (1964)

A Passage to India (1984)

Rabbit-Proof Fence (2002)

Schindler's List (1993)

To Kill a Mockingbird (1962)

White Man's Burden (1995)

Documentaries

"Frontline: A Class Divided" (1985)

Race and Ethnicity

American History X (1998)

Babel (2006)

Bamboozled (2000)

Bread & Roses (2000)

A Bronx Tale (1993)

The Constant Gardener (2005)

Crash (2004)

Cry, the Beloved Country (1995)

Dangerous Minds (1995)

A Day Without a Mexican (2004)

Dirty Pretty Things (2002)

Do the Right Thing (1989)

A Dry White Season (1989)

Europa, Europa (1990)

Finding Forrester (2000)

Get on the Bus (1996)

Hotel Rwanda (2004)

In My Country (2004)

La Misma Luna (Under the Same Moon) (2007)

Land of Plenty (2004)

Maid in America (1982)

Maria Full of Grace (2004)

Miss Evers' Boys (1997)

Mississippi Burning (1988)

Mississippi Masala (1991)

My Family (1995)

Once We Were Warriors (1994)

187 (1997)

Panther (1995)

Rabbit-Proof Fence (2002)

School Daze (1988)

Secrets & Lies (1996)

Stand and Deliver (1988)

Strictly Ballroom (1992)

Taxi Driver (1976)

The Wedding (1998)

A World Apart (1988)

Zebrahead (1992)

Documentaries

Crossing Arizona (2006)

Ethnic Notions (1986)

Farmingville (2004)

Incident at Ogalala (1992)

Los Trabajadores/The Workers (2003)

"Mexico: A Death in the Desert" (2004)

"Surviving the Good Times: A Moyers Report" (2000)

Troubled Harvest (1990)

Understanding Race (1999)

Walking the Line (2005)

What's Race Got to Do With It? (2006)

Race/Class/Gender Intersectionality

Bend It Like Beckham (2002)

Casa de los Babys (2003)

The Color Purple (1985)

Higher Learning (1995)

Jungle Fever (1991)

Just Another Girl on the I.R.T. (1992)

The Long Walk Home (1990)

Maria Full of Grace (2004)

Menace II Society (1993)

Mississippi Masala (1991)

Monster's Ball (2001)

187 (1997)

Set It Off (1996)

Skin Deep (1989)

Whale Rider (2002)

White Man's Burden (1995)

Documentaries

The Letter: An American Town and "The Somali Invasion" (2003)

Mickey Mouse Monopoly (2001)

What's Race Got to Do With It? (2006)

Religion

Agnes of God (1985)

Bruce Almighty (2003)

The Chosen (1981)

The Da Vinci Code (2006)

Der Himmel Über Berlin (Wings of Desire) (1987)

Dogma (1999)

Doubt (2008)

The Exorcist (1973)

The Last Temptation of Christ (1988)

The Mission (1986)

Not Without My Daughter (1991)

The Prince of Egypt (1998)

Sister Act (1992)

Sister Act 2: Back in the Habit (1993)

Documentaries

Jesus Camp (2006)

Religulous (2008)

Trembling Before G-d (2001)

Science Fiction

Alien (1979)

Aliens (1986)

Back to the Future Part II (1989)

The Beast From 20,000 Fathoms (1953)

Beneath the Planet of the Apes (1970)

Black Hole (1980)

Blade Runner (1982)

Born in Flames (1983)

A Clockwork Orange (1972)

Demolition Man (1993)

Destination Moon (1950)

Fantastic Voyage (1966)

Forbidden Planet (1956)

The Incredible Shrinking Man (1957)

Jurassic Park (1993)

The Matrix (1999)

On the Beach (1960)

Planet of the Apes (1968)

Star Trek: The Motion Picture (1979)

Star Trek II: The Wrath of Khan (1982)

Star Trek IV: The Voyage Home (1986)

Star Wars (1977)

Star Wars: The Empire Strikes Back (1980)

Star Wars: Return of the Jedi (1983)

Them! (1954)

Total Recall (1990)

2001: A Space Odyssey (1968)

Voyage to the Bottom of the Sea (1961)

Waterworld (1995)

Sexuality and Gender

The Adventures of Priscilla, Queen of the Desert (1994)

Before Night Falls (2000)

Boys Don't Cry (1999)

Brokeback Mountain (2005)

The Crying Game (1992)

If These Walls Could Talk 2 (2000)

In & Out (1997)

The Incredibly True Adventure of Two Girls in Love (1995)

Kinsey (2004)

Ma Vie en Rose (1997)

Milk (2008)

Normal (2003)

Stonewall (1995)

Transamerica (2005)

Documentaries

The Celluloid Closet (1995)

Last Call at Maud's (1993)

Out of the Past: The Struggle for Gay and Lesbian Rights in America (1997)

Paris Is Burning (1990)

Small Town Gay Bar (2006)

The Times of Harvey Milk (1984)

Trembling Before G-d (2001)

Sex Work(ers)

Irma La Douce (1963)

Sweet Charity (1969)

Working Girls (1986)

Documentaries

Born into Brothels: Calcutta's Red Light Kids (2004)

The Good Woman of Bangkok (1991)

Live Nude Girls Unite! (2000)

Not a Love Story: A Film about Pornography (1981)

Sacrifice (2000)

Shinjuku Boys (1995)

Sisters and Daughters Betrayed (2002)

Trading Women (2003)

Social Class and Inequality

Amar Te Duele (2002)

Attica (1980)

Born Yesterday (1950)

Bread & Roses (2000)

The Color Purple (1985)

Dirty Dancing (1987)

Educating Rita (1983)

8 Mile (2002)

Gosford Park (2001)

Grapes of Wrath (1940)

The Greatest Game Ever Played (2005)

The Great Gatsby (1974)

Hey Hey It's Esther Blueburger (2008)

The Ice Storm (1997)

Inventing the Abbotts (1997)

La Règle du Jeu (The Rules of the Game) (1939)

Le Charme Discret de la Bourgeoisie (The Discreet Charm of the Bourgeoisie) (1973)

Matewan (1987)

Miss Pettigrew Lives for a Day (2008)

My Fair Lady (1964)

The Nanny Diaries (2007)

Native Son (1986)

Norma Rae (1979)

Pride & Prejudice (2005)

The Pursuit of Happyness (2006)

Pygmalion (1938)

Reds (1981)

The Remains of the Day (1993)

Room at the Top (1959)

Save the Last Dance (2001)

Six Degrees of Separation (1993)

Step Up (2006)

The Swimmer (1968)

Trading Places (1983)

The War (1994)

Documentaries

Morristown: In the Air and Sun (2007)

People Like Us: Social Class in America (2001)

Social Class—Private Schools

The Bells of St. Mary's (1945)

Chasing Holden (2001)

Dead Poets Society (1989)

The Emperor's Club (2002)

Finding Forrester (2000)

Goodbye, Mr. Chips (1969)

The History Boys (2006)

The Lords of Discipline (1983)

Outside Providence (1999)

School Ties (1992)

Taps (1981)

Social Interaction

Annie Hall (1977)

The Big Chill (1983)

The Break-Up (2006)

Clerks (1994)

The Graduate (1967)

Harold and Maude (1971)

Open Water (2003)

Pretty Woman (1990)

Sisterhood of the Traveling Pants (2005)

Six Degrees of Separation (1993)

12 Angry Men (1957)

When Harry Met Sally . . . (1989)

Who's Afraid of Virginia Woolf? (1966)

Social Movements and Social Change

Bloody Sunday (2002)

Butterfly (1982)

Deadly Deception (1987)

Gandhi (1982)

In the Name of the People (2000)

Iron-Jawed Angels (2004)

Kilomètre Zéro (Kilometer Zero) (2003)

Love and Anarchy (1974)

Lumumba (2000)

Making a Killing (2002)

Matewan (1987)

Medium Cool (1969)

Romero (1989)

Salt of the Earth (1954)

The Take (2007)

Documentaries

Amandla! A Revolution in Four Part Harmony (2002)

At the River I Stand (1993)

Before Stonewall (1984)

Berkeley in the Sixties (1990)

Eyes on the Prize: America's Civil Rights Movement 1954–1985 (1987)

The FBI's War on Black America (1990)

The Fight in the Fields: Cesar Chavez and the Farmworkers' Struggle (1997)

Freedom on My Mind (1994)

The Greening of Cuba (1995)

Kent State: The Day the War Came Home (2000)

Made in India (2000)

Made in Thailand (2003)

Making Sense of the Sixties (1991)

Out of the Past: The Struggle for Gay and Lesbian Rights in America (1997)

A Place Called Chiapas (1998)

Rebels With a Cause (2000)

Scout's Honor (2001)

Showdown in Seattle: Five Days That Shook the WTO (2000)

30 Frames a Second: The WTO in Seattle (2000)

This Is What Democracy Looks Like (2000)

Union Maids (1976)

Uprising of '34 (1995)

Weather Underground (2002)

With Babies and Banners: Story of the Women's Emergency Brigade (1979)

Women Organize! (2000)

You Got to Move (1985)

Zapatista (1999)

War and Gender

A Few Good Men (1992)

Born on the Fourth of July (1989)

Casualties of War (1989)

Courage Under Fire (1996)

Dr. Strangelove or: How I Learned to Stop Worrying and Love the Bomb (1964)

Full Metal Jacket (1987)

G. I. Jane (1997)

Un Long Dimanche de Fiançailles (A Very Long Engagement) (2004)

War and Violence

All Quiet on the Western Front (1930)

Apocalypse Now (1979)

Big Red One (1980)

Black Hawk Down (2001)

Born on the Fourth of July (1989)

Braveheart (1995)

Breaker Morant (1980)

The Bridge on the River Kwai (1957)

A Bridge Too Far (1977)

Casablanca (1942)

Casualties of War (1989)

Catch-22 (1970)

Cold Mountain (2003)

Das Boot (1981)

The Deer Hunter (1978)

The Dirty Dozen (1967)

Dr. Strangelove or: How I Learned to Stop Worrying and Love the Bomb (1964)

Enemy at the Gates (2001)

The English Patient (1996)

Europa, Europa (1990)

Full Metal Jacket (1987)

Gladiator (2000)

Glory (1989)

Gone With the Wind (1939)

Good Morning, Vietnam (1987)

Hope and Glory (1987)

The Killing Fields (1984)

Lacombe Lucien (1974)

The Last of the Mohicans (1992)

La Vita è Bella (Life Is Beautiful) (1997)

The Longest Day (1962)

MASH (1974)

Master and Commander: The Far Side of the World (2003)

No Man's Land (2001)

Oh! What a Lovely War (1969)

Paths of Glory (1957)

Patton (1970)

The Pianist (2002)

Platoon (1986)

Regeneration (1997)

Ride With the Devil (1999)

Salvador (1986)

Saving Private Ryan (1998)

Schindler's List (1993)

The Thin Red Line (1998)

Three Kings (1999)

Welcome to Sarajevo (1997)

Women and Gender

Anchorman: The Legend of Ron Burgundy (2004)

Antonia's Line (1995)

Bagdad Cafe (1990)

Because I Said So (2007)

Bend It Like Beckham (2002)

Bhaji on the Beach (1993)

The Birdcage (1996)

The Break-Up (2006)

Calendar Girls (2003)

Chocolat (1989, 2000)

The Circle (2004)

Girlfight (2000)

The Handmaid's Tale (1990)

Ice Princess (2005)

The Incredibly True Adventure of Two Girls in Love (1995)

In Her Shoes (2005)

I've Heard the Mermaids Singing (1987)

Just Like a Woman (1992)

The Magdalene Sisters (2002)

Material Girls (2006)

Mean Girls (2004)

Mrs. Doubtfire (1993)

Mujeres al Borde de un Ataque de Nervios (Women on the Verge of a Nervous Breakdown) (1988)

Mulan (1998)

Muriel's Wedding (1994)

North Country (2005)

Personal Velocity: Three Portraits (2002)

Real Women Have Curves (2002)

Shirley Valentine (1989)

The Stepford Wives (1975, 2004)

Strangers in Good Company (1990)

Thelma & Louise (1991)

Whale Rider (2002)

What Women Want (2000)

White Chicks (2004)

Documentaries

Michael Kimmel on Gender (2008)

Women and Work

Bread & Roses (2000)

Cheaper by the Dozen (2003)

The Devil Wears Prada (2006)

Fast Food Fast Women (2000)

Nine to Five (1980)

Norma Rae (1979)

North Country (2005)

Silkwood (1983)

Women Buddy Films

Bagdad Cafe (1990)

Beaches (1988)

Coup de Foudre (Entre Nous) (1983)

Crimes of the Heart (1986)

Desert Hearts (1985)

Divine Secrets of the Ya-Ya Sisterhood (2002)

The First Wives Club (1996)

A League of Their Own (1992)

Leaving Normal (1992)

Mystic Pizza (1988)

Nine to Five (1980)

Passion Fish (1992)

Personal Best (1982)

Sisterhood of the Traveling Pants (2005)

The Turning Point (1977)

The Women (2008)

Women's Sexuality

About Schmidt (2002)

Ad Ogni Costo (Grand Slam) (1968)

American Gigolo (1980)

Antonia's Line (1995)

As Good As It Gets (1997)

The Banger Sisters (2002)

Being Julia (2004)

The Bliss of Mrs. Blossom (1968)

Boomerang (1992)

Bull Durham (1988)

Calendar Girls (2003)

A Change of Seasons (1980)

Cocoon (1985)

Daughters of the Dust (1991)

Desert Hearts (1985)

A Dirty Shame (2004)

The Evening Star (1996)

The Graduate (1967)

Grandma's Boy (2006)

Grumpy Old Men (1993)

Harold and Maude (1971)

How Stella Got Her Groove Back (1998)

Innocence (2004)

Ladies in Lavender (2004)

La Pianiste (The Piano Player) (2001)

La Vie Devant Soi (Madame Rosa) (1977)

The Madness of King George (1994)

Meet the Fockers (2004)

Moonstruck (1987)

Must Love Dogs (2005)

The Opposite of Sex (1998)

The Piano (1993)

Roman Spring of Mrs. Stone (2003)

Sammy and Rosie Get Laid (1987)

Saving Face (2004)

Shirley Valentine (1989)

Something's Gotta Give (2003)

Songcatcher (2000)

The Squid and the Whale (2005)

Steel Magnolias (1989)

A Streetcar Named Desire (1951)

A Summer Place (1959)

Sunset Blvd. (1950)

Terms of Endearment (1983)

Thomas Crown Affair (1999)

A Touch of Class (1973)

Under the Tuscan Sun (2003)

Unfaithful (2002)

An Unmarried Woman (1978)

Wedding Crashers (2005)

Witches of Eastwick (1987)

Documentaries

Still Doing It: The Intimate Lives of Women Over 65 (2004)

Work and the Workplace

Bread & Roses (2000)

Coyote Ugly (2000)

Daddy Day Care (2003)

The Devil Wears Prada (2006)

Employee of the Month (2006)

The Firm (1993)

Freedom Writers (2007)

Glengarry Glen Ross (1992)

Michael Clayton (2007)

Modern Times (1936)

Nine to Five (1980)

Norma Rae (1979)

On the Waterfront (1954)

The Pursuit of Happyness (2006)

10,000 Black Men Named George (2002)

Documentaries

Life and Times of Rosie the Riveter (1980)

Roger & Me (1989)

Work in Restaurants

Alice Doesn't Live Here Anymore (1974)

As Good As It Gets (1997)

Cocktail (1988)

Five Easy Pieces (1970)

Frankie and Johnny (1991)

Heavy (1995)

Home Fries (1998)

It Could Happen to You (1994)

Little Man Tate (1991)

Mystic Pizza (1988)

Nurse Betty (2000)

Office Space (2000)

Pay It Forward (2000)

Pleasantville (1998)

Return to Me (2000)

Thelma & Louise (1991)

Untamed Heart (1993)

Waitress (2007)

The Wedding Singer (1998)

Thanks to the following for contributing to this list of films:

For the constant flow of film recommendations from our colleagues on the Sociologist for Women in Society (SWS) listserv.

American Society of Criminology

Introduction to Criminology Syllabi Collection

Coeditors: Denise Paquette Boots (University of Texas at Dallas) and William Reese (Augusta State University): Recommended Media List

Rick J. Scheidt, Ph.D.

http://apadiv20.phhp.ufl.edu/cinema.htm

War films:

http://www.channel4.com/film/newsfeatures/microsites/W/greatest_warfilms/results/5-1.html

Guide to documentary and independent films:

http://www.bullfrogfilms.com/index.html

And to the following individuals for their contributions of film titles: Michelle Bemiller, Kim Cook, Mark Cox, Erin Farley, Katie Gay, Wendy Grove, Michael Kimmel, Christina Lanier, Jess MacDonald, Mike Maume, Abigail Reiter, Miranda Reiter, John Rice, and Rachel Schneider.

About the Editors

Jean-Anne Sutherland is an assistant professor at the University of North Carolina, Wilmington, in the Department of Sociology and Criminology, where she teaches Introduction to Sociology, Gender and Society, Social Psychology, and Sociology Through Film. Her areas of research include sociology of mothering, with a focus on the experiences of guilt and shame in the mothering role. Her chapter titled "Ideal Mama, Ideal Worker: Negotiating Guilt and Shame in Academe" appears in *Mama PhD: Women Write About Motherhood and Academic Life,* edited by Caroline Grant and Elrena Evans (Rutgers University Press, 2008). Additionally, her research includes sociology through film, specifically the construction of pedagogical frameworks for teaching with film.

Kathryn Feltey is an associate professor at The University of Akron in the Department of Sociology, where she teaches Qualitative Methods, Sociology of Gender, College Teaching of Sociology, and Sociology Through Film. Her research focuses on women's experiences of dislocation as a result of poverty, domestic violence, or disaster. She is coeditor of a special issue of the *National Women's Studies Association Journal,* "New Orleans: A Special Issue on Gender, the Meaning of Place, and the Politics of Displacement," as well as editor of the Gender Section of the online journal *Sociology Compass.*

About the Contributors

Carleen R. Basler is an Assistant Professor of Sociology and American Studies at Amherst College. She completed her BA at the University of California, Los Angeles, and her PhD at Yale University. Her teaching and research are primarily concerned with race and ethnicity, political identity, social stratification, and social movements. Basler is a Mexican American who splits her time between residences in Amherst, Massachusetts, and Los Angeles, California.

Robert C. Bulman is a Professor of Sociology at Saint Mary's College of California. He received his BA in sociology from the University of California at Santa Cruz in 1989 and his PhD in sociology from the University of California at Berkeley in 1999. His areas of expertise include the sociology of education, the sociology of adolescence, the sociology of culture, and the sociology of film. He has published a book on the depiction of high schools and adolescents in Hollywood films—*Hollywood Goes to High School: Cinema, Schools, and American Culture* (Worth Publishers, 2005).

Janet Cosbey is an associate professor at Eastern Illinois University, Charleston, in the Department of Sociology/Anthropology, where she teaches family and society, gender roles and social change, and social gerontology. Her research interests include the scholarship of teaching and learning, retirement, and caregiving. She has published articles in *Teaching Sociology* and several gerontology journals.

James J. Dowd is a Professor of Sociology at the University of Georgia. He studied sociology at St. Peter's College (BS), the University of Maryland (MA), and the University of Southern California (PhD). He regularly teaches Introduction to Sociology, as well as courses on culture, social theory, and the military. He has taught Sociology in Film for more than 15 years and has published a number of papers having to do with films and the film industry in journals such as *Current Perspectives in Social Theory, Sociological Perspectives,* and *Teaching Sociology.*

Karla A. Erickson is an Assistant Professor of Sociology at Grinnell College. As a feminist ethnographer of work in the new economy, she is primarily interested in the social connections that emerge out of service interactions. From the exchanges between a waitress and customers in the booth of a restaurant to bedside in a hospice ward, Erickson explores how workers manage their identities, and what consumers want and do or do not receive in commercialized care settings. Her work has appeared in *Ethnography, Qualitative Sociology,* and *Space and Culture.* Her research has been anthologized in *The Gendered Society Reader,* edited by Michael Kimmel, and *Restaurants: The Book,* edited by David Sutton and David Berriss. Her first single-authored book, *The Hungry Cowboy: Service and Community in a Neighborhood Restaurant,* was published in March 2009 by the University Press of Mississippi. With Hokulani Aikau and Jennifer Pierce, Erickson coedited and wrote several chapters in the recently published *Feminist Waves, Feminist Generations: Life Stories From the Movement* (University of Minnesota Press). Erickson received her PhD in American Studies and Feminist Studies from the University of Minnesota in 2004, an MA from Hamline University in 1998, and a BA in English and Women's Studies from Illinois Wesleyan University in 1995.

Henry A. Giroux holds the Global TV Network Chair in English and Cultural Studies at McMaster University in Canada. His most recent books include *America on the Edge* (Palgrave Macmillan, 2006), *Beyond the Spectacle of Terrorism* (Paradigm, 2006), *Stormy Weather: Katrina and the Politics of Disposability* (Paradigm, 2006), *The University in Chains: Confronting the Military-Industrial-Academic Complex* (Paradigm, 2007), and *Against the Terror of Neoliberalism: Politics Beyond the Age of Greed* (Paradigm, 2008). His newest book is *Youth in a Suspect Society: Democracy and the Politics of Disposability* (Palgrave Macmillan, 2009).

Susan Searls Giroux is Associate Professor of English and Cultural Studies at McMaster University in Hamilton, Ontario. She is the author, with Henry A. Giroux, of *Take Back Higher Education: Race, Youth and the Crisis of Democracy in the Post–Civil Rights Era* (Palgrave, 2004) and, with Jeffrey T. Nealon, *The Theory Toolbox: Critical Concepts for the Humanities, Arts and Social Sciences* (Rowman & Littlefield, 2003). She is also managing editor of *Review of Education, Pedagogy, and Cultural Studies,* and recently coedited a special issue of *College Literature* titled "The Assault on Higher Education" (Fall 2006). She has published numerous articles, which have appeared in *The CLR James Journal, College Literature, Cultural Critique, JAC, Social Identities, Third Text, Tikkun,* and *Works and Days.*

Roberto G. Gonzales is an assistant professor at the University of Washington. His current work examines how the laws of the state, acculturation processes, and the cultural requirements of adulthood collide to shape the realities of unauthorized immigrant youth. It also pays close attention to the ways in which these youth experience, navigate, and critique the many contradictions inherent in becoming "American." Gonzales is also the cofounder of Video Machete, a Chicago-based, intergenerational collective, committed to transforming communities and addressing social change in communities and society through multimedia and video production.

Ed Guerrero is Professor of Cinema Studies and Africana Studies at New York University. His influential books, *Framing Blackness* (Temple University Press, 1993) and *Do the Right Thing* in the Modern Classics series (British Film Institute, 2008), explore black cinema, its critical discourse, and political economy. Guerrero has also written extensively on black cinema, culture, and politics for such journals as *Callaloo, CINEASTE, Discourse, Ethnic and Racial Studies, Film Quarterly, Journal* *of Popular Film and Television,* and *Sight & Sound.* Ed Guerrero has served on numerous editorial and professional boards, including *Cinema Journal, Quarterly Review of Film and Video, Race/Ethnicity,* and the National Film Preservation Board of the Library of Congress.

Michael Kimmel is Professor of Sociology at SUNY at Stony Brook. His books include *Men Confront Pornography* (Crown, 1990), *Against the Tide: Profeminist Men in the United States, 1776–1990* (Beacon, 1992), *The Politics of Manhood* (Temple University Press, 1996), *The Gendered Society* (Oxford University Press, 2003), and *Manhood: A Cultural History* (Oxford University Press, 2006). He coedited *The Encyclopedia on Men and Masculinities* (2004) and *Handbook of Studies on Men and Masculinities* (2004). He is the founder and editor of *Men and Masculinities,* the field's premier scholarly journal, a book series on Gender and Sexuality at New York University Press, and edited the Sage Series on Men and Masculinities. He is the Spokesperson for the National Organization for Men Against Sexism (NOMAS) and lectures extensively in corporations and on campuses in the United States and abroad. His new book, *Guyland: The Perilous World Where Boys Become Men* (HarperCollins, 2008), has just been optioned by Dreamworks for a feature film.

Travis W. Linnemann is a PhD candidate in Sociology at Kansas State University. His research areas are criminology and sociology of gender. Linnemann has published in peer-reviewed journals focusing on the influences of race on criminal court outcomes and alternative treatments for court-involved youth. Prior to his return to graduate school, Mr. Linnemann held a variety of positions in juvenile justice, concluding with a multiyear project to reduce minority overrepresentation in the Kansas Juvenile Justice System.

Betsy Lucal is Associate Professor of Sociology at Indiana University South Bend, where she also teaches in the women's studies program. She teaches courses on gender and sexuality, sociological theory, social movements, and sociology of food. She is a member of Indiana University's Faculty Colloquium on Excellence in Teaching, and winner of the IU South Bend Distinguished Teaching Award and the all-IU Sylvia Bowman Award for Excellence in Teaching. She currently serves as Chair of the American Sociological Association's Section on Teaching and Learning in Sociology. She also is a deputy editor of *Gender & Society,* journal of sociologists for women in society.

Michael A. Messner is Professor of Sociology and Gender Studies at the University of Southern California, where he teaches courses on sex and gender. His most recent book is *It's All for the Kids: Gender, Families and Youth Sports* (University of California Press, 2009).

Andrea Miller is the Director for the Center of the Study of Human Rights at Webster University. As director she is currently coordinating the Year of International Human Rights for the College of Arts and Sciences at Webster University. Miller led Webster University's inaugural Human Rights Summer Institute for high school students in the St. Louis area. She is a faculty member in the Behavioral and Social Sciences Department, where she teaches courses on gender, sexuality, and bisexuality. Miller holds an MA and a doctoral degree in sociology from American University. Her work focuses on gender and sexuality at the local and global level. She is currently working on a manuscript titled "Are Bisexual Men Lying: Men Doing Bisexuality."

Christopher W. Podeschi is an Assistant Professor of Sociology at Bloomsburg University of Pennsylvania in Bloomsburg. He obtained his PhD in 2002 from the University of Nebraska—Lincoln. His research, including his dissertation, focuses primarily on culture and the natural environment. A project nearing completion quantitatively examines trends in the depiction of nature in children's books. Other work looks at meaning of place and its relationship to concern for the environment and development/sprawl. He has also developed a strong interest in applied environmental sociology after administering a survey in support of a community-based water-quality program when he worked for Coastal Carolina University. Recent courses Podeschi has taught include Sociological Theory, Race and Ethnic Inequality, and Environmental Sociology.

L. Susan Williams, Associate Professor of Sociology, specializes in gender, violence, and inequality and is considered highly qualified in both quantitative and qualitative research methodologies. Her study of adolescent girls in Connecticut broke ground in terms of empirically documenting the effect of local community characteristics on life decisions of individual girls. More recently, her research has focused on youth and gendered violence, resulting in a comprehensive study of incarcerated girls and boys; her most recent paper is "Bad Girls and Rural Pathways: Girls' Deviance and Local Social Control." Williams teaches courses on gender, criminology, and diversity and has been recognized for excellence in teaching. Williams holds several national positions of service, including membership in a Sister to Sister Task Force designed to cultivate relationships across disciplinary and cultural boundaries. She serves Kansas State University at several levels, including the Honor Council and the Institutional Review Board.